Items should be returned on or before the last date shown below. Items not already requested by other borrowers may be renewed in person, in writing or by telephone. To renew, please quote the number on the barcode label. To renew online a PIN is required. This can be requested at your local library.
Renew online @ **www.dublincitypubliclibraries.ie**
Fines charged for overdue items will include postage incurred in recovery. Damage to or loss of items will be charged to the borrower.

nt with *The*
reporting on
assignments
on platform
ed 167 lives.
lling, on the
ook Power in
y Robinson;
Irish waters.

ONCE UPON A TIME IN THE WEST

The Corrib Gas Controversy

Lorna Siggins

TRANSWORLD IRELAND

TRANSWORLD IRELAND
an imprint of The Random House Group Limited
20 Vauxhall Bridge Road, London SW1V 2SA
www.rbooks.co.uk

First published in 2010 by Transworld Ireland,
a division of Transworld Publishers

A CIP catalogue record for this book
is available from the British Library.

ISBN 9781848270947

Addresses for Random House Group Ltd companies outside the UK
can be found at: www.randomhouse.co.uk
The Random House Group Ltd Reg. No. 954009

The Random House Group Limited supports the Forest Stewardship
Council (FSC), the leading international forest-certification organization. All our
titles that are printed on Greenpeace-approved FSC-certified paper carry the FSC logo.
Our paper procurement policy can be found at
www.rbooks.co.uk/environment

Typeset in 12.25/15pt Ehrhardt by
Falcon Oast Graphic Art Ltd.
Printed and bound in Great Britain by
Clays Ltd, Bungay, Suffolk

2 4 6 8 10 9 7 5 3 1

Mixed Sources
Product group from well-managed
forests and other controlled sources
www.fsc.org Cert no. TF-COC-2139
© 1996 Forest Stewardship Council
FSC

CONTENTS

ACKNOWLEDGEMENTS

Many, many people helped with the research and compilation of this book, and it wouldn't be between covers now without the tenacity and enthusiasm of Eoin McHugh of Transworld Ireland, the patience and keen eye of Transworld editor Brian Langan and a mixture of both shown by literary agent Jonathan Williams. I am also very grateful to Katrina Whone, managing editor, and Kate Tolley, editorial co-ordinator, both of Transworld Publishers in London.

I wish to thank Geraldine Kennedy, editor of *The Irish Times*, for her support and for the use of the newspaper's material; *Irish Times* assistant editor Fintan O'Toole for his generous introduction; *Irish Times* managing editor Eoin McVey; news editor Kevin O'Sullivan and picture editor Frank Miller; editor of *The Western People* James Laffey; Andrew O'Rorke of Hayes and Sons; Dr Mark Garavan, Conor O'Clery, Arthur Reynolds and Dermot Somers for much textual advice and counsel.

I am most grateful for the co-operation of staff with Shell E&P Ireland, Enterprise Energy Ireland and contractors; staff with An Bord Pleanála, the Department of Energy, Communications and Resources, the Department of the Environment, the Garda Síochána and the Garda Síochána Ombudsman Commission, and the Environmental Protection Agency; non-governmental organizations Afri, Front Line and Table; and the many residents of the Kilcommon parish communities in Rossport, Glengad, Pollathomas, Lenamore and Ballinaboy, along with members of the Rossport Solidarity Camp.

All sources are identified in the text, except for those who requested anonymity – bearing in mind that the Corrib gas project controversy continues. I wish to thank: Dr Simon Berrow, Roisín Boyd, Brian Barrington, BL, Gerardine Boyle, Des Branigan of Marine Research & Associates, Stein Bredal, Fergus Cahill of the Irish Offshore Operators' Association, Padhraig Campbell, Sarah Clancy, Ed Collins, Terence Conway, Leo Corcoran, Mary and Willie Corduff, Páraic Cosgrove, Áine and Gregory Daly, Anne Daly, Eddie Diver, Teresa Diskin, Lelia Doolan, Alfred and John Donovan, Terry Dunne, Eamonn Farrell, Tadhg Foley, Peter Flynn of Clarke and Flynn Solicitors, Alan Gannon of Claffey Gannon & Co, Solicitors, Pauline Garavan, Maura Harrington, Jacinta Healy, Colm and Gabrielle Henry, Michael D Higgins TD and Labour Party president, Rita Ann Higgins, Mary Horan Moran, Colin Joyce and Christy Loftus of Shell E&P Ireland, Anthony Irwin, Enda Kenny TD and Fine Gael leader, Richard Kuprewicz of Accufacts Inc, Mary Lawlor of Front Line, Borghild Lieng, Dr Ronan Lynch, Benny McCabe, Sr Majella McCarron, Simon McGarr of McGarr Solicitors, Bríd McGarry, Bríd McGrath, Vincent McGrath, Philip McGrath, Ruairí McKiernan, Liamy MacNally, Chief Supt (retd) Tony McNamara, John Monaghan, Imelda Moran, PJ Moran, Fidelma Mullane, Monica Muller, Breeda Murphy, Joe Murray, Treasa Ní Ghearraigh, Bríd Ní Sheighin, Brian Ó Catháin, Risteard Ó Dómhnaill, Pat and Mary O'Donnell, Donncha Ó Faoláin and Sinéad Ó Nualláin, Eoin Ó Leidhin, Dónall Ó Mearáin, Ciarán Ó Murchú, Micheál Ó Seighin and Caitlín Uí Sheighin, Brendan Philbin, Michael Ring TD, Áine Ryan, Minister for Energy Eamon Ryan TD, Betty Schult, Sheila Sullivan and Brent 'Jack Bauer' Parker.

ACKNOWLEDGEMENTS

I am indebted to Seamus Heaney for permission to reproduce
Belderg. For secondary source material, I am indebted to *The Irish
Times*, *The Irish Independent*, *The Irish Daily Mail*, *The Irish
Examiner*, *The Guardian*, *The Daily Telegraph*, *Bloomberg*, *The
Western People*, *The Mayo News*, *The Connaught Telegraph*, *The Irish
Mail on Sunday*, *The Sunday Business Post*, *The Sunday Times*, *The
Sunday Tribune*, *The Sunday World*, *The Phoenix*, *Upstream*, *Village*
magazine; BBC Radio, Channel 4 television, MidWest Radio,
Newstalk, RTÉ radio and television, Raidió na Gaeltachta, TG4,
Today FM, TV3, and website royaldutchshellplc.com.

Family and close friends have demonstrated the utmost patience
while this work was in progress; you know who you are, and thank
you.

FOREWORD

At one level – the level at which it is experienced by the people of the Erris peninsula in County Mayo – the conflict over the Corrib gas field is a small and intimate affair. It concerns their own lives, their farms, their homes, their communities. It is played out on and against a landscape that may be exotic and remotely beautiful to most of us, but that is utterly familiar to them.

Indeed, in some respects the conflict may have enhanced that sense of smallness. At times the communities themselves have been swamped by outsiders – by police and security guards, and by organized protesters. People who were happy to live in relative obscurity have become objects of scrutiny, of surveillance, of curiosity and sometimes of contempt. Above all, in the unfolding of the story over the last decade, these people have been treated, again and again, as irritating irrelevancies. In the grand scheme of things, on the scale at which governments and global corporations operate, they are grains of sand, significant only in their ability to clog up the wheels of progress.

At another level, though, the Corrib story is very big indeed. Contained within it are many of the large themes of twenty-first-century politics – the relationship between governments and transnational corporations; the tension between environmental protection and our collective hunger for fossil fuels; the balance between democracy and development; the need for a local identity in a globalized economy. The very intimacy of a story unfolding in a remote, sparsely populated corner of Ireland, on the edge of Europe – the very name of Erris means 'western promontory' – allows us to see these large forces not as abstractions, but as immediate human dilemmas.

What has been happening in Erris has a very powerful resonance around the world. As events in Ireland have a way of doing, it subverts the simple and static opposition between the developed and developing worlds. Ireland, as the most globalized economy in the world, as a member of the European Union and as a favoured base for US-owned corporations, is emphatically a part of the developed world. But the plight of the Erris communities is in many ways more reminiscent of that of an indigenous society in the developing world. Though theoretically citizens in a liberal democracy, those who have stood in the way of the exploitation of the Corrib gas field by a consortium led by Shell found themselves with very little protection from their own government. Instead of seeking to negotiate a settlement on behalf of these citizens, Irish governments aligned themselves to an overwhelming extent with Shell, putting the resources of the state behind the acquisition of land and, when locals objected, mounting a policing operation that at one point included the deployment of the navy.

A central part of this process has been the configuration of the problem as one of law and order rather than of democracy and fairness. Having generated conflict through high-handedness, Shell and the state have ended up effectively criminalizing local dissent. Therefore the very idea that people have a right to be consulted about, and to have an effect on, the decisions that shape their lives is at stake.

What has often been lost in this process is that, time and again, objective assessments have concluded that local anxieties about health, safety and the environment are neither irrational nor unreasonable. The vast majority of those in Erris who have ended up as objectors to the Corrib project never imagined themselves as the kind of people who would engage in a decade-long conflict with one of the world's largest companies and with their own government. They simply did what sentient citizens are supposed to do in a functioning democracy – tried to understand what was happening around them and the implications it might have for their families and neighbours. If people can be criminalized for being alert to their own immediate world, democracy itself is in deep trouble.

In the context of such a concerted campaign to define Corrib as essentially a law and order issue, the value of consistent, even-handed and accurate reporting is immeasurable. Events in Erris have often been emotive, summoning up on the one side deeply rooted memories of nineteenth-century exploitation and on the other fears of everything from Luddism to terrorism. The story has dragged on to a point where most people who do not have a stake in it are inclined to pay attention only when there is a flashpoint or a dramatic moment, usually one involving physical confrontation or violence. The day-to-day texture of life in what has become an embattled part of the world has tended to be lost, as has the mundane reality that people have rational fears and that neither the state nor Shell has ever entered into a genuine process of consultation and compromise.

Lost, that is, except for the outstanding reportage of Lorna Siggins for *The Irish Times*. Siggins has always treated the reporting of the Corrib conflict as being in the public interest even during the times when the public has not been interested in it. She has stuck to the basic but crucial task of recording what people from all sides have to say, of checking it against known facts and of ignoring hysteria and propaganda. She has brought calm insight, independent observation and a superb capacity for elucidating complexities to bear on that task. With so many axes being ground and so many agendas being served, Siggins has been consistently and unshakeably trustworthy.

In the context of a decade-long story, however, even such exemplary reporting has its limits. As with any important set of events, context is everything – or, in this case, the multiple contexts that Lorna Siggins explores so skilfully here, from the traditions of Erris to the global energy system and from environmental impacts to governmental attitudes. Sometimes a book simply has to be written, and sometimes there is only one writer with the authority to write it. This is such a book and Siggins is such a writer. When two such imperatives come together they create a third – a book that demands to be read by anyone interested in the workings of contemporary power.

Fintan O'Toole

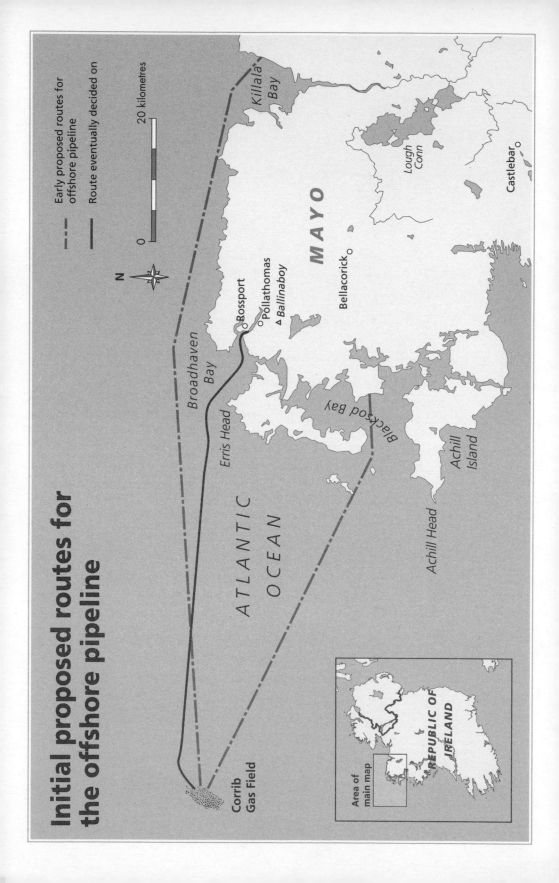

Initial proposed routes for the offshore pipeline

Early proposed routes for offshore pipeline

Route eventually decided on

20 kilometres

N

ATLANTIC OCEAN

Corrib Gas Field

Broadhaven Bay

Erris Head

Rossport
Pollathomas
Ballinaboy

Blacksod Bay

Achill Head

Achill Island

Bellacorick

MAYO

Killala Bay

Lough Conn

Castlebar

Area of main map

REPUBLIC OF IRELAND

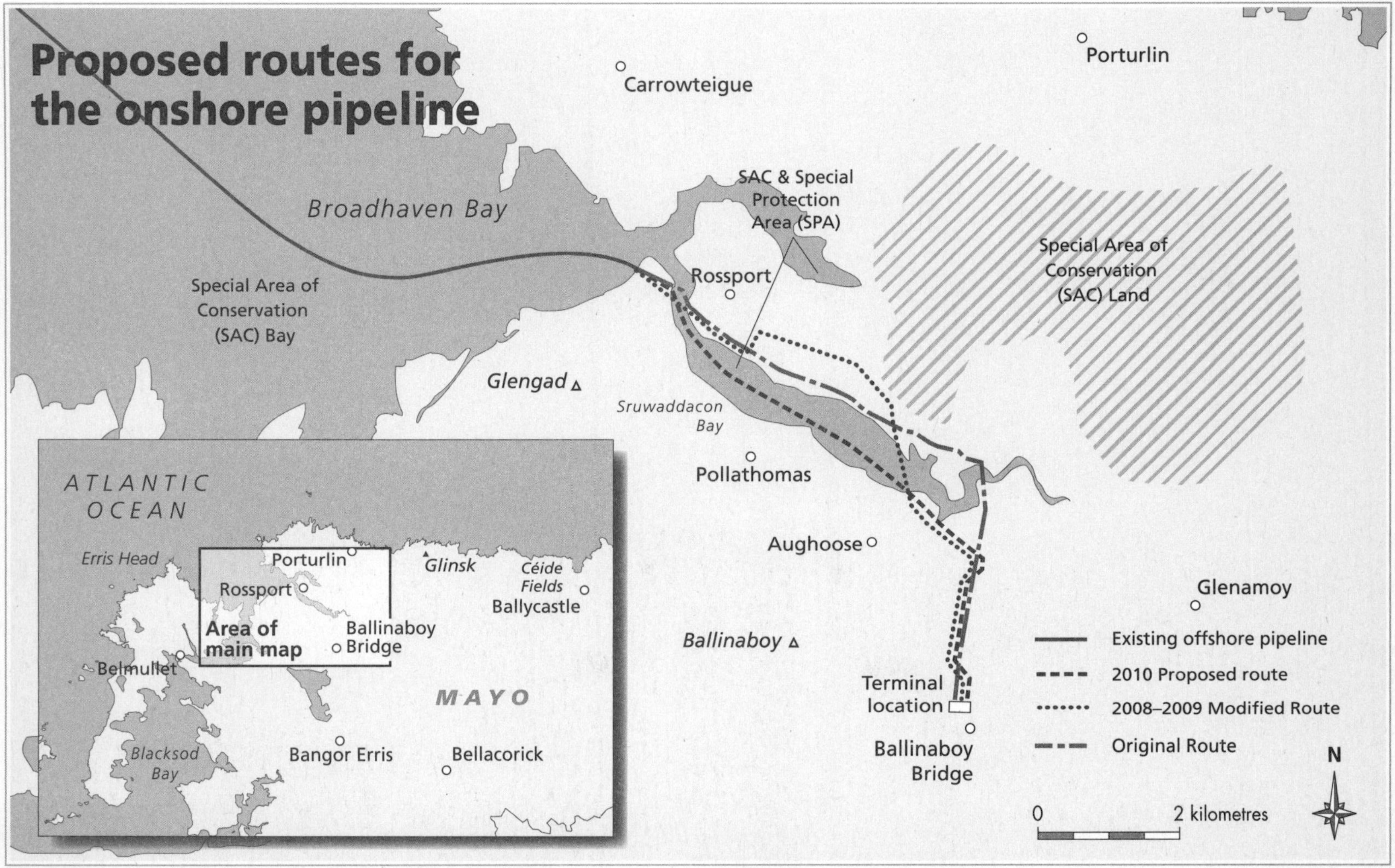
Proposed routes for the onshore pipeline
Porturlin
Carrowteigue
SAC & Special Protection Area (SPA)
Broadhaven Bay
Rossport
Special Area of Conservation (SAC) Land
Special Area of Conservation (SAC) Bay
Glengad
Sruwaddacon Bay
Pollathomas
Aughoose
Glenamoy
Ballinaboy
Terminal location
Ballinaboy Bridge
Existing offshore pipeline
2010 Proposed route
2008–2009 Modified Route
Original Route
N
0
2 kilometres
ATLANTIC OCEAN
Erris Head
Porturlin
Rossport
Glinsk
Céide Fields
Ballycastle
Area of main map
Ballinaboy Bridge
Belmullet
MAYO
Blacksod Bay
Bangor Erris
Bellacorick

What gets us into trouble is not what we don't know, it's what we know for sure that just ain't so.

MARK TWAIN

Prologue

FROM SRUWADDACON
TO SANTA CRUZ

Even in the inky light before dawn, there is a spectacular quality to the bog plain extending over much of the barony of Erris in north Mayo. Simultaneously bleak and beautiful, it rolls out, mile upon mile, under a big wide Atlantic sky. The brown canvas is framed by hills and mountains and hectares of conifers, while fragile powerlines threaded on wind-strained poles straddle the margins. Momentarily desolate, all it takes is a swarm of swifts, a curlew's cry, a bounding hare to bring the view to life.

Any vehicle can be seen for miles: even at speed, an ambulance hardly needs a siren. Traffic was sparse in the early hours of 23 April 2009 as an emergency vehicle sped south to Mayo General Hospital in Castlebar. The patient was fifty-five-year-old Rossport farmer Willie Corduff, a father of six from a small mixed holding on the shoreline of Sruwaddacon estuary in north Mayo. With him was his youngest son, Liam. Willie's wife, Mary, followed in her sister-in-law's car.

The sequence of events had begun almost twenty-four hours earlier – some might say ten or eleven years earlier – when the phone rang in the Corduff household. Shortly after 7 a.m. on 22 April, heavy machinery had begun moving on to coastal land at Glengad, across the estuary from Rossport. The hamlet had been identified as the landfall location for gas due to be piped ashore from the Corrib gas field some eighty kilometres off the Mayo coast. Willie Corduff and a neighbour, Gerry Bourke, took the fifteen-minute round-estuary trip by road to meet a number of other residents. Watching the contractors move in, the residents shared their concerns about authorization for this work. Glengad beach, at the mouth of the Sruwaddacon estuary in north Mayo's Broadhaven Bay, is bordered by the Glenamoy bog complex. Both the estuary and the bog had been designated as protected habitats under a European Union (EU) directive.

Farmers like the Corduffs were well accustomed to the constraints that the Brussels directive imposed on their land and commonage. Shell E&P Ireland, the company that had hired the contractor, was well aware of the constraints also – just eighteen months before, it had been directed by a government minister to restore part of a special area of conservation at Glengad to its original state, following unauthorized test drilling by contractors there. Frustrated at the response now to their queries about authorization for further work, and driven by long-held concerns about the health and safety aspects of the entire gas project, Corduff and Bourke spotted a delivery truck, parked in the Shell 'compound'. They crossed a sheepwire fence and dived in under the chassis, pledging that they would not move until they were shown the necessary paperwork.

The gardaí arrived from Belmullet, the nearest station, and ordered the two men to leave. When both refused, gardaí managed to pull Bourke out. Corduff, though, a big, broad man, had wedged himself in tight, and he wasn't budging. During the various efforts to remove him, caught on video-camera, Corduff roared at the officers.

Years of frustration were reflected in his choice of language. The month of April had particular resonance for him. It was in April 2002 that a government minister had approved the plan of development for the Corrib gas field, signing away access by the developers to private land for an onshore pipeline the following month. It was in April 2003 that a planning appeals inspector had turned down permission for the onshore gas terminal for the project, describing it as 'the wrong site' from the perspective of strategic planning, government policy on balanced regional development, minimizing environmental impact and sustainable development.[1] And it was in April 2005, after a new planning application had been approved, that Corduff was cited in a High Court injunction taken by the gas developer, Shell – defiance of which led to him spending one of the best summers in years in jail, along with four neighbours.

Gravel was thrown at him under the truck, he recalled later; he said that a garda dug the sharp end of a stone into his left ankle. Several gardaí tried to crawl under on hands and knees, and he warned them that he had boots with toecaps on that would 'bust them'. They said he wouldn't dare. Security guards employed at the Glengad site stood back, on watching brief. One of their superiors periodically asked Corduff if there was anything he needed, including food. All that Corduff wanted was some evidence, and he said he would be happy to speak to the garda superintendent.

His wife, Mary, drove to Belmullet Garda Station with a friend and neighbour, but she was told the superintendent couldn't speak to her. She was handed a 'fact sheet' outlining the relevant authorizations. The local authority had maintained that 'ancillary works' didn't require planning permission; residents, and An Taisce, the Dublin-based environmental organization, maintained that they did and that there had been a precedent for the same.

As night fell, Corduff was given a mobile phone to talk to journalists, sitting at desks in warm offices while he continued his vigil under the vehicle. He described to reporters how he had been harassed by the gardaí during his protest. As for the security, 'They've been perfectly civil,' he said. 'I've no complaint with them

5

at all.' That was to change in the early hours of the following morning. His brother-in-law, Peter Lavelle, had opted to stay close by; the night was cold, wet and windy. At one point, Corduff remembered hearing 'banging, and noise'. Several hours later, estimated at about 2 a.m., he crawled from under the truck to stretch his legs and to talk to Lavelle.

The two believed at this stage that they were on their own: there was no sign of any gardaí or security staff. Minutes later, according to Corduff, he was seized by several strong arms, hit on the side of the head, forced to the ground. He said his head was compressed into his chest. Lavelle was also surrounded by figures, said to be wearing dark clothing and night-vision gear. He was separated from Corduff and unable to help.

Liam, Corduff's son, had arrived minutes earlier with Glenamoy farmer P. J. Moran and his sister Bridie. They heard shouts and cries for help, then met Lavelle, who was extremely upset and told them that Corduff was in trouble. They encountered a line of security guards shining torches in their faces. Bridie Moran could see Corduff on the ground, and it appeared that several men in balaclavas were 'sitting on him'.

Liam demanded access to his father. He said that the senior security officer present refused to let him through. He recalled that two gardaí asked unsuccessfully if Liam could be let into Shell property. When he was eventually allowed in about thirty minutes later, Liam was shocked to find his father moaning, unable to open his eyes. Stone chips covered his face. Ambulance staff placed Corduff on a spinal board with a cervical collar. They noted in their patient care report that he was 'confused'.

The medical report at Mayo General Hospital recorded an 'alleged assault by security guards'. It said that the patient had been 'kicked all over the body and had LOC [loss of consciousness]', along with 'headaches, nausea and vomiting'. A CT scan proved normal, as did a series of X-rays. Corduff was discharged the following day, having spoken to several reporters from his hospital bed. 'I thought they would kill me,' he said.

By then, however, a statement from the gardaí had been broadcast on radio news bulletins. It said it was investigating an incident that had occurred at approximately 11.30 p.m. at the Shell compound at Glengad the night before. Up to fifteen men armed with iron bars and chains had 'entered the compound at Glengad and threatened security staff'. One intruder had started a digger truck, which was 'used to cause damage', and one staff member had been injured on the arm and had received medical attention.

The Garda statement added that a 'protester', whom it did not identify, and who was 'present on the site since yesterday' was 'this morning removed from the compound by security staff'. He was 'taken to hospital by ambulance as a precaution'. Gardaí later said that the ambulance had been called as the 'protester' was 'feeling unwell'.

The incursion that the gardaí described may have been linked to the 'banging, and noise' that Corduff said he had heard when still under the truck. Other parts of the Garda statement did not tally with Corduff's account, and with what local people believed to have occurred. However, to a confused general public, truth had long become a casualty in a protracted and bitter dispute.

An expert analysis of medical records for the incident, conducted subsequently at the request of the human rights organization Front Line, confirmed that Corduff's injuries were consistent with 'a history of assault'. On 27 April 2010, Front Line, based in Dublin but with an international profile for working with human rights defenders, published a report by barrister Brian Barrington on the situation in north Mayo. Barrington made a series of recommend-ations: one key suggestion was that gardaí outside Mayo should reinvestigate the alleged assault. Several weeks before the report came out, a garda called to Willie Corduff's farm to tell him that there would be no criminal prosecutions. Corduff was upset, but not surprised.

Security policies and the personnel employed to implement them had long been aggravating factors in the resistance of local inhabitants to the energy project. Over a period of several years, there had been claims of 'shady agents' working on both sides. The behaviour of uniformed security operatives protecting Shell's interests had antagonized the local community when, for instance, unidentified security staff filmed a man taking his grandchildren to Glengad beach, a priest going about his parish business, and a local hostel owner accommodating guests. Supporters of the project, on the other hand, appeared convinced that protesters would take the law into their own hands at any juncture. Trust had broken down; attitudes had hardened into irreconcilable difference.

In the days that followed Corduff's hospitalisation, there would be statements of support for Corduff from a former United Nations assistant secretary general, Denis Halliday, and from South African Archbishop Desmond Tutu, based on Corduff's profile as a small farmer who had put his life on hold over health, safety and environmental concerns. A week later, two government ministers attended a highly charged meeting held by the local community. 'For decades, two or three gardaí policed this community without any trouble,' they were told by one tourism operator. 'And now three hundred gardaí can't control it.'

1

BEDROCK

It began some 230 million years ago, or so geologists estimate. Around that time, a reservoir began to form deep below the bottom of the sea. Generations of Irish schoolchildren had been brought up during the early years of the republic to believe that Ireland had no mineral resources of note, but the real map of Ireland is at least ten times its land mass, extending across the continental shelf.[1] The 430-million-year-old island was once part of a plate that moved north on the surface of the planet. When Europe and North America began drifting apart, three-dimensional troughs or 'basins' were created, comprising thick piles of sediment with potentially valuable deposits of gas and oil.

Under a kilometre of chalk, sandstone deposits contained what would be identified many millions of years later as the Kinsale gas field, while thick seams off the Dublin coastline would harbour coal. Deeper waters on the Atlantic margin, running from the Slyne basin off the west up to the Donegal basin off the north-west, represented Ireland's potential 'new frontier'. [2]

The Kinsale gas field was declared commercial in 1973 by its developer, Marathon, and in 1975 the company signed an agreement with Bord Gáis to supply 125 million cubic feet of gas daily for a twenty-year period, starting from 1979. A production platform that would process the gas offshore, stripping it of water and impurities, was commissioned that year. It was anchored fifty kilometres off Kinsale, and a subsea pipe was constructed to channel the gas ashore.

However, the government's deal with Marathon came into sharp focus when a coalition involving Fine Gael and Labour was formed and the oil crisis hit world economies in the early 1970s. The new minister for industry and commerce was Justin Keating, an academic vet and farmer who was best known for his television work. Keating had initially opposed Ireland's application to join the European Economic Community before having a change of heart. He was aware of the arguments against membership, which had dominated debate in Norway. The Scandinavian state had serious concerns about the impact of such membership on its marine resources, both oil and fish. The year before Ireland's accession in 1973, Norway had established a state-owned oil and gas company, and a separate directorate to manage and control its newly found resources, following the discovery of oil on the Ekofisk field on its continental shelf.

Statoil, as it was named then, aimed to ensure Norwegian state participation in the North Sea, and to build up a pool of expertise independently of the oil and gas majors. This induction began with the deal that ensured that Statoil was a 50 per cent partner with Mobil. The training agreement was the first in a series which ensured that Statoil – and Norway – would become a world leader in deep-sea technology, exploration and production.

Keating looked to the Norwegian model, and he wasn't alone. In spring 1973, a former Naval Service officer, Fergus Cahill, who was head of communications at the Institute for Industrial Research and Standards, had proposed that the government set up a study group to examine the entire area of offshore development. It should be as widely based as possible, he said, in a paper to the Cork Scientific Council, because no single existing agency had expertise in this area. Cahill quoted Britain's Baron Balogh, a junior energy minister in the 1970s, who had said there had been a 'silly and dangerous muddle of gigantic proportions' in relation to making the most of British offshore opportunities. Cahill contrasted Britain's approach with that of Norway, which had adopted a tough bargaining position with the oil companies. This had ensured that discoveries in the Norwegian sector of the North Sea could provide an annual income of around US$100 for every man, woman and child in Norway, the equivalent of several thousand euro annually in today's terms.

'The effect of a like sum on the Irish economy would be, for instance, to reduce income tax by roughly two-thirds,' Cahill said in his paper.[3] In December 1973, a conference was held in Jury's Hotel, Dublin, which laid down a blueprint for developing oil and gas resources off Ireland. Norway sent representation, and state agencies, including the Industrial Development Authority and Cahill's employer, the Institute for Industrial Research and Standards, participated.

The Ireland Exclusive Offshore Licensing Terms for oil and gas, drawn up by Keating, was intended to ensure Irish state participation on its continental shelf, and to ensure that companies exploiting the resource were subject to royalties and taxation. The licensing terms would govern prospecting and exploration licences and petroleum leases. Under Keating's terms, the state or its agencies could take a maximum 50 per cent stake in development and exploitation of any commercial discovery. Companies had to commit to a programme of drilling wells, surrendering 50 per cent of the licensed area after four years. The state would not be subject to exploration costs, but would have full access to exploration data.

Keating's plan included establishment of a state company, the Irish National Petroleum Corporation (INPC). 'I wanted this company for more reasons than just for money,' Keating explained, thirty-five years later. 'We had to understand the industry from end to end, the experts, the geologists and people who interpret seismic data. We had to know what other companies got up to by having our own.'

Keating had 'good, dedicated civil servants', who had carried out research with his support, and had contrasted how Britain and Norway approached their largesse in the North Sea. Britain on one side was 'wasting its resource as the big oil companies had such influence, and Norway was using it brilliantly to secure its future', Keating recalled.

'Norway was not very different in size from us, but with poor farmland. It's really not a country but a sea route – it depended on sea and on fish and hydroelectricity, but it was not a naturally rich place, and it was very poor at the beginning of the twentieth century.

'But the Norwegians had a far-sighted government, made up of very patriotic, independent-minded people, who were determined to do the best for themselves. And the government was just wise enough in advance, when it realized the extent of the North Sea resource and that it could copperfasten part of its future. It realized how other countries much bigger than it had managed the resource badly; and it introduced what I think is still a brilliant policy.'[4]

That Keating and company managed to introduce a similar policy in Ireland – with a 50 per cent 'state take', as he called it – was all the more extraordinary, given that the world economy had 'walked off a cliff' with the 1973 oil crisis. The price of oil multiplied by five, there was a worldwide recession, and many products associated with oil-based chemicals disappeared off world markets. 'It was a very sharp crisis which we survived without people knowing how desperate it was in 1973,' he recalled.

'We had no clout on world markets; oil countries had all. That weakened me as I wanted to make conditions that were much harder than the oil and gas companies expected, much harder than they had in Britain . . . When we ultimately produced the conditions that we did, they were a little taken aback as they hadn't appreciated this was coming.'[5]

The Cosgrave Fine Gael–Labour coalition government fell in 1977 before the impact of the legislation could be gauged accurately. Keating's successor in the new Fianna Fáil-led administration was Des O'Malley, who was not enthusiastic about the Irish National Petroleum Corporation (INPC). As O'Malley said in a government debate in 2001 on its privatization, he had only formed it 'because I had to'. He did not 'altogether agree with it', but events in Iran, including OPEC's dispute with the 'seven sisters', as the major oil companies were known, and the Shah's overthrow in 1979, had forced his hand. Producer countries like Iraq would only sell to a state-owned company.

Significantly, Norway had actually offered Ireland a block in the North Sea, but would only deal with a state company. The invitation was not taken up, and opportunities for working with Norway to build up a pool of independent expertise were also lost. It transpired later that the particular Norwegian block contained the Gulfaks field, one of the largest on its continental shelf.[6]

The INPC's brief was much more limited than that envisaged by Keating, when it was established in 1979. Its focus was on securing strategic oil supplies; it would not be involved in drilling, exploration or development of resources. A petroleum affairs division was established, initially located in the Department of Industry and Commerce. When specific expertise was required, oil and gas industry consultants were hired. In 1979 O'Malley rejected a request to name such consultants in the Dáil 'in view' of 'national interest'.

Successive ministers continued to adopt this 'national interest' approach. Labour Party energy minister, and Tánaiste, Dick Spring, declined to name consultants in March 1986, and his successor in energy, Ray Burke, declined in October 1987. Spring and Burke were also involved in significant changes to the licensing terms for the oil and gas sector in Irish waters, set by Keating back in 1975. From 1975 to 1985, the oil majors had drilled around a hundred wells, but had reported nothing that was of commercial value. By the mid-1980s, the sector was telling the business press that any finds were 'marginal'.

Bowing to this pressure, in April 1985 Spring reduced state royalties and introduced a sliding scale for state participation in marginally profitable fields. The following year, he abolished state participation in marginal fields altogether. Burke's turn came in 1987, when Fianna Fáil returned to power. From 1985, Marathon had been seeking to change its agreement with Bord Gáis on the Kinsale gas field, and secured a subsequent court agreement that allowed it to restrict its annual supply from Kinsale and to fix a new price for gas beyond a certain limit.

Burke began discussions with the exploration companies, as he told the Dáil in April 1985, and said he would take 'whatever additional steps' he deemed necessary to 'accelerate exploration activity'.[7] He announced his steps in the Dáil on 30 September 1987. Oil and gas production would be exempt in future from all royalty payments, there would be 100 per cent tax write-off against profits on capital expenditure for exploration, development and production extending back twenty-five years. Labour Party leader Dick Spring, who had initiated the first review of Keating's fiscal terms in 1985, was outraged. He described Burke's changes as 'an act of economic treason'.

Burke's defence was that he was 'gravely concerned about exploration prospects'. In January 1988, Burke confirmed that he had consulted, or had been effectively lobbied by, the industry. He told the Dáil that he had met representatives of more than twenty companies 'from around the world', some of which had been working off Ireland before. It emerged that a senior civil servant had advised him against the direct meeting.[8] On 30 May 1991, a new energy minister, Bobby Molloy, Progressive Democrat founder and TD for Galway West, reviewed the terms again, and the following April Minister for Finance Bertie Ahern included

Burke's oil and gas fiscal changes in the 1992 Finance Act, and reduced corporation tax to 25 per cent as an 'incentive'. The reaction from opposition parties was muted.

By June 1992, just after the Act was passed and the new licensing terms were in place, it appeared that state opportunities for any accruing benefit were minimal. There was no longer any obligation by developers to drill at an early stage if a discovery was made, and it appeared that any oil and gas found could be mothballed, and delivered at a time of optimum market prices. The government's effective ceding of all control over large tracts of offshore territory had a precedent. After Ireland's accession to the EU, then known as the European Economic Community, access to biologically rich Irish waters on the continental shelf was effectively ceded in return for concessions to the agricultural sector. By 1995, the World Bank would note that Ireland was listed among the top seven countries in the world with 'very favourable' terms for exploration.[9]

British company Enterprise Oil had established a base in Ireland in 1984. It owed its origins to privatization of the British government's stake in the North Sea. In the mid-1990s, a US company called Santa Fe had applied for a block off the Irish west coast which Enterprise subsequently took a 60 per cent stake in. Corrib was the second well in a slow drilling sequence, and was 'almost missed' as the weather window was closing in. Enterprise geophysicist Nick Dancer christened it thus; all his prospects bore the names of Irish lakes and rivers. Under a complex series of deals, Statoil then acquired shares – initially 15 per cent in a deal with Enterprise, and then up to 42.5 per cent when it acquired Norwegian listed company Saga, which had in turn bought out Santa Fe.

Enterprise Energy Ireland (EEI), formally incorporated with a registered address in the Bahamas, moved to appraisal stage of the Corrib field, but there were delays in moving a rig out to conduct this – due in part to a reported reluctance by the company to employ Irish staff offshore. Irish men had been employed on Marathon projects and had worked on rigs in the Irish and Celtic Seas, and all over the world, and the practice of using some non-skilled 'roustabouts' from Foynes in County Limerick had been introduced by the Total energy company in 1972 when it was drilling in Irish waters.

However, staff hired by Enterprise via the SIPTU trade union in Foynes for the Petrolia rig on Corrib in 1996 proved problematical: according to the company, the union had insisted that everyone on a reserve list should be employed, which the company felt could cause problems for health and safety. Some union representatives suspected they were not wanted, in case they might pass back to shore sensitive information relating to the size of a well drilled. When the company mobilized a second rig in 1998, it decided not to offer any jobs at Foynes, resulting in a picket at the County Limerick port by dock workers in sympathy with their offshore colleagues. At this stage, Enterprise was headed by a straight-talking Scot, John McGoldrick. Fianna Fáil was back in government, and the then minister for natural resources, Dr Michael Woods, cited EU regulations on free movement of labour to defend Enterprise's decision to move its supply base to Ayr in Scotland.

In late July 1998, McGoldrick was a guest in the Fianna Fáil fundraising tent at the Galway races. The concept of such a 'tent' or marquee had been initiated by the party's fundraiser, Des Richardson, four years before, and it had proved to be a very attractive informal lobbying opportunity. With McGoldrick was one of his public relations advisers. They were in ebullient form at the Galway race week, but it was an ebullience that was to spill out into the streets in a reported confrontation with several former offshore workers.

While a first well had been marginal, tests on a second in the northern element of the Corrib field had established gas flow rates of up to 60 million standard cubic feet a day. It was regarded as phenomenal – one of the biggest well tests that Enterprise had ever completed. Reserves were estimated to be about a trillion cubic feet – two-thirds the size of Marathon's Kinsale Head field. However, that estimate was to fall slightly, to about 850 billion cubic feet, in a subsequent three-dimensional seismic survey and further well drilling.

The company's chief executive, Pierre Jungels, was quoted as stating that a further well would be drilled that summer before proceeding with full-scale commercial development. The partners would seek project sanction in December, Jungels said, 'where there will be a clearer picture of the engineering and commercial aspects of the project'. He also confirmed that one Brian Ó Catháin would succeed John McGoldrick, then head of the Dublin subsidiary, who was leaving to manage Enterprise Oil's North American assets in the Gulf of Mexico.

Ó Catháin, a Belfast-born geologist, was based in London with Enterprise; when he had examined logs for Corrib, he had wondered if this might provide his ticket home. And so it did. Educated at St Malachy's College, Ó Catháin was a fluent Irish speaker who had studied at Bristol University and joined Shell in 1984, where he was trained as a drilling and petroleum engineer. He worked for the multinational in the North Sea, New Zealand and Holland, leaving in 1990 to develop his business expertise in the oil and gas industry in London where he was recruited by Enterprise.

Significantly, the western seaboard had no supply of natural gas; Kinsale was feeding the greater Dublin market via a pipeline built from Cork to the east. With no transmission link to the west, the Corrib partners initially thought they might sell the gas directly to a power station. McGoldrick had done much of the necessary lobbying with his team for political support in advance of formal applications for state approval. His company had held presentations locally, including one in McGrath's pub in the small village of Pollathomas in north Mayo. It had also identified, and finally secured, an onshore site for a 'reception' or processing terminal, where methane and water would be separated, and from which gas would be sold.

Locations for the terminal had been examined up and down the coast, from Clare to Donegal. It was 'always' going to be onshore, according to the company.

An accident just over a decade before had marked a watershed for the oil and gas industry in Europe, which had already been plagued by a number of safety incidents. In early July 1988, some 165 people working on the Piper Alpha rig 120 miles (193 kilometres) north-east of Aberdeen in the North Sea had lost their lives in an explosion; two rescuers also died, bringing the toll to 167. The platform, dating back to 1976, had been deployed by Occidental Petroleum (Caledonia) Ltd for oil drilling and was then converted to gas production – breaking some key safety rules by merging some sensitive areas. One of just over sixty survivors described to this author, who reported for *The Irish Times* from Aberdeen directly after the series of explosions, how the gas leak screamed 'like a banshee wail' before its full impact was felt under living quarters.

An inquiry established by the British government in November 1988 determined that the initial condensate leak was caused by maintenance work being conducted simultaneously on a pump and related safety valve. Occidental was found guilty of inadequate maintenance and safety procedures, but faced no criminal charges.

The spotlight was now on safety standards generally offshore. In April 1989, five workers were killed when an oil platform leased from a California oil services company exploded off the Nigerian coast. Shell and its partner, Nigerian National Petroleum Corporation, disagreed on how reporting of the incident was to be handled. In the same month, but 'half a world away', an explosion occurred on Shell's Cormorant Alpha rig in the North Sea as it was being evacuated due to a gas leak. It was a fortunate 'near miss'.

One of the recommendations in the subsequent inquiry into

Piper Alpha advised that enforcement of safety should move from the Department of Energy to the British Health and Safety Executive (HSE) to avoid a conflict of interest.

At the outset, there was considerable concern in Mayo that the county would not benefit from Corrib – not only with provision of gas, but also provision of jobs. A series of articles in the *Western People* by Mike Cunningham, ex-director of the Norwegian subsidiary, Statoil Exploration (Ireland) Ltd, raised some serious questions about benefits to the county and to the state. Oil and gas licensing terms should be reviewed, he said. Cunningham noted that some two thousand 'downstream' jobs had been generated by the Kinsale gas project. On 18 August 1999, he wrote,

> The reticence of the industry to communicate and discuss their plans makes it particularly difficult to ascertain what is planned for Mayo's Corrib gas. Leaving aside commercial constraints, this tight-lipped stance in respect of the development of Corrib gas and, indeed, all planned exploration activities off our coastline is in stark contrast to the manner in which the Norwegian and British oil and gas industry developed. Norwegian communities along their western seaboard demanded that they be kept informed at all stages from licence awards, seismic operations, through to exploration and development. Their demands were accommodated . . .

Cunningham's views were echoed by politicians, including Fine Gael Mayo TD Jim Higgins, who was concerned about the benefits to the state and the county: the Irish taxpayer could end up 'buying back our natural resources without any tangible commitment to an agreed number of jobs either offshore or onshore'.[10] There was even concern that the gas would not be pumped ashore to Mayo at all – a concern that led the *Western People* to mount a public campaign in November 1999 to ensure that the 'Mayo gas find benefits the people of the region in a tangible way'.

Initial mapping for a pipeline linking the wellhead offshore to the coast identified a number of 'preferred' routes, with several possible landing points. The map showed the gas linking into the national network of transmission pipelines, running to Dublin and on to the Irish Sea interconnector to Britain. It was only latterly that Broadhaven Bay was selected, with a landfall at Glengad.

Successful negotiations by Enterprise with the semi-state forestry company Coillte secured the substantial inland location at Ballinaboy Bridge, where an agricultural research station had undertaken work in the 1950s. The site was selected by New Zealander Kelvin Wallace, an Enterprise engineer in the spring–summer of 1999, according to Brian Ó Catháin. From a geotechnical point of view it was regarded as ideal, as its long use for research into peatlands and grass cultivation meant it was no longer 'virgin bog'. From an environmental point of view, it would have a modest visual impact in a sparsely populated area. It was not in a special area of conservation, and was a single parcel of land, reducing legal complexities. However, the eventual landfall and part of the pipeline route would run through fragile habitats and private land.

Ó Catháin arrived in Dublin in the spring of 2000 and he was delighted to be home, as it was the first time he had ever worked in Ireland. Enterprise had commissioned several consultancy studies to support Corrib's project sanction by various state bodies, and a government petroleum lease would be secured the following year on 15 November 2001. Significantly, a European Union directive on gas aimed to open up the market to competition; this could simplify authorization of the production pipeline, which would run some nine kilometres inland through private property. Significantly, it also emerged that the semi-state company Bord Gáis intended to extend to the west.

In August 2000, a report by Christy Loftus in the *Western People* suggested for the first time that all was not as it seemed in relation to promises of benefits for north Mayo and the wider region. It referred to the findings of a consultancy study commissioned by the government from a North American company, the Brattle Group.

The Brattle report was 'generally positive' towards the Corrib project, but ruled out the possibility that Mayo towns would benefit – in spite of an assertion to this effect by Minister for the Marine and Natural Resources Frank Fahey. In Loftus's words, it also 'knocked on the head' a proposal to use gas to replace the turf-fired power station at Bellacorick in north Mayo, which was due to be decommissioned by the ESB in 2004 with the loss of three hundred jobs.

It did suggest that entry charges on Corrib gas developers to the Bord Gáis Éireann ring main should be designed to recover in full the cost of the 'spur' from the ring main to the proposed Glengad landfall. It also suggested that the spur would offer 'other benefits', such as bringing gas to 'some local populations', but affirmed that the gas company should develop estimates of gas consumption by those towns that did not currently have a supply.

Enterprise's public affairs manager, Rosemary Steen, gave the report a 'cautious' welcome, Loftus reported, but said that her company was concerned that Brattle was 'sending out a number of wrong signals' in terms of access to the national network, especially in areas such as the north-west, which 'did not currently have access' to gas. Mayo County Council chairwoman Annie May Reape had not seen the report, but told the newspaper that if it recommended that the Corrib find would not benefit Mayo, then she would urge the government to reject the report 'out of hand'. 'The issue of how Mayo benefits from the Corrib gas will be top of my agenda while I am *cathaoirleach* [chair],' Councillor Reape said. If this required 'changes to legislation, then so be it'.[11]

The hazel saplings may seem a little fragile, but their roots in Dooncarton, otherwise known as Glengad Hill or Caipín, above the small coastal village of Pollathomas are sturdy. Patches of scree mark the impact of landslide. There are some barbed-wire intervals to navigate. As the undergrowth surrenders to furze, the summit and an aircraft radar dome seem deceptively close. Turning around to catch a breath, one is struck by the vista opening out – to the north, the headland of Glinsk and two knolls, Tanaí Mór and Tanaí Beag , or the 'Dolly Partons', as they are locally known. To the west, the silver and indigo Atlantic running into Broadhaven Bay and the winding Sruwaddacon estuary (*Sruth Mhada Conn* – the long narrow channel or current, named after the dog of Conn). In the stillness, you can hear your own breath; a hammer tapping on a roof across the water echoes for miles.

Directly below is the 'ferry', where the estuary narrows close to the village of Rossport. The ferry pier was built by the Congested Districts Board in 1891 for small rowing boats, navigating what is regarded as the second strongest current in Ireland. To the north-east is the Glenamoy river, meandering like a creek into the tide. Beyond it again, the Muing na Bó river, named after the marsh of the cows, rising to the east near Glinsk and wending its way to the sea by Sruwaddacon. Spanning it, Annie Brady's bridge, named after the wife of Sir Thomas Brady, inspector of fisheries, who was so grateful for the help local people gave her when her horse and carriage separated in a flood that she promised to pay for a structure – though she died shortly after, her heartbroken husband kept her promise.

Rossport, Pollathomas and Glengad are constituent villages of an area of Erris known as Dún Chaocháin, within the parish of Cill Chomáin or Kilcommon, comprising about two thousand people. The Irish-speaking area is renowned for its music and '*seanchas*'; the folk-lore associated with the coastline drew the late Scottish writer David Thomson to the area, recording the stories of 'Michael the Ferry' in *The People of the Sea*. And the north Mayo coastline is a stunning seascape, bordered by high cliffs and sea stacks colonized by flora, fauna and sea birds in largely undisturbed habitats.

The endangered corncrake is still found here, along with the rare red-necked phalarope, while the Inishkea islands, where the Norwegians established a whaling station in the early twentieth century, and Broadhaven Bay are home to wintering barnacle and Brent geese.

It was along this remote sea border that one of the last golden eagles naturally resident in Ireland in the last century nested and hunted. Many of the breeding birds are now classified as rare and protected species, such as the Leache's petrel, with its dramatic Latin name *Oceanodroma leucorhoa*, selecting the Stags of Broadhaven, the four imposing conical peaks, as its only nesting site in Ireland, and the little egret, becoming a recent overwinter visitor in Broadhaven's tidal inlet at Sruwaddacon.

The barony of Erris, as it is still known, takes its name from the Irish '*iar ros*', or western promontory. One of its most famous legends is that of the Children of Lir, turned to swans by Aoife, third wife of Lir of the Tuatha Dé Danann. When the four swans eventually arrived in north Mayo, they lived their lonely lives for three centuries on Inis Glora – from the Irish 'Inis Gluaire' or the Island of Purity – until a disciple of St Patrick, Caemhoc, came out by sea and rang the bell, which freed them.[12] A monastery on Inis Glora, now in ruin, is said to have been founded in the sixth century by St Brendan the Navigator, the monk reputed to have discovered America. The thirteenth-century Book of Ballymote said that the bodies of the dead buried on the island 'do not rot'.

It was on the mainland close to the outcrop where the swans found their redemption that the evidence of the oldest known farming in Ireland, by Neolithic people some five thousand years ago, was unearthed. The excavations took place in Behy, close to Belderg, an area now known as the Céide fields at Ballycastle. The archaeologist principally responsible, Dr Seamus Caulfield, was the son of a locally based schoolteacher, who had long been fascinated by the area's heritage. The significance of the Belderg discovery was such that it inspired Nobel laureate Seamus Heaney to write of how:

When he stripped off blanket bog
The soft-piled centuries

Fell open like a glib;
There were the first plough-marks,
The stone-age fields, the tomb
Corbelled, turfed and chambered,
Floored with dry turf-coomb.

A landscape fossilized,
Its stone-wall patternings
Repeated before our eyes
In the stone walls of Mayo.[13]

The community extended far west, leaving its traces with court tombs in Rossport. The communities farmed on soil above and around some of the oldest rocks in the two islands, equivalent in age to the Scottish Outer Hebrides, with shales and gneisses formed by pre-Cambrian activity 600 million years past.[14] The last ice age never reached this part of Connacht, and the late naturalist Professor Frank Mitchell remarked that one would require many hats for full appreciation of the landscape – an archaeological hat, for instance, to study the Bronze Age activity, a quaternary cap for inshore marine life, and a geomorphological one for exploration of the cliffs carved by wave erosion running continuously from Belderg to Benwee Head.

One of the best known sea stacks is at Dún Briste, or 'broken fort', close to Downpatrick Head some three miles north of Ballycastle. The *Annals of the Four Masters* has indicated that it was not always a free-standing structure, and that in 1393 a land bridge or arch there collapsed during a great storm. People were rescued, it was said, by *'caolaighe cnáibe'* or ropes made of hemp. In 1991, it was climbed by English mountaineer Mick Fowler and American Steve Sustad.

Local legend had it that the stack was once home to a pagan chieftain named Crom Dubh, who tried to throw the patron saint into his everlasting furnace. However, Patrick cast a stone carved with a cross, creating a blowhole named 'Poll a Sean Tine' or the Hole of the Ancient Fire. Patrick then struck the ground with his crozier, the link with the mainland broke, and Crom Dubh was left to his fate where he was devoured by midges.

Mayo or '*muigh eó*', the Plain of the Yew, was named as a single entity after the sacred plant by Saxon monks; but the northern fastness, bounded by the sea and by the Ox and Nephin mountains, would always maintain its distinct identity where it was once known as the kingdom of the Uí Fiachrach, a powerful dynasty that once controlled large parts of Sligo and Mayo, including the kingship of Connacht during the seventh and eighth centuries.

When the Normans arrived in Ireland in the early twelfth century, they left their mark, with the Norman Barrett family granting township status to Dún Chaocháin. The Barrett title was transferred during Cromwellian times, when Protestant settlers, and refugees who had been dispossessed of their land elsewhere, began making Erris their home. It is no accident that the motto for another Mayo dynasty, the O'Malley seafarers, proclaimed its dominance as *terra marique potens* or 'powerful by land and sea'. During pirate queen Grace O'Malley's time in the sixteenth century, the fishing grounds off the Mayo coast were particularly lucrative. Erris was drawn into, and suffered the consequences of, opposition to the English Crown during the dying days of the Gaelic clan system.

Queen Elizabeth I's appointee, Sir Richard Bingham, resumed his governorship of Connacht after the ill-fated wreck of much of the Spanish Armada fleet in 1588. He ordered his deputy to seek out and kill any Spanish survivors, and in a subsequent rebellion by the Bourke clan and confederates, Bingham marched into Erris to exact retribution.[15]

Kilcummin strand, near Benwee Head and close to the north Mayo village of Killala, was the accidental location for a significant event in Irish history, the 1798 insurrection. It wasn't the chosen landing point for French forces under General Humbert, who had intended to sail into Lough Swilly, County Donegal, with three ships carrying about a thousand French troops in support of a rebellion planned by the United Irishmen. A sea cave near Downpatrick Head was the location for one tragic episode associated with that year. Some twenty-five local men who were hiding from the English 'Redcoats' in the aftermath of the '98 rebellion were lowered by their womenfolk on to a ledge at the foot of the cave through a blowhole, but were caught by the tide and drowned.

Biologically rich marine life, richer than the banks of Newfoundland, sustained fishing and farming communities in isolated areas like Porturlin, Carrowteigue, Pollathomas, almost sixty miles from the centre of county administration in Castlebar, and twenty-four miles from the nearest county council office in Belmullet, the town marking the isthmus across to the Mullet peninsula. Further inland, however, acres of blanket bog, formed by woodland clearance, provided the main employment for those who could stay in the first decades of post-independence Ireland. They worked with the state company Bord na Móna on peat cultivation, or in the associated peat-powered electricity generation station at Bellacorick.

The peat mantle may have provided fuel and feeding for resourceful sheep, and a habitat for unusual plant species, such as the Mediterranean heather (*Erica erigena*), but it was never able to sustain extensive food production or protect its population from some of the worst effects of landlordism and famine in the nineteenth century. Almost 90 per cent of Mayo's population is said to have been dependent on the potato when the blighting fungus, *Phytopthora infestans*, began destroying crops from August 1845.

Within two years, workhouses built earlier that decade to relieve poverty were unable to cope with numbers. Small tenant farmers requiring relief for their families were forced to quit their land if they owned more than a quarter of an acre. This was dispossession in exchange for food.

Between 1841 and 1851, Mayo's population dropped dramatically from almost 389,000 to just over 274,000, according to census figures, and the impact on Ireland's population overall has been well recorded, with approximately one million people dying and a million more emigrating, some of whom never made it across the Atlantic, on so-called 'famine' or 'coffin' ships. The soup kitchen set up in 1848 by local Quaker landlords, the Bournes at Rossport House, was one of many across a county that was severely hit by further catastrophe – the subsequent 'great clearances' – when almost a quarter of a million people were evicted from the land due to rent arrears between 1849 and 1854.

The house at Rossport had been built around 1820, and the Máistir Mór, head of their household during the famine, was regarded as a 'good man in bad times'.

Among the tens of thousands of families affected were the Davitts of Straide in County Mayo. Michael, second child of Martin and Catherine Davitt, was born during the early part of the great famine in 1846, and was four and a half years old when the landlord's agents forced the door of their house with a battering ram, ordered the family out, chucked their furniture on to the road and set fire to the thatched dwelling. The family travelled to a textile town in Lancashire where Martin Davitt hoped to get work. Michael was nine when he sought to supplement the family income in a local cotton mill. He was eleven when his right arm was badly injured by a machine, and he had to be chloroformed for amputation.

Michael Davitt joined the Irish Republican Brotherhood (IRB), or Fenians, in Haslingden in 1865, with his parents' knowledge, and was to become its organizing secretary and arms agent for England

and Scotland. Shortly after he had persuaded his parents to emigrate to North America, he was arrested for his IRB activities, charged with treason and felony, and imprisoned for fifteen years at the age of twenty-four. He spent much of his sentence in one of the toughest jail environments, Dartmoor in Devon. He served seven years. During a return to Ireland, he was invited back to Mayo by James Daly, a large farmer, hotelier, and proprietor and editor of the *Connaught Telegraph*. The poverty witnessed by Davitt, his continued involvement in the IRB and his meeting with Fenian leader John Devoy on a trip to North America, was to lead to a 'new departure' in Irish politics involving land law reform.

Davitt's Land League was formally inaugurated in James Daly's hotel, subsequently the Imperial on Castlebar's Mall, on 16 August 1879. The league became a shared platform for nationalists, moderates and revolutionaries. The movement contributed a new word, 'boycott', to the lexicon, which transferred into many languages, after the ostracization of a land agent, Captain Charles Boycott, in the Lough Mask area of south Mayo.

However, Gladstone's legislation, which led to a return of land to tenants who then became landowners – a group later described as 'twenty-acre capitalists' – was not what Davitt had envisaged. As biographer Bernard O'Hara has noted, his radical views on social policies had 'little effect in the new state', which his land reform movement had helped to forge.[16]

The race memory of poverty and death was such in Erris – described by Victorian naturalist Robert Lloyd Praeger as 'inspiriting' but also 'the wildest loneliest stretch of country to be found in all Ireland' – that it was a magnet for South Dubliner and author J. M. Synge. He had travelled the west coast for a series of articles in 1905 for the *Manchester Guardian*, which analysed the progress of the socio-economic schemes run by the former Congested Districts Board. Erris yielded one of his best-known works, *The Playboy of the Western World*.

Staying in the Royal Hotel in Belmullet, Synge heard of, and wrote a poem about, Danny, a rate collector, a less than popular member of the community. Incorporated in the plot of the subsequent play was the story of Lynchehaun, in hiding on Achill Island to the south for committing a terrible deed. He was smuggled away on a ferry, which used to run between Doohoma in Erris and Achill, and was taken by many emigrants across the Atlantic.

Synge's text is rich in references to Erris, be it the strand at Doolough where horse racing still takes place, the 'Owen' river, drawn from the Owenmore, or the crossroads at Gaoth Sáile (Geesala) where Pegeen Mike's father Michael James was due to meet Philly Cullen and a couple of friends en route to Kate Cassidy's wake. The author said he felt the need to pay tribute to the 'folk-imagination of these fine people', who had not shut their lips on poetry, even though they had lost so much else.[17]

The famous late eighteenth, early nineteenth-century poet and last of the 'wandering bards' Antoine Ó Raifteirí had been born in Kiltimagh and one of his most famous verses, '*Cill Aodáin*', spoke of his longing to be in the 'centre of County Mayo'. Erris's place in literary history was rooted in Irish, rather than English, as recorded by scholar Tomás A Búrc, who undertook much work on its Irish annalists for the state's Folklore Commission, and Professor Séamus Ó Catháin, who worked with residents on stories and place-names and ensured that the collections of Micheál Ó Corrduibh and Máirtín Ó Conghaile, among others, were housed in an archive in University College Dublin (UCD).[18]

It was this rich seam that drew an east Limerick graduate, Micheál Ó Seighin, to the area in June 1962, when the Irish language organization Gael Linn was finding it difficult to get Irish-speaking teachers for a local secondary school. Ó Seighin was the youngest born of a family of nine. His parents were married on the day that civil war broke out, and an aunt was involved in Cumann na mBan, the women's organization founded in Dublin in 1913 to support moves towards an independent republic. Although reared with only school Irish, Ó Seighin majored in history and Irish, with English as a subsidiary, at University College Galway (now NUI Galway), where he became committed to the first language.

He acted in the national Irish-language theatre An Taibhdhearc in Galway, was an enthusiastic participant in the university's Literary and Debating Society, and became head of the Cumann Éigse and president of the students' union. In his last year at college, he spoke at a Union of Students in Ireland conference in Galway attended by a fellow Limerick man, Gerry Collins, who was past president of University College Dublin students' union. Collins approached him afterwards to congratulate him on his address, telling him it was 'great, but to ease off a bit on the Irish. "Don't overdo it!"' Ó Seighin remembers Collins, a future Fianna Fáil minister, advising him.[19]

Ó Seighin was a member of An Comhchaidreamh, a group of Irish-speaking graduates who had earlier established Gael Linn, so it seemed only natural that he would offer to teach at a new Irish-speaking secondary school in Rossport. As he noted afterwards, if he had been living in Finland he would have done the same – supporting the Finns at a time of Swedish cultural domination.

The post was for two years, but he never left, having met a young woman named Caitlín Ní Dhochartaigh who was from the village of Ceathrú Thaidhg (Carrowteigue), and was researching local phonetics with a priest, an tAthair Colmán Ó hÚalláchain. On his first evening in Rossport, Ó Seighin had stayed at the house of Anthony John Ó Gionnáin, whom he remembers as a 'wonderful, remarkable' individual. 'He took me for a walk, and I didn't know why at the time,' Ó Seighin recalled years later. 'He took me along to the shore and around the sweep that is Rossport and back along the other side. He was pointing out the houses to me and saying, "My uncle is there," or "My first cousin is there" and "That man up there is my relative." I was young but by the time I had ended the walk I had known exactly what Anthony was doing. What Anthony was saying to me was, keep your mouth shut until you know where you're going and everything will be okay . . .'[20]

The young teacher, not used to keeping quiet, was also to learn that this area, Dún Chaocháin in the sprawling parish of Cill Chomáin (Kilcommon), looked seawards, rather than landwards, and that there were subtle but distinct differences between areas. Sometimes it related to use of the Irish language, but the distinctions also reflected different populations, and 'different colonizations'.

At first, the landscape seemed bleak and unforgiving, reflecting a history of suffering within the post-famine sub-consciousness of those still living there, but he learned to love the contrasts afforded by the changing western light. He learned to know the area through its Irish placenames, which recorded indicators, events, milestones of centuries before. In a changing world, such placenames had the effect of bonding people together, in his view.

The Ó Seighins settled, reared a family, became involved in drama, *sean nós* singing, traditional music in the community. Micheál's political sympathies would have been with Sinn Féin, in an area dominated by support for either Fianna Fáil or Fine Gael. In 1995, the year before gas was discovered off the coast, the couple were instrumental in the formation of a community development co-operative, Dún Chaocháin Teoranta. Its aim was to co-ordinate development, integrate material and human resources, and one of its first publications was a survey of the needs of its young residents.

A youth plan was initiated to provide sport, arts, crafts and other social activities. It received funding from the Allied Irish Bank's Better Ireland programme and state agencies such as Údarás na Gaeltachta. The co-op became a model for others around the country, through its range of community and language-based projects and classes, training courses, translation and childcare, and adoption of information technology. Some fifteen hundred placenames were recorded, walking routes developed, guidebooks published, along with bilingual CD-ROMs. It was sustainable tourism long before the phrase came into fashion.

There were other initiatives. In June 1993, artists and sculptors signed up to a *meitheal*, or community effort, based in Ballycastle and Belmullet, while students at St Patrick's secondary school in Lacken worked with students from Dun Laoghaire College of Art and Design. Placenames in both Irish and English and flora and fauna were recorded, and an existing ogham stone was incorporated in the route catalogued by Alan Lonergan. Dealbhóireacht 5000 Teoranta, or Sculpture 5000, was established to manage the project with the support of state agencies, chaired by local authority official Peter Hynes. Its directors represented a number of groups and private organizations, including the Sculptors Society of Ireland, Mayo 5000 and Meitheal Mhaigh Eo, the latter chaired by Ó Seighin. In July 2002, a group of US architectural students and architect Travis Price placed a number of works on the sculpture trail, named Tír Sáile, where the sea and the land meet. The expedition was recorded by *National Geographic* magazine.

In September 2005, there was an empty chair at the Dún Chaocháin Teo annual general meeting in the Seanscoil Ceathrú Thaidhg. The *Western People* noted that it was the first time since the foundation of the organization just over ten years before that Micheál Ó Seighin 'could not be in attendance'. He was in prison. 'Micheál's contribution to the development of the local community over the last twenty years, as well as his love of the Irish language, drama, *sean nós* singing and traditional music was widely acknowledged by those present,' the newspaper reported. 'It is his high regard for the area which prompted him to take a stand against the proposed unprecedented gas pipeline and its potential damage to the environment and way of life in this close-knit community.'[21]

2

THE WAKE

From a developer's perspective, Broadhaven Bay seemed ideal for a gas pipeline landfall. A small and scattered population lived in an area that successive central administrations had almost given up on. The proximity of past to present, which made Erris such a distinctively independent region, was not a factor in the business equation. A strong influencing factor was a potential gas terminal site inland, owned by Coillte, the semi-state company that had been established to manage state forests.

Two years before, Enterprise and its partners had employed consultants to survey the Connacht coastline to identify the most suitable landfall for a pipeline running in from the gas field. A shortlist included Killala and Blacksod bays, Emlagh point (west of Westport), Liscannor and Doughmore bays in County Clare. However, final selection would be determined by a processing site, as the unprocessed gas could only be piped a certain distance. Much later, this cut the shortlist down to Killala and Broadhaven bays in north Mayo, and four sites were located within these coastal areas – Sruwaddacon estuary, Bunatrahir Bay, Ross Point near Killala, and Rathlee Head. Killala would require a 145-kilometre link, while the

Broadhaven Bay landfall point was around 80 kilometres from the gas field.

The inland location for the terminal would require a deal with Coillte. It had secured title to a 165-hectare area of forest at Ballinaboy, in what appeared to be a largely unpopulated area close to an animal-testing laboratory, the Charles River Bio Laboratory, at Glenamoy. The land had been registered in two parts – one part under the old Department of Lands, as An Foras Talúntais had run a research unit there on the bog in the 1950s and 1960s, and the other in the name of a Mrs McDonnell. During the summer of 1999, Coillte had placed a simple notice in the *Western People* announcing that it was registering this second part under its ownership. By November 1999, it was all registered in Coillte's name.

The uncomplicated ownership was very attractive. Too big for Corrib gas needs alone, the site had potential for expansion if other resource finds should be made off the west coast. Enterprise engaged consultants to prepare an environmental impact statement (EIS), and the developers began negotiating with Coillte to purchase it, subject to planning permission. Details of the deal, signed in July 2001 for a reported sum of just over €2.7 million, were not made public at the time.

Caitlín and Micheál Ó Seighin had first heard about this pipeline in 2000. There had been reports of a gas discovery off Mayo the year before. In August 1999, the *Mayo News* wondered if the county would benefit from the 'gas bonanza', while rival newspaper the *Western People* initiated a readers' campaign to ensure that gas was piped ashore to Mayo.

Caitlín Uí Sheighin recalled that the first notification was an announcement in April 2000 in the parish newsletter 'that we were all going to be rich. I said to Micheál, "*Bhuel, dá mbéadh aon mhaith ann, ní thiocfadh sé anseo* [if there's any good in it, it won't come here]. That was my reaction. At the time I thought it would be done like Kinsale. So, fine, I had no problem in the world with it.'[1]

Some time after that, the Bishop of Killala, Dr Thomas Finnegan, and the parish priest of Kilcommon, Father Declan Caulfield, were flown by helicopter to bless the rig – 'dressed up in the benediction robes, with a television crew', Micheál Ó Seighin recalled. 'This was a source of huge scandal for people,' he said. 'People thought straight away, What are they up to? This is a reversion to an abuse of power by the clergy. People of my age had seen this before . . .'

At the time, Ó Seighin was still teaching, and told his geography students that he had no interest in it. 'I said to them that either this crowd don't know what they're doing or they know more than I know. It seemed that they were intending to sink this huge pipeline up the side of the hill which everyone knows will draw all the water into the drain and will take the side of the mountain down.'

At that stage, the proposed route onshore from Broadhaven Bay was via the fishing village of Pollathomas below Dooncarton Hill and overlooking Sruwaddacon estuary, with a landfall in the neighbouring village of Glengad. The route was subsequently changed by the developers to run across the northern estuary. The push by the Church was followed up by the initiation of an organization called One Voice for Erris, which held meetings in the parish priest's house. It was supported by the Council for the West, which had been set up by the western bishops to lobby for development for the western seaboard after successive reports had highlighted economic disparities between east and west. The invitees to the One Voice meetings tended to be people with business interests, the Ó Seighins recall, but several residents, including Inver schoolteacher Maura Harrington and Willie and Mary Corduff of Rossport, kept up attendance even when they felt they were not welcome.

They had reason to be so. North Mayo had experienced more than its fair share of fantastical promises, particularly in the years leading up to the 2004 closure of a major employer, the Bellacorick power station. During what was known as the 'Lemass era', when

Seán Lemass was Taoiseach from 1959 to 1966, state industries were developed to harness resources like peat, and to provide air services and shipping. A power-generating station fuelled by milled peat drawn off the top of the blanket bog began pumping power from Erris into the national grid in 1962.

Some of the large acreage of bogland acquired by the state in Erris was also earmarked for other projects. In the early 1960s a company named Min Fhéair Teo was established in Muingmore, Geesala, to produce grass meal and cubes for cattle feed, and there was a subsequent venture to make briquettes out of peat, which ran into problems due to moisture content. The backers of this latter project were Norwegian company Norsk Hydro. Ó Seighin heard from a guesthouse owner, who had accommodated some of the Norsk Hydro management, that they had withdrawn finally because they 'had not got the contract for the pipeline'. The man didn't know what they were referring to. Ó Seighin felt that he did.

Bellacorick in Erris became one of a network of peat-powered generating stations. Ironically, this is perceived today as one of Ireland's most environmentally destructive technologies. The imposing cooling tower, all three hundred feet of it, dominated the unyielding landscape for several decades, while the hamlet where it was located became known for its 'musical bridge' – visitors who took a fist-sized stone and ran it along the capstones would hear a series of notes. When the tower was demolished in a controlled explosion in October 2007, there were some lumps in throats among an estimated three thousand who travelled to watch.

'Santa's workshop' was how one resident remembered his mother describing it, while for another it was a home for elves. 'I watched it go with a tear in my eye,' another resident told the *Irish Independent*. 'We lived in the shadow of the station. It kept bread and butter on our tables in the sixties and seventies when there was hardly any work available locally and emigration seemed the only option.'[2] Even before its closure, however, the landmark had been replaced by slightly smaller, but similarly active, wind turbines, as Bord na Móna developed its first wind farm.

There were to be more job promises, and more job losses, in the area, even during the economic boom of the late 1990s, which appeared to pass Erris by. In mid-1997 Asahi, a Japanese-run synthetic-fibre plant established in Killala, announced its closure with the loss of 315 jobs. Asahi never achieved its planned targets, mainly because its commissioning coincided with the oil crisis of the early 1970s, which upset the market for the raw material, and because of competition from spinning mills thereafter in Taiwan and China. The knock-on effect was significant. The state rail company, Iarnród Éireann, had transported its raw materials along the Ballina rail link. The British multinational Marks & Spencer was reported to have relied on it for some 80 per cent of its thread requirements.

In August 2000, Bishop of Killala Dr Thomas Finnegan travelled to Belmullet to sympathize with the 114 staff of a lingerie factory, run by Warner's, which had shut down. At one point in its ten-year history, the plant had employed almost the entire population of the town, including returned emigrants. Just ten years before, the bishop had been on hand at the same plant in Belmullet industrial estate when the Babygro factory closed. 'The way we see it is that if the whole of Erris broke off the mainland of Ireland and fell into the Atlantic, there would be a sigh of relief in Dublin,' one employee told the press.[3]

It appeared that the state's heavy dependence on foreign direct investment did not sit well with Erris at a time when Ireland was in the middle of an economic boom. Cranes dominated the Dublin land-scape, as construction of new retail space to feed demand reached fever pitch, and in 1998 the securing of the historic Good Friday agreement in the North also augured well for the future. The fact that the economy's powerhouse had never been felt in any sustained way in north Mayo influenced the initial public enthusiasm for the proposed gas project. It was only in early autumn of 2000 that Micheál Ó Seighin began to get more concerned.

He spotted a small report in newspapers about financing and construction by state company Bord Gáis of a pipeline from north Mayo to Galway. This was a result of negotiations with Enterprise, which had initially intended to build a link to the grid – but Bord Gáis was reluctant to allow a private operator into its territory. The announcement was made by Taoiseach Bertie Ahern on 4 October, and in the same week his marine minister, Frank Fahey, declared that a number of towns in the west would benefit directly.

The minister had to withdraw his initial press release when it emerged that not all of the towns he had named were within a sniff of securing a supply, given their population and location. Planning permission had not even been submitted for the terminal or refinery yet; and nothing of any significance relating to the project had been approved. Ó Seighin's antennae were up. There had been no application for planning permission for 'anything', as he recalled. 'Yet the deal with Bord Gáis was up and running.'[4]

There were also reports that the British car and aviation engine manufacturer Rolls-Royce had plans to build a 64-kilowatt power station at Bellacorick when the peat station finally closed. Ó Seighin felt this was 'ridiculous', as 'no developers in their right mind' would build a power station in such an isolated area. In fact, nothing ever transpired as a result of the Rolls-Royce 'plans', which were later described as a 'fairytale' by Harry Barrett of the Labour Party in Mayo.

Considerable groundwork had been done on the Corrib project by the developers. During the summer of 2000, Enterprise sought to win over local support for it, with promises of up to five hundred jobs in an area with unemployment estimated as high as 30 per cent. As the second largest inward investment of any project in Ireland for some years, this was the west's equivalent of Intel, the computer multinational in County Kildare, the company said.

Fianna Fáil councillor and secondary school teacher Paraic Cosgrove was among those who welcomed the plan. The forestry at

Ballinaboy was just ten kilometres from his home, a fine bungalow built on a height just outside Bangor Erris. Cosgrove was one of a family of eleven, five of whom did not survive. He knew the reality of economic hardship. Most of his siblings emigrated from Bangor Erris at a time when it was the only option. He studied at University College Cork (UCC), spending summers on temporary jobs, such as welder's assistant on the Cork–Dublin gas pipeline, at Tynagh mines in County Galway and in the agricultural testing centre at Ballinaboy.

Cosgrove secured his first teaching job in Geesala in 1968, and moved in 1981 to Belmullet Vocational Educational College, where he taught science, mathematics, French, Irish, 'even religion', as he remembered. In 1997, he retired on health grounds, and two years later he lost his seat on Mayo County Council – saying afterwards that 'In politics, you are only as good as your last favour.' As far as he was concerned, the gas project deserved every bit of local support it could get.

Several kilometres down the road from Cosgrove's in Ballinaboy, Jacinta Healy wasn't so sure. She still remembers the knock on the door. His name was Dukes – Brian Barry Dukes – and he was involved in site investigations for a terminal that would be built on Coillte land in the locality, he told her. She had noticed activity around the forestry weeks before, but hadn't thought too much about it. At mass the previous Sunday, Father Declan Caulfield had spoken enthusiastically of a gas project – Jacinta Healy's neighbours had mentioned it afterwards. However, she was busy rearing four children with her husband James, and trying to hold down a home-care health-service job. She had her youngest, nine-month-old Matthew, in her arms when Dukes asked if he could come in to chat.

'The house was upside down, as we were building an extension,' she said. 'He was in his late fifties – very nice – and spoke of site investigations, giving me a bit of description about what was planned. I remember I asked him if there was much gas out there. "Do you see that young fellow you have in your arms?" he told me. "He will be a grandfather before it's finished." I always said he was the only one among the developers who ever spoke the truth.'

Over in the village of Rossport, Mary and Willie Corduff had heard the parish priest tell them, 'Your poverty is over,' and that there were going to be 'unreal' benefits from the gas project. 'For rural Ireland, the Church was always right,' Corduff explained. 'It could be shining down sun and the hay in the field and if tomorrow it would be raining they'd still go to the church. So it was the news about the gas coming from the Church that convinced a lot of people.'

The Corduffs had endured their own type of hardship in one of the most beautiful corners of one of the most beautiful areas in the Erris region. Willie's father Francis bought the family farm in 1947 – 'The place was then pure bog with a fallen-down house,'[5], his son said. Corduff senior had been born in Rossport but had gone to England for work. He returned to marry in 1946, borrowing the money from his brother, Druie, to buy the farm, reclaiming land from the sea to make it viable.

One of Willie's relatives, Rossport teacher Harry Corduff, had made national news headlines in 1957 when he spent his Easter school holidays in Mountjoy jail in Dublin. In 1955 he had refused to tax his car in protest over the bad roads and poor facilities in the area, then refused to pay a fine of £4 18 shillings. 'They have the jet age in Shannon now, but we are still in the stone age here,' said a neighbour, Thomas Burns, when he and neighbours celebrated Harry's return home with a victory calvacade from Ballina in late April 1957. 'Before this, even the Ballina people just thought of us as the wild men from the mountains,' another neighbour, John Naughton from Carrowteigue, told the newspapers.[6]

Willie Corduff was one of a family of five. He left school early, earned his keep by cutting seaweed and turf, and at one point decided to go to Dublin to find employment in construction. His mother kept writing to him. Twice a week he would receive her letters, entreating him to come back home. 'They were heart-broken,' he recalled. 'So I felt I'd go home and that was it. I never left . . .' He married Mary Lavelle, who came from the same village as his mother, about twelve miles across the bay. The couple had six young children when a fire broke out on the farm. It started in a shed, and the couple's older boys, Francis and Liam, then fifteen and sixteen, had to restrain their father from going in to rescue the animals, most of whom smothered in the smoke. A calf that escaped died of smoke inhalation further up the road. The fire brigade from Belmullet, a half-hour away, had to use seawater and fought the blaze all through the night – staying on even when it appeared quenched to ensure that the hay did not reignite.

It was a terrible but defining period in the couple's lives. They had bills to pay, and no income, and no obvious alternative source of one since they had no means of transport. They cut turf and seaweed, grew their potatoes, lived on milk from their herd. Willie went fishing for three years, and quipped later that they were 'probably the worst three years that ever came in the fishing' as the 'catches were bad'. Neighbours and friends helped out in any way they could. It was a type of support that the Corduffs could never forget. As they explained time and again, 'That community spirit was there,' and always would be, even during the 'boom' years of the 1990s when the so-called Celtic Tiger passed parts of the west coast by without so much as a lash of its tail. For the Corduffs, this was something that oil and gas developers never quite understood, as the couple began to find out more about what would be one of the largest infrastructural projects in the state at the time, planned for their little locality.

In 2000, the Enterprise team examined the ground tests undertaken in and around Pollathomas for the onshore pipeline route, but the results pointed to erosion and stability risks – in fact, in September 2003 the village was to suffer a devastating landslide from Dooncarton Hill. The developers decided to transfer the onshore pipeline route to the other side of Sruwaddacon estuary. Much of the land was protected under the EU Habitats Directive, as was the estuary itself. The 'optimum' route would run through Rossport village, almost directly opposite Pollathomas. Up to forty landowners could be affected.

Brian Ó Catháin travelled up to discuss it with Rossport residents in a local pub. Company representatives checked out the premises on a Tuesday, and it was empty – this would be ideal for a public meeting. The team notified the community, extended invitations, and arrived that evening as planned. Jaws dropped as they entered. The premises was packed, but for an entirely different reason.

Anthony Cox had died in Manchester and his body had been brought home to be buried, following a wake in Doherty's licensed premises near Corrán Buí (Corraunboy). Relatives and friends had travelled from England to attend. At least half of the visitors for the funeral had worked on pipeline construction in Britain – among thousands of Irish emigrants engaged in constructing underground railway lines, such as the London Tube, motorways, bridges and even the Channel Tunnel a number of decades previously. With drink flowing freely, the questions from the floor were thick and fast. How was this pipeline going to be welded? How thick was the line? There were all sorts of technical questions. 'They knew more about pipelines than I did,' Ó Catháin recalled later. 'It turned out to be a total fiasco. It was just an accident. We had thought it was going to be a nice quiet place to have a meeting. It got us off to a very bad start.'

It was there that Ó Catháin met Rossport residents who were already very concerned about where the pipeline was running. Among them were Gerald and Monica Muller, a German couple living in Rossport, and Willie and Mary Corduff. In the late 1960s Gerald, a blacksmith and silversmith, had bought a smallholding, and within months Monica, from the highly industrialized Ruhr valley, had joined him. The couple converted an old outhouse into a craft workshop, planted a vegetable garden and fertilized it with seaweed from the shoreline, and built a conservatory on to their cottage. They attended céilís in the local community centre, and became very much part of the community.

As Mary Corduff remembered, the Enterprise team in the pub were 'arguing among themselves' over answers to questions from the floor. It was also the first time that she discovered that the pipeline was not running through Pollathomas but crossing the estuary to run through Rossport and close to her home.

In August 2000, farmer Brendan Philbin had a caller to his home at Rossport South. It was a representative of Enterprise, and he wanted to explain to him that a pipeline system would be going through his land. Philbin had heard about the project, but understood that the pipeline route was on the Pollathomas side of the Sruwaddacon estuary. He was 'very surprised and shocked' at the news that it was otherwise, and began to realize that the project 'was not being handled correctly as too many changes were being made in a very rapid fashion'.

Philbin recalled that he put forward some 'basic questions' to Enterprise representatives at several meetings, and 'they individually and collectively did not seem to have answers or understand our position'. He was particularly concerned about the issue of handling peat – how it would be removed and stored for the refinery footprint; given its inherent instability, how stable the pipeline could be when buried in a metre of peat; the possibility of corrosion in acidic soil; and the possibility of buckling and fatigue with the pressures involved. Full wellhead pressure had been cited at 345 bar; to put this in context, Philbin established that the pressure in a car tyre was two bar.

As a farmer who had relied on turf as a main source of fuel, Philbin knew the material intimately and recognized the risks involved in removing large quantities in an area surrounded by rivers and streams. 'The engineers . . . kept confusing our argument about peat with the characteristics of gravel material . . . I felt very concerned and disillusioned.'[7]

Philbin was not alone. In October 2000, he received a phone call from Bríd McGarry, who managed a sheep farm with her widowed mother Teresa in the next village of Gortacragher, across the back of Willie Corduff's farm. Teresa was worried about the pollution from the proposed refinery. Her daughter Bríd, who held a degree in food science and food chemistry at University College Dublin, was also concerned. She had first heard about the gas project in 1999, when there was, as she said, 'great excitement' about the project and a 'perception that it involved bringing in clean gas'.

When Bríd began to research the environmental impact of gas refineries, she had no idea that the pipeline route was to traverse her family's land. Even when the route was switched across the Sruwaddacon estuary to Rossport, she and her mother still had no real indication. So when Enterprise staff came on to their land, the McGarrys asked who had given them authorization. They were told it 'came from Coillte', but the McGarrys were never forwarded any letter to confirm this in spite of a promise of same.

In November 2000, a month after Bord Gáis committed to taking gas from Corrib and building the spur to Galway, Enterprise applied on behalf of the Corrib gas partners for planning permission from Mayo County Council to build a gas terminal at Ballinaboy. An environmental impact statement to accompany the application was available for inspection in the local authority offices in Castlebar, and in the Garda stations in Ballina and Belmullet.

Micheál Ó Seighin received a worried call, also from a landowner, and agreed to go and look at the environmental impact statement, along with his daughter, also named Bríd, and colleagues from Dún Chaocháin Teo, Treasa Ní Ghearraigh and Uinsionn MacGraith. Their visit to Belmullet coincided with a public meeting hosted that same night by the Council for the West, supported by One Voice for Erris, in the town's convent. Ó Seighin was 'horrified' by what he read in the two ring binders in the Garda station. The group were perusing them in the hallway, and after half an hour a garda took pity on them and invited them into the relative comfort of an inter-rogation room. Examining some of the drawings and maps, Ó Seighin noted that, apart from a pipeline landfall at Glengad, there was also a proposed outfall pipe taking waste from the terminal back to the middle of Broadhaven Bay.

'There was, of course, no understanding at all that the bay has a tow inwards instead of out,' he remembered. 'These were spatial realities that struck me immediately. The chemical mix was horrifying.' Ó Seighin, Ní Ghearraigh and MacGraith returned home but Bríd stayed on for the convent meeting. 'There had been a number of these meetings, although I don't know why they were trying to push the project so strongly then,' her father recalled. 'There was no real opposition to it yet at all.'

Bríd Ní Sheighin 'couldn't get over the stupidity of it' when she arrived home afterwards. 'No one obviously knew what they were talking about . . .' She recalled being 'very very upset'. There was, as her father gathered, a sense of pervasive unreality and a 'suspension of disbelief'.[8]

One of those who did ask questions from the floor was Sister Majella McCarron of the Sisters of Our Lady of Apostles. A native of County Fermanagh, she had spent thirty years in Nigeria and taught education at the University of Lagos. In 1993, she had met writer and activist Ken Saro-Wiwa, who was leader of the Movement for the Survival of the Ogoni People (MOSOP) in the Niger delta, after she had been invited to become a contact point for an international missionary movement, the Africa-Europe Faith and Justice Network. The network's simple objective was to identify any business with a base in Europe that had a negative impact on Africa.

She worked alongside Saro-Wiwa for the best part of a year, and founded the Ogoni Solidarity group in Ireland in early 1995, some months after she returned home on a sabbatical. She never saw the writer alive again and had to quit her life in Nigeria after the events of November 1995: on 10 November, Saro-Wiwa was one of nine Ogoni leaders, all of whom had campaigned against the activities of Shell in the Niger delta, who were hanged by the Nigerian dictator, Sani Abach. Saro-Wiwa had been falsely accused of murder and tried by a military tribunal in which proper legal representation and appeal were denied.[9]

Sister Majella had travelled west by public transport, and was brought from the railway station in Ballina to Belmullet by the parish priest, whom she thought she had come to help. It was only when she walked into the meeting that she realized she was definitely not preaching to the converted. So she spoke of her experience abroad and warned of the risks, aware that her church had been recruited as a local advocate for the project.

Ó Seighin began researching the health impacts on the Internet. His daughter Bríd remembers spending Christmas in Uinsionn MacGraith and Treasa Ní Ghearraigh's house, perusing the environmental impact statement. 'That Christmas,' Bríd said after-wards, 'was when the project began to control our lives.'

Much of the local focus then was on the impact of the planned terminal at Ballinaboy, but a series of meetings was set up by the Irish Farmers' Association (IFA) to discuss the upstream production pipeline, which would be coming ashore from the gas field to Ballinaboy, and the separate Bord Gáis transmission pipeline, running at much lower pressure from Ballinaboy to Craughwell in County Galway. The discussions focused mainly on compensation. The IFA had invited a consultant to help local farmers deal with the project promoters. It transpired that the consultant worked for the same company that advised Enterprise on agricultural matters. The IFA came up with a proposed package for affected landowners, involving compensation of €11 per linear metre for permanent way-leave over lands, €13.50 per linear metre for crop loss and disturbance and €3 a linear metre for 'early sign up' – totalling €27.50 per linear metre.

The Enterprise figures and those on offer from Bord Gáis were similar, Brendan Philbin noted, and the offer was later modified to €18 for a forty-metre working strip for the upstream pipeline. 'I knew at this stage that the IFA were instructed to impress on the landowners present that this was basically a Bord Gáis pipeline with a few minor modifications and that there was no real difference between them. We were told that we were getting "money for old jam" by an employee of the company.

'Needless to say, myself and other landowners lost total respect for the IFA after the events which had unravelled before us.' In Philbin's opinion, 'We were to be sold out for a pittance by an organization which was supposed to represent us. Michael Davitt . . . would be turning in his grave in disgust.'

Philbin delved further and found 'major differences' between the terms for the Kinsale gas field off Cork and Corrib, following changes in state licensing and lease terms. 'In effect, the people of Mayo and Ireland would gain practically nothing from these riches in our oceans . . . The indigenous people of Erris were also to be sacrificed into the oil bargain with the environmental repercussions and the unanswered health and safety aspects . . . To add even further insult to injury, the Irish people were denied this colossal revenue, which would have contributed enormously to the upkeep and maintenance of our hospitals, schools, roads and other relevant infrastructure.'[10]

Brian Ó Catháin was endeavouring to pursue a diplomatic route. Carne golf club near Belmullet was regarded as one of the most beautiful links in the region, and a location for movers and shakers, so Enterprise arranged golf days there for staff and contractors. Still, Ó Catháin was aware that a certain resistance had already built up, due, he believed, to an unfortunate decision by Enterprise management before his time with Corrib. The first environmental impact statement on aspects of the project had been put out to tender under EU procedures, and was awarded to a British-based company. He believed in hindsight it should have been given to an Irish firm that would have appreciated local sensitivities. 'I think it was a big mistake,' he conceded later. 'It effectively meant that for many landowners in the area, the first people with whom they had contact on behalf of the Corrib project were environmental scientists and engineers, some of whom had English accents.'

Speaking at the Humbert Summer School in Ballina, north Mayo's administrative centre, Ó Catháin predicted that the gas field would provide up to half of Ireland's requirements, and that the project could be a fillip for local development. The terminal would be involved in a simple procedure of separating water from what was 95 per cent methane gas, he explained. For geologists and engineers like him, it seemed very straightforward. Terminals like this had been constructed all over the world, in Britain, in France, in built-up areas of the Netherlands. Many such European terminals had been set on the coastline, where visual impact was not such an imperative. The north Mayo coastline was 'phenomenal', he acknowledged, and therefore the terminal should have minimal visual impact at Ballinaboy.

In meetings with residents, he explained that the pipeline had to be designed to withstand high pressure, but would not be carrying gas at high pressure. Major European trunk lines were running under people's houses in Rotterdam, for instance, through Vienna and Amsterdam without any issue. This was 'routine'. The Bishop of Killala, Dr Thomas Finnegan, was very supportive. The gas find represented a 'new ray of hope' with great potential, he felt, particularly when the peat-fired Electricity Supply Board (ESB) power station at Bellacorick was due to close in 2004.

Dr Finnegan had a problem, though: he was worried about recent reports which suggested that Mayo towns might not benefit directly from this gas find. They also ruled out proposals to replace the Bellacorick peat-powered station with a gas-fuelled one. 'The government must now ensure that the infrastructure, both legal and physical, needed to make Mayo gas available in Mayo is put in place immediately,' Dr Finnegan said.[11]

3

SEASCAPE

On the evening of 28 October 1927, a retired doctor in Cleggan, County Galway, picked up a bad-weather warning on his magnetic radio. He called his farmhand and asked him to ride to Rossadillisk village immediately, and warn the fishermen not to put to sea. He was too late. The fleet from Cleggan and its neighbouring island of Inishbofin had left, as had currach crews from Inishkea Island and Lacken and other parts of the Mayo coastline. The Cleggan disaster, as immortalized in poet Richard Murphy's verse of that title, was one of the worst marine accidents of its type on the west of Ireland coastline in the early twentieth century. Forty-five men from Inishkea, Lacken, Inishbofin and Cleggan perished. One of the survivors, James Cloherty from Inishbofin, recalled that the storm came on almost without warning. He heard 'terrible screams and shouts in the darkness'. It was believed that the death toll would have been much lower if some of the fishermen had not been so anxious to save precious equipment.

John O'Donnell, known as John 'a Cladach', had gone out with his brother Jim in a currach from Porturlin on the north Mayo coastline. Miraculously, a currach skipper who was swept back into the harbour was able to tell neighbours that the O'Donnell brothers had been fishing east of there. A search party was sent out, and the pair were located by their shouts in a cove, several hundred feet below a sheer cliff-face. Ropes hauled them up to safety in the darkness.

The Cladachs, as the late John O'Donnell's family are still known, continue to fish one of the most treacherous parts of the western seaboard out beyond the Stags of Broadhaven. In 1991 John's son, Pat, started a processing business to handle the port's crab landings. His earnings were augmented in the summer by the lucrative wild salmon fishery until the introduction of a ban on driftnetting in Irish waters in 2007. The factory, located at the back of Pat O'Donnell's house on a cliff edge overlooking Oileáin na Muice or Pig Island, provided seasonal employment for up to twenty people during the peak years, with O'Donnell selling crab claws at home and exporting to a European market hungry for Irish shellfish. He named one of his boats *James Collins*, after a young man he hadn't known who had died from injuries sustained when he became caught in machinery in a factory in Ballina. O'Donnell's boatbuilder had witnessed the accident. 'The Chief', as O'Donnell was known, was sufficiently moved to commemorate a lost life in this way.

Historical maps of this part of the western coastline record it as a graveyard for lost vessels. It was off the village of Rossport that the Air Corps had conducted its first night helicopter mission in July 1988, rescuing three fishermen. Ballyglass pier beyond Belmullet, used by the O'Donnells and other Porturlin fishermen, was the focus for another less fortunate rescue attempt in 1988, when John Oglesby, one of a generation of successful young Donegal-based fishing skippers, had his leg severed by a sheave pin while lifting and loosening a trawl warp on the deck of his vessel, *Neptune*.

Crew member Tom Rawdon was also injured. The nearest Royal National Lifeboat Institution (RNLI) stations were many hours' steaming to the north at Arranmore Island, County Donegal, and to the south the Aran Islands, off the Galway coast. The Air Corps rescue squadron was on the east coast at Baldonnel. A call was put through to Britain for assistance, but the Royal Air Force calculated that the vessel would have reached port before it could fly to its assistance. John Oglesby died within sight of land. His son, Martin, was among the crew battling to keep him alive. Joan McGinley, a friend and member of a fishing family in Donegal, was so moved by the event that she initiated a highly successful campaign to improve west coast rescue services, which led to the establishment of the Irish Coast Guard several years later.[1]

As part of this increased west coast emergency cover, the RNLI opened a new lifeboat station at Ballyglass in north Mayo. The lifeboat crew, along with Pat O'Donnell and a number of fishing skippers, was called to a very challenging rescue in October 1997, when emergency services received a report that four people had failed to return from a boat trip to Belderrig pier.

Retired German businessman Will Ernst von Below had taken Tony Murphy, his wife Carmel and their eleven-year-old daughter Emma out in a currach. In the heavy swell, they ran into trouble and became trapped in one of the larger caves when their currach capsized. It was 5.30 p.m. and near dusk when the Ballyglass RNLI lifeboat, the Irish Coast Guard unit from Killala and various fishing vessels began searching the coastline.

Pat had his son Jonathan on board *Blath Bán*, while Pat's brothers Tony and Martin also steamed out to assist. At about 7 p.m., shouts and whistles were heard coming from a tideline-level cave on Horse Island. Lights trained on the cave mouth could pick out the reflective strips on lifejackets.

The rescue was co-ordinated on land by local garda officer and second coxswain of the Ballyglass lifeboat Tony McNamara. The Ballina-based Grainne Uaile diving club offered to help, and the Irish Coast Guard Killala unit took the divers to the cave mouth, standing by as two of them, Josie Barrett and Michael

Heffernan, swam in with a light line. The club divers became separated from one another in a three-metre swell, and Barrett was
eventually picked up in a state of exhaustion. Garda divers were
fully kitted out when they were dropped by the Irish Coast Guard's
Shannon-based Sikorsky rescue helicopter in a field close to
Ballyglass pier.

Sean McHale and Martin Kavanagh of the Irish Coast Guard
made repeated efforts to get the Garda divers into the cave,
negotiating an enormous swell. The rescuers found the Murphy
family huddled together in a crevice, just a metre above sea level,
but neither currach owner von Below nor diver Michael Heffernan
had survived. The Irish Coast Guard's inflatable engine had been
damaged. It looked as if both the rescuers and the survivors were
trapped.

In an act of pure faith, Garda Ciarán Doyle volunteered to swim
the thousand metres out of the cave with the 250-metre line
attached to the inflatable. Out at the cave mouth, Pat O'Donnell on
Bláth Bán spotted the exhausted diver emerging and recovered
him, with the essential line, from the water. Working with his
brother Martin, in sister craft *Sinéad*, O'Donnell hauled out the
Irish Coast Guard inflatable with the survivors on board. It was
widely acknowledged as an act of experienced seamanship.

Pat O'Donnell returned to recover the bodies of von Below and
Michael Heffernan. At the state's first marine rescue awards
eighteen months later, the late Michael Heffernan was post-
humously awarded a gold medal, which was accepted by his wife,
Annamarie, and two children Leigh Anne and Michelle. Silver
medals were given to Garda Ciarán Doyle, Sean McHale of the
Irish Coast Guard and Josie Barrett of the Grainne Uaile diving
club, and bronze medals to Garda Dave Mulhall, Garda Sean
O'Connell and Martin Kavanagh of the Irish Coast Guard. Pat
O'Donnell and his brothers received letters of thanks. O'Donnell
remembered that his son, Jonathan, 'worked like a man that night'.

*

When word began to spread in Kilcommon parish about plans for processing gas that had been discovered off the Mayo coastline, Pat O'Donnell recollected thinking this could be good news for the area. West coast fleets were weathering some of the worst effects of the EU Common Fisheries Policy, which made it very difficult to compete with favourable quotas secured by France and Spain. It was when some of the detail began to emerge that he grew concerned. There was talk of a discharge pipe from a refinery that would be built somewhere onshore. His son Jonathan would bring stories home from school in Rossport, where Micheál Ó Seighin was teaching. Would heavy metals in effluent affect Broadhaven Bay? O'Donnell wondered. He had to meet EU regulations on seafood quality, and he did not want to lose his business.

Keen to have some independent evaluation of the information then provided by Enterprise, he commissioned a study from an expert at Southampton University on the effects that the project might have on the marine environment.

In a preliminary report completed for O'Donnell's company, Porturlin Shellfish Ltd, in February 2001, Dr Alex Rogers of Southampton University's Ocean and Earth Science Department expressed serious concerns about the environmental impact of the gas landfall. He said that the development would 'appear to offer very little to the local community'.

Discharge of heavy metals, including mercury and quantities of methanol, through an outfall pipe releasing water from the planned gas refinery into Broadhaven Bay could contaminate the marine environment, Rogers warned. He noted that the project's own environmental impact assessment, which he had examined as part of his study, was 'contradictory' on the behaviour of contaminants in the marine environment. He cited many instances of the potentially negative impact of bioaccumulation of heavy metals, referring to the Minamata mercury deaths in Japan in the late 1950s, for instance.

Rogers warned that the release of such materials in even small quantities was of great concern to local fishermen, and to the protected habitats of Broadhaven Bay and Sruwaddacon estuary. He also questioned the decision to use methanol, which would be released with the discharged water. The scientist did not confine his assessment to the marine environment. He questioned the impact of releasing nitrogen oxide from the proposed refinery, and noted that little attention appeared to have been paid to sensitive land habitats – such as the Glenamoy bog complex. The Mullet peninsula was one of four areas in Ireland where the threatened corncrake could be found, he noted.

He dismissed as 'superficial' the developers' environmental impact assessment onshore survey; it was, he said, mirrored by an 'uninformative' and 'poorly executed' study of marine fauna in the intertidal zone, shallow subtidal zone and offshore areas. Given that the cold-water coral *Lophelia pertusa* had been recorded close to the Corrib field, towed camera surveys or submersible images should have formed part of the offshore work, in his opinion. 'Likewise, a few day grabs [samples] of the inshore area are of no use in assessing the diversity of the shallow water fauna and flora in the Belmullet area,' Rogers wrote.

Ballinaboy had been chosen for economic reasons, and for access to the sea as a dumping site for waste water, Rogers contended. He questioned whether other options had been considered for processing the gas, such as Bellacorick where there was already a peat-fired power station, or even Galway, where there was a proposed link-up with the natural gas system.

The project's environmental impact assessments 'discuss difficulties of placing a platform at sea over 350m depth of water, but BP [British Petroleum] has floating facilities in deeper water off the west coast of Scotland', he observed. 'Whilst Ireland has an urgent requirement for expansion of the offshore hydrocarbon industry, care should be taken not to destroy the country's natural heritage that, in the case of the west coast, is amongst the most beautiful and unspoilt in Europe.'[2]

Pat O'Donnell submitted a copy of the study to Minister for the Marine and Natural Resources Frank Fahey. 'We are employing twelve people here in crab processing and have done since 1992,' O'Donnell told *The Irish Times* in February 2001. 'I come from a family of eleven, and we made it through the hard times here. Enterprise Oil wasn't around then, and we don't need it now.'[3]

The Rogers report was quickly rejected by Enterprise managing director Brian Ó Catháin.

'We are not Asahi Mark Two . . . We are concerned about the environment and we are here to stay,' Ó Catháin told *The Irish Times* in the same month, referring to the abandoned Japanese textile factory based in Killala. A month before, the Corrib gas field had been declared commercial and Fahey had informed the Dáil several weeks after the confirmation. His department also asked Enterprise for a study of alternatives to an onshore processing plant. Consultants David Fox and Associates advised that 'the developers . . . should not be required to change or consider changing the Corrib development scheme'.[4]

However, that same month, Mayo County Council's senior planner wasn't happy with some of the detail in the terminal planning application and requested further information. Ó Catháin said that a response was being prepared, which would not cause undue delay. 'The last thing we want to do is to put any pollutants into the bay, and the emissions from the factory are similar to those you'd get from a gas central heater in your house,' he said. Some 99 per cent would be 'carbon dioxide and water', and it would be less polluting than Bellacorick power station, he said. The pipeline route was proving 'contentious', he acknowledged, with sixty parcels of land affected and forty-five landowners involved – the majority of whom were 'supportive' of the project. The objectors wanted the pipeline to be routed through bog commonage, rather than through their land, he contended.

'The problem is that Dúchas [the state body responsible for the environment and heritage] doesn't want us going through the

commonage, where there are turbary rights, or through virgin
blanket bog which is part of a Special Area of Conservation,' he
said. 'And we can't go through the estuary as that is also protected.
We honestly feel we can't have consulted more. We aren't going
away. We are going to be there for the next twenty years. We want
to be seen to be a good part of the community, like Coca-Cola in
Ballina, employing fifty as operators at the [Ballinaboy] plant, and
additional staff for services like catering.'

Donegal to the north was already benefiting, he acknowledged.
Killybegs in south Donegal was developing as its supply
base, handling '75 per cent' of its vessel movements associated
with the rig, and Carrickfin airport had become its air transport
base.[5]

The terminal planning application was withdrawn from Mayo
County Council, and resubmitted to the local authority in April
2001. It was a significant development, but there was still little trust
among local people in the council, at a time when they were also
being frustrated by Coillte in their efforts to glean more information on the Ballinaboy land sale.

Coillte had agreed initially to a request to meet residents, and an
announcement was made on the local MidWest Radio. However, on
the Friday before, Coillte cancelled 'due to unforeseen circumstances', but in an email to one local resident on 25 May 2001 it
offered to meet 'representatives' of 'your group'. It also said that it
was 'surprised to learn that it was your intention to hold a public
consultation meeting on Monday next as we had stated that
consultation on the development in question is being carried out by
Enterprise Energy and we are not qualified to carry out
consultation on such a development'.

'Coillte is a state-owned body, we are all stakeholders in it,'
residents responded on the same date. 'There can only be a public
meeting and not a private consultation under these circumstances,
except of course if Coillte has anything to hide or is very anxious to

keep proceedings in regard to the sale of land to Enterprise Energy Ireland secretive.'

On MidWest Radio, Coillte's chief executive Martin Lowery defended the decision to sell the land. Coillte didn't respond to requests to invest the purchase price of the Ballinaboy site for the benefit of the local community. In mid-May 2001, the company, which was by far the largest landholder in the state, announced profits of over €25 million for the previous year.[6]

Residents registered a formal complaint with the Forest Stewardship Council about Coillte's failure to consult. The council, established in London by a number of international forestry non-governmental organizations in 1993 to support environmentally appropriate, socially beneficial and economically viable forestry management, had awarded Coillte certification in 2001. It sent a representative to Rossport to talk to residents, but two years later it had still not withdrawn certification on the basis of the community's complaints. The actual sale of Ballinaboy was predicated on planning permission, and this would take longer than the developers anticipated.

Within weeks of the Rogers report on the marine risks, the Erris Inshore Fishermen's Association (EIFA) was formed, and community representative Eddie Diver (Eamon Ó Duibhir) was appointed to the chair. The EIFA sought a meeting with Enterprise on the issue of the outfall pipe. On 6 June 2001, a group of twelve fishermen faced Enterprise representatives Michael Murray, Agnes McLaverty and RSK environmental consultant David Taylor, who had carried out the environmental impact assessments examined by Rogers, in a hotel in Geesala.

Diver made clear that his members were not opposed to the development of the Corrib field, but would not, under any circumstances, entertain the concept of an effluent outfall pipe from the Ballinaboy plant into Broadhaven Bay. Fishermen knew well that, in spite of assurances from Enterprise about dispersal, the currents at the outfall location ran inwards to Broadhaven Bay and not out to sea. And Diver understood the fishermen, having come from the Gaoth Dobhair area of north-west Donegal. He had moved to Erris to work with Roinn na Gaeltachta, the government's Gaeltacht department in 1967, and never left, for the isolated coastal environment was 'very close to home'.

The Enterprise team referred to best industrial practice in the design of the effluent treatment plant at the proposed terminal. The fishermen were adamant. This was their livelihood at stake. Enterprise then agreed to a further redrafting of the environmental impact statement for the offshore element, with additional computer modelling on bathymetry, as in measurement of depths of bodies of water, and tidal movements in Broadhaven Bay, as well as offshore site investigation off Dooncarton Hill, which was scheduled to take place that summer. The developers also gave assurances that the field development proper was 'unlikely' to begin 'until such time as the fishermen's concerns were fully addressed and resolved'.[7] As the association was to find out over the passage of time, this promise was to prove rather hard to keep.

Early in 2001, when Pat O'Donnell was posting Dr Rogers's study to the minister, leading *Irish Times* environmental writer Michael Viney raised some further issues of concern to the local community, including the fishermen. Analysing the first in what would be an 'armful of environmental impact statements in glossy ring binders' on the gas project, he said that the operations in Sruwaddacon Bay in particular could 'expose the folly of compromising on industrial intrusion into a remote and undisturbed estuary granted one of the EU's top badges of protection'. He noted that there was no baseline study in the environmental impact statement to date of bottlenose dolphins, which frequent Broadhaven Bay where the 'terminal effluent will be released at some four kilometres out'.[8]

The fishermen weren't the only group with concerns about marine impact. In early June 2001, the Irish Whale and Dolphin

Group (IWDG) marked the tenth anniversary of Ireland's whale and dolphin sanctuary, as declared by former Taoiseach Charles J. Haughey, with a call for a proper environmental assessment of the Corrib gas field development. One of its members, photographer Shay Fennelly, had also made a complaint to the EU environment directorate over proposed blasting in Enterprise's 2001 environmental impact statement during offshore pipeline laying. The non-governmental organization, based in Kilrush, County Clare, appealed to the government to apply the EU's Environmental Impact Assessment Directive to seismic surveys and exploration drilling in Ireland's marine territory. It called on the minister for arts, heritage, the Gaeltacht and the islands to stop further development of the onshore aspects of the Corrib gas terminal until a survey of cetaceans had been carried out.

'At present, there is another seismic survey under way off northwest Mayo as part of the Corrib gas field development, and there is no attempt to assess or mitigate this impact on cetaceans,' IWDG director Dr Simon Berrow told *The Irish Times* of 6 June 2001. 'We know this area is used by a wide range of cetacean species, including pilot and minke whales and white-sided, common and bottlenose dolphins, and that seismic activity may displace whales and dolphins from their preferred habitat.' There was only 'one brief reference' to cetaceans in the developers' revised environmental impact statement submitted to Mayo County Council, Dr Berrow pointed out.

In spring 2001, an outbreak of foot-and-mouth disease in Britain prompted a series of government measures to prevent the spread of the disease to the Republic. During the height of that alert, in April, Teresa McGarry and her daughter Bríd recorded another incident of trespass on their land by developers' agents. An Enterprise representative informed them, when they confronted him, that the pipeline route had been further modified and would be going through 'all of their fields'. It was the first they knew of it. No one had ever written to or otherwise approached them.

'This was the manner in which we were consulted,' Bríd McGarry recalled afterwards. 'We were dictated to on our own private property in a very arrogant and demeaning manner as if we really did not matter in the bigger scheme of things. We felt isolated and alone in our plight, as the big push was on by the developers and the state authorities to rush this project through, regardless of how people were to be treated on the ground.'

Down in Galway, Padhraig Campbell and some colleagues from the offshore oil and gas industry were generating publicity over their concern about Irish employment – and benefits to the state – following their dispute with Brian Ó Catháin's predecessor, John McGoldrick. As spokesman for the SIPTU trade union's offshore oil and gas committee, Campbell called for withdrawal of drilling rights for companies that refused to hire Irish-based oil-rig workers. An initiative by the Irish Offshore Operators' Association (IOOA) to work with the state training agency, Fás, on providing Irish jobs was dismissed by Campbell as 'mere window-dressing' to placate the minister and public opinion.

Campbell had also come across a report by consultants Wood MacKenzie two years before, which had claimed that there could be as much as five trillion cubic feet of gas lying in structures around the Corrib field. The area was geologically 'quite simple' and could prove to be very productive and economic to develop, the report had said.

In April 2001, Minister for the Marine and Natural Resources Frank Fahey visited the West Navion drill ship hired by Enterprise Oil for the Errigal prospect off Donegal, and told a reporter that he would review the 1992 exploration terms awarded to oil and gas companies if there was a successful outcome to that year's drilling programme. It caused some concern within the IOOA, which sought to put the idea to bed.[9]

In June 2001, the *Western People* quoted Gerald Keane, principal engineer with the state industrial development wing, Enterprise Ireland, as expressing concern about the industry's use of Irish services and expertise. He had maintained a register of some two hundred suitably qualified Irish companies that were capable of supporting and servicing the activities of companies such as Enterprise and Statoil, but the industry was 'bypassing them' by insisting that they first register with an Aberdeen company that maintained a similar international database on a fee-paying basis.[10]

In May 2001 Enterprise hosted a press visit to the Sedco 711 rig, drilling what the company described as possibly the last well before production on the Corrib gas field. The journalists, including this writer for *The Irish Times*, flew by helicopter from Knock airport with Enterprise's Brian Ó Catháin and team on a stunningly beautiful day, with the Atlantic resembling a tranquil turquoise pond from above. On landing, Ciarán Nolan, Enterprise geo-physicist, described how the five wells drilled on the field would look if superimposed on a map of Dublin. They would extend from Croke Park on the northside to Stephen's Green, King's Inns, Temple Bar and Dolphin's Barn across the river Liffey. 'In 2003, we hook these up to a manifold and run the pipeline ashore,' he said.

The press group were told how technology had revolutionized the work in water three times the depth of the Kinsale gas field, using a semi-submersible drilling rig that had roamed the world's exploration fields from the North Sea to west Africa. The rig cost £110,000 sterling a day to run at that time, including helicopter transport costs for crews to and from Carrickfin airport, County Donegal. Most of the staff on board at that time had Scottish accents, with connections ranging from Aberdeen to Amsterdam. Information was so sensitive that mobile phones were not permitted on the rig, but Ó Catháin was keen to stress that real-time data was fed immediately to a restricted access website in the Department of the Marine's petroleum affairs division. It was 'nonsense' to suggest that information was tampered with in any way, as implied by SIPTU, Ó Catháin said.

He spoke about Ireland's relatively small but growing gas market, which would dictate the pace of Corrib's development, and about whether more wells would be drilled after production. The previous February, the government had sanctioned construction by Bord Gáis of a second interconnector, parallel to its existing link with Scotland, in spite of lobbying to the contrary by Marathon and 'grave concern' expressed by the Irish Business and Employers' Confederation (IBEC).[11] Bord Gáis was also constructing a distribution pipeline from north Mayo to Craughwell, County Galway, he explained to the journalists, connecting Corrib to the national grid. It would run past Bellacorick power station, which would be developed for further production beyond its current life. 'Bord Gáis talks about bringing gas to the west,' Ó Catháin said. 'But the gas will be coming from the west, not the other way.'[12]

4

THE BOGONI

The small group of active objectors pooled their resources. Bríd McGarry agreed to research emissions, as did Jacinta Healy and fellow residents of Ballinaboy. Gerard Muller undertook a study of bridges to see how they would cope with increased traffic. Two rooms of the Ó Seighin house were taken over with books, reports, papers, as Micheál Ó Seighin read and researched copious amounts of information. Caitlín remembers that they were so busy that there were times when they barely spoke from dawn till dusk.

Ó Seighin's caustic assessment made headlines in Mayo. 'The entire community here now realizes the scale and toxicity of the effluents and emissions about to be imposed on this area – land, air and sea – by the refining activities of Corrib Gas,' his submission to the revised planning application said. The developers' plan to strip some sixty acres of peat, to a depth of up to fifteen feet, and redistribute it in the adjoining forestry 'seems to have come out of the teddy bears' picnic. Apart from the sheer bulk and viscosity of the mass, the logistics rival those of NATO in Kosovo. Enterprise Oil have not in any way shown (a) that they understand the problem and (b) that they have any idea how to cope with it.'[1]

As concern gained momentum, Church figures including Dr Finnegan unsuccessfully proposed a representative committee for the area at a meeting in Glenamoy. Still, public bodies did not appear to be listening – or distanced themselves altogether, Ó Seighin recalled. There was a perception that the function of these organizations was to 'do what they were told'. He found it shocking, having been reared in a time and place where the politicians were those who had emerged from the war of independence and the civil war, and had 'put their lives on the line and believed in what they were doing'. [2]

The *Sunday Business Post* noted in late May 2001 that the community was 'divided about the offer' of compensation for way-leave to lands on the proposed pipeline route.[3] In June, Enterprise declined an invitation to attend an annual Ken Saro-Wiwa seminar in memory of the executed Nigerian writer and activist, organized by Sister Majella McCarron of Ogoni Solidarity Ireland, with Maura Harrington and others in Geesala. Ogoni Solidarity, which received support from the Catholic Church's development aid organization Trócaire and from Afri, had been holding the seminars in various parts of the country since 1996, a year after the execution of Saro-Wiwa and his eight colleagues.

Enterprise said it didn't consider it appropriate to attend when planning applications were still in train, and believed that it was 'outrageous' to link the company to the Ken Saro-Wiwa affair, which had been 'horrific'. The organizers should have invited someone from Shell, Enterprise told *The Irish Times*.[4] In fact, as Sister Majella recollected afterwards, Shell had been invited and had declined.

On the Friday evening of the seminar, Komene Famaa from the Ogoni people spoke about the devastation wrought on tribal lands in Nigeria by Royal Dutch Shell, where up to half a million of his people lived. Sister Majella knew Famaa from her time in Nigeria as he had been an active member of the Movement for the Survival of the Ogoni People. He had left to do a short course in Rome, wasn't able to return and continued living under refugee status in Italy. He had come to Ireland to take several post-graduate courses.

In spite of vast revenues generated by mineral production, the Ogoni had just one hospital and the services of one doctor for seventy thousand people, he explained. Access to clean water and rampant poverty were still issues, in spite of the oil wealth. He believed there were many similarities between the people of Erris and his own people, in relation to language, fishing and farming. 'The land is our life, and our life is the land,' Famaa said. His testimony was to prompt Maura Harrington to suggest that, in view of what the people of Erris were facing, they should consider calling themselves the 'Bogoni'.

The Department of the Marine did accept an invitation to attend, and was represented at the seminar by Michael Daly, petroleum affairs division principal officer. He explained that all assessments of the Corrib gas project were in accordance with Irish and EU law and included 'modern' and 'public' safeguards. Daly confirmed work on legislative change would shortly enable the minister to give Enterprise access to land linking the landfall at Rossport to the Ballinaboy terminal.

The Irish Times elaborated on this:

> A spokesman for the Minister confirmed that the 1960 Petroleum and Other Minerals Development Act allowed him to acquire compulsory ancillary rights, including way-leave, but the Act did not allow for oral hearings. The Minister preferred to rely on the 1976 Gas Act, which allowed him to make acquisition orders for compulsory purchase, but which also provided for oral hearings.
>
> 'The decision to make compulsory purchase orders was still a matter for the Minister, and it was for the company to apply to administer it,' the department spokesman said.[5]

Also speaking at the Saro-Wiwa seminar, economist Richard Douthwaite questioned the wisdom of bringing gas ashore now from Corrib, and asked if it was worth sacrificing a sensitive site in Broadhaven Bay for the equivalent of twenty days' gas supply in Europe when there were many applications for wind farms in the area. He also questioned the idea of burning gas for electricity, as proposed for Bellacorick. And he warned that gas production would peak in 2020, then fall off dramatically. 'What is this state doing about global warming?' he asked.

Padhraig Campbell of SIPTU focused on licensing terms. There was no firm evidence that towns in the west would get gas, as Ireland would be 'buying it back' from the developer. Very little of a forecast IR£1 billion, which would be spent developing the gas find, would come back to Ireland, he contended, and he called for a 'Norwegian mentality' if Ireland was to 'avoid becoming the filling station of Europe. The oil industry now has the raw material, means of production, distribution and markets under their control and Ireland is out of the equation,' Campbell said. The *Mayo News* reported that a 'lively' question and answer session was chaired by Páraic Breathnach, founder of the Macnas street theatre company in Galway, who 'reiterated the disregard shown to the people of Erris by Enterprise Oil and by those local and central government representatives' who had failed to accept invitations to attend.[6]

Ian McAndrew, who worked with Údarás na Gaeltachta and was public relations officer for One Voice for Erris, had travelled with Micheál Ó Seighin and several journalists, including Liamy MacNally of MidWest Radio and photographer Shay Fennelly, to the Shetlands in the spring of 2001. Their three-day fact-finding mission aimed to learn how the Scottish islands had coped with the discovery of 'diamonds' – as in the North Sea oil some hundred miles offshore – and the construction of one of the largest oil terminals in Europe, Sullom Voe, on the archipelago.

The group met members of the Shetland Islands Council, who explained how they had passed legislation in 1974 that gave their body considerable control over development in and around the islands. The council took ownership of the land earmarked for the terminal site and developed 'multiple roles', including majority shareholder of the tug company, planning authority and environmental regulator, as Fennelly explained in an *Irish Times* report of the visit.[7] It negotiated disturbance money and a stipend for every tonne of oil brought ashore, charged port duties on every ship, and channelled the monies into charitable trusts for the benefit of Shetland Islands inhabitants. The benefits had included social welfare, leisure and recreational projects, environmental improvements, arts funding, and economic development, and by 2000 one of the trusts had had a reserve fund of £230 million sterling.

The Shetland Islands Fishermen's Trust had a reserve fund of £3 million sterling, Fennelly noted, and fishermen had a direct role in environmental monitoring due to the potential adverse consequences of a terminal on their inshore waters. 'We don't pay individual compensation, we pay it into a Shetland Fishermen's reserve fund,' the association's chief executive, John Goodlad, told the Mayo group. 'When you are dealing with compensation from oil companies, you must all hang together or you will hang separately. Speak with one voice . . .'

Speaking with several voices, residents tried to glean more information at a public meeting that Minister for the Marine and Natural Resources Frank Fahey held in Erris on 25 July 2001.

Fahey, Fianna Fáil TD for Galway West, had been granted his first cabinet post as minister for the marine and natural resources in January 2000, and was keen to assert himself as minister for the entire west of the country. He was also keen to assure the public about the 'strength of the environmental assessment procedures associated with the development of the Corrib field', as articulated in a press statement of 22 June 2001, outlining details of the session he planned in Geesala.

'The environmental impact assessments in respect of the development for the Corrib gas field, which will address the subsea structure, pipeline and gas terminal, are required by law to be comprehensive, informative and public. The public should have the utmost confidence in the impartiality and professionalism of the services within my department and the local authority in ensuring that the proposed development fully respects legitimate environmental concerns,' he said.

Fahey was accompanied to Geesala by senior officials from his department's petroleum affairs division, including geologist Dr Keith Robinson, staff from the state heritage agency Dúchas, and Sean O'Riordan, representing consultancy ERM, which had been hired by the department to examine the environmental impact statement commissioned by Enterprise for Corrib's terminal. ERM had already made clear in its bid for the contract that it understood the ministerial requirements. The bid explained that one of its key tasks would be to 'identify key representatives in the various stakeholder groups, for example, fishermen's organizations and conservation groups'.

In its 'overall approach' section, ERM said it understood that the Corrib project needed to be 'accessed in a tight timeframe', and said it was also 'aware that the NGO [non-governmental organization] movement is mobilizing itself in response to renewed current activity in the Irish offshore sector'. As Ó Seighin noted afterwards, when he read the bid, 'They were hired!'

The 'seminar' was recorded on videotape by Fahey's department – which caused some upset among those attending as their permission had not been sought in advance. Subsequently the department acknowledged this as an 'oversight'. It also confirmed that a copy had been given to the developer. Over time, the video record, which some anxious residents believed was intended to form a profile of objectors, would serve another purpose. The minister insisted that each speaker should identify themselves and where they came from – not just by county, but by locality.

Winifred Macklin didn't wear a Mayo accent, but her grandparents would have some time before. She spoke passionately about her strong attachment to an area she had moved to in her bid to find a clean, safe environment. On 21 October 1971, the Glasgow suburb of Clarkston where she lived had been rocked by a gas explosion, killing twenty-two and injuring a hundred. Most of the victims were young women working as retail assistants or mothers out doing family shopping.

Her grandmother had looked out on Sruwaddacon Bay, as had her grandmother before her. 'I want my grandchildren, my great-grandchildren and all the children in this area to be able to breathe clean air, swim in clean waters and look at the healthiest environment in Europe that we have in the west coast of Ireland,' she said. 'I think the people from Skibbereen right up to Donegal should raise objections.' Surely there were other options for Erris, she suggested – solar power, wind power, wave power, power from 'the Atlantic crashing in'.

Monica Muller of Rossport spoke of a 'sell-off' of Irish gas and oil by 'several government ministers', with 'no royalties and no tax benefits'. Any gas would not be coming to Mayo, but would be exported to the highest bidder with no benefit to the state. Dr Keith Robinson disputed this: Ireland had a total of 142 wells drilled, compared with 6,000 in the British sector of the North Sea, he said. 'My slide showed you that, at the end of 1992, there was no interest in exploring in Ireland, there was no future for exploration in Ireland, so it was deemed necessary at that time to bring in terms which would be more attractive. It is true that there is no royalty and there is no state participation. There is a tax on profits, a corporation tax at a special rate of 25 per cent. And that is a fact which everybody knows.'

Significantly, Robinson continued, the department had done its own economic evaluation of the Corrib development 'if it goes ahead. And it is not seen by us as being unduly profitable at all . . .'

'We are not here today to discuss the [licensing] terms,' the minister said firmly, in response to Monica Muller.

'It is very much to the point,' Muller contended. 'Because by "we", I presume you meant the government and the Irish people . . . we have lost control over what is happening. Enterprise Oil is in total control. Once the proposed project goes ahead, it will set a marker for any other company to follow. You have given it away. It is gone.'

This was a 'totally wrong and inaccurate statement', the minister countered. Muller reminded him that his own official had said, in an earlier presentation at the meeting, that the government was 'in no position to dictate to the oil companies'.

Maura Harrington, then vice-principal of the primary school in Inver, was next to her feet. From Doohoma on the tip of Erris's southernmost peninsula, overlooking Tullaghan Bay, she had taught at Inver since 1973. Her family background was Fianna Fáil, and in the early 1980s, she had collected IR£500 locally to help families of striking miners in Britain – many other people in Ireland of every political hue had contributed to similar collections. As far as she was concerned, the project appeared to be a *fait accompli*, and she was critical of the department's dealings with ERM to assess the environmental impact assessment.

Dr Mark Garavan, sociology lecturer at Galway-Mayo Institute of Technology (GMIT), introduced himself as from Castlebar. He went to the heart of the matter. The minister's role was 'incompatible', he said, in representing both regulator and developer. Had the company submitted a plan of development to government, which would give a clearer picture of the overall project? he asked. Two recent press releases had stated that such a plan had been submitted.[8]

Dr Terry McMahon of the Marine Institute explained how a marine licence vetting committee (MLVC), specially convened by the minister for this purpose, would assess an environmental impact statement and make a recommendation. If the company did not follow conditions it 'will be stopped', he asserted. Civil servant Michael Daly of the department's petroleum affairs division confirmed that the 'process at the moment is at a very advanced stage of finalizing of a [petroleum] lease', and once this was complete, the Corrib partners could apply for a plan of development, with an accompanying environmental impact statement. As of that moment, there was no formal plan with the department, Daly said.

Sean O'Riordan of ERM sought to defend his consultancy. It was independent, he stressed. In fact, as it was to emerge later, there were several recorded contacts between ERM and Mayo County Council's planning office, even though the consultants had been commissioned by the department.[9]

Fritz Schult, Pollathomas hostel owner, was worried about the impact on local tourism, and wanted to know if the plan of development, as referred to by Dr Mark Garavan, would be published. 'Everything will be published,' the minister said. Later it would transpire that even court actions could not force key documents into the public domain.

There was a series of further questions, each pressing. Ian McAndrew of Údarás na Gaeltachta, who had been on the recent fact-finding trip to the Shetlands, said he would like to see as much benefit accruing to the area from the gas development as it had in the Shetlands. The 'one glaring omission' from the minister's presentation related to details of the deal for the Ballinaboy land, concluded with Coillte. Land had been sold 'behind closed doors', he said, without it going to public tender, and there were 'tribunals in Dublin sitting on a lot less'.

Jacinta Healy, resident at Ballinaboy and 'neighbour of the proposed terminal', wanted to know why there was no Environmental Protection Agency representative present to take questions on residents' concerns about emissions. Would there be a regular monitoring regime? Another resident asked about the impact on the area's main water supply, Carrowmore lake. Eddie Diver, of the Erris Inshore Fishermen's Association, expressed concern on behalf of his members about the potential pollution of Broadhaven Bay. The relevant part of the environmental impact statement on this was 'totally inadequate', he said.

In each case, the minister's answers were confident and assured. There would be 'no difficulty' with monitoring emissions, Fahey said, and Coillte didn't require the department's approval for the daily running of its business. He would have 'no difficulty' with putting the land deal into the public domain, he said. He spoke of the need to 'create competition' through this new supply of gas, with gas coming from two interconnectors via Britain. 'The big advantage of Corrib is that we are now going to get gas into the west of Ireland. The ring main, which is being built from Dublin to Galway, would not be built were it not for the fact that the Corrib gas is coming ashore,' he said. The towns of Claremorris, Castlebar, Ballina and Sligo would 'immediately benefit'. What was more, Bellacorick peat-fired power station would reopen as a gas-fired power station, thanks to Corrib, he continued.

Fine Gael TD for Mayo Jim Higgins intervened. He shared 'people's anxieties' on licensing terms for oil and gas, he said. Of course there should be incentives for developers, but Ireland's sole take of a 25 per cent corporation tax with no royalties contrasted sharply with the terms in Norway and Britain. An approach that treated the state and its people as 'joint stakeholders' should be considered.

Higgins said he hoped that environmental concerns in relation to Corrib could be 'overcome' and that the target date of 2003 could be met. However, it would be a 'travesty', he said, if the only jobs accruing to the region were 'on the rig itself or on the terminal'. He also referred to the several departments involved in various approval roles for the project, and posed a key question – one that was to shadow the project for years after: 'Is there anybody actually pulling the whole lot together? What we need really is a definite cohesive plan in place to deliver jobs, direct jobs, sustainable jobs to the region,' he said. People feared what would happen when the gas expired in twenty years' time – when Erris might be left with 'an empty trough, an empty pipeline, a damaged environment and no real sustainable jobs for the region'.

Fine Gael councillor for Erris, Gerry Coyle, wanted the minister and his panel to know that the people present were 'very normal people' who were 'not nuisance makers', but whose fears had 'not been allayed by what they had heard so far. They are farmers, they are fishermen, they are schoolteachers, they are from all walks of life,' he said. 'They have fears, genuine fears . . .' Why had the minister and his panel not visited the area six months ago, or even a year ago, to explain the situation?

Micheál Ó Seighin, and several others present, wanted to know why the minister did not insist on gas being cleaned 'outside', as in offshore. And why were the health authorities not involved? Was it not correct that Ireland was coming from a very weak regulatory position? And how was the issue of access to land for the pipeline going to be handled? 'This land is agricultural now, but as soon as the process begins, or however it begins, it becomes commercial land,' Ó Seighin added.

Responding, Sean O'Riordan of ERM admitted that strategic planning was lacking. In Australia, North America and the Nordic states, a 'very highly developed strategic assessment' took place before any of the planning or policy-making. However, the EU had only in the last month adopted a directive that would require the strategic assessment of policies, such as in transport or the oil and gas sector.

It was approaching seven o'clock in the evening, and the minister didn't want to 'rush anyone', but appealed to those with their hands up to confine their queries to one question. Mary Corduff wanted some details on the legislation applying to access to land, while John Moloney from Glenamoy asked the Dúchas expert how that organization could be 'happy siting this development in the middle of a huge Special Area of Conservation [SAC], totally surrounded by SACs'.

Chris Tallott of the North West Mayo Action Group, representing some three hundred staff working with Bord na Móna and the ESB who faced losing their jobs at Bellacorick, wanted some details on the proposed gas station. He welcomed it, but felt that it should be a 'much bigger gas station'. The minister said it was his 'wish' that agreement would be reached by Enterprise with the local community. 'It is not my intention if at all possible to get into compulsory anything unless and until all other efforts have failed, and in that situation it would be a government decision as to whether there was a need for compulsory acquisition or anything else.'

Monica Muller intervened: 'Enterprise has already threatened us with compulsory acquisition procedures, so the process is already up and running,' she pointed out.

She was right. The minister would sign off on the acquisition of private land in less than a year, the first steps being taken within three months of the Geesala gathering, when notices were published in the *Western People* before the end of November 2001.

When he first threw the floor open to questions at that meeting, the minister had promised that 'We will be back here in several ways on several occasions to look at the various concerns and issues you have.'[10] This didn't quite happen, according to many of those present at the meeting, and some documents were not published during his tenure, as he had promised. And just days after that gathering, shortly before 5 p.m. on a bank holiday, Friday, 3 August 2001, Mayo County Council gave planning approval for the terminal.

The minister did release a report by his marine licence vetting committee in early 2002, which effectively approved the project but with significant conditions. There was no 'major impetus' or action plan for Erris, as he had promised at the meeting. There were no immediate plans for gas in that part of the region, or a power station. Neither would the developers have reached agreement with landowners when the minister authorized compulsory access to land by a private company – the first time that this had been done by the state since independence, but permitted, according to the developers, under an EU directive liberalizing the market.[11]

Fahey has declined to comment publicly on this, but sources close to him say that his intentions to report back, to publish key documents, were 'genuine'. He had convened the Geesala meeting in a bid to try and win over the support of objectors like Maura Harrington and Micheál Ó Seighin. Afterwards, he believed that this was not possible. He had 'followed his department's advice rigorously' it was suggested, given the enormous significance of the project to the west – a region where industrial development had been held back by insufficient energy sources. During his time as a TD, his ear had been 'bent by industrialists' who had repeatedly said that they wouldn't locate in the west of Ireland for this reason. Corrib would, in his view, have a transformative effect.

Just weeks after Geesala, in August 2001, Fahey rebuked those people with 'legitimate environmental concerns' whom he had tried to mollify just a month before – and who now appeared to be exercising their democratic right to appeal the planning decision approving the terminal. Development of the west was not being held up by the government but by objectors to infrastructural developments, the minister told the Humbert Summer School in Ballina, which had taken the theme 'Corrib gas: flagship for developing the west'. If there was a general election in the morning, there would be six independent and Sinn Féin politicians elected to seats on the western seaboard, he said. The local Irish Farmers' Association chairman and a potential election candidate, Sean Clarke, responded that he found Fahey's remarks about independents 'insulting'.

That same month, developers Enterprise concluded a deal, advertised as 60 per cent of the Corrib field production, with Bord Gáis. Given that Enterprise shared the field with partners Statoil and Marathon, 60 per cent of its share represented just 27 per cent of the field overall. There was no clear picture to the future of the balance.

5

THE BERTIE BYPASS

The bulletin came in the post, heralding a 'boost for the entire region'. Shortly after receiving Mayo County Council's approval for its gas terminal in August 2001, Enterprise published its first edition of *Corrib Natural Gas News*. The four-page publication forecast five hundred jobs in construction at various stages, and some '50–65 jobs' on completion. It described how fuel from the 230-million-year-old gas reservoir would be extracted. It was 'a very clean, natural fuel and is composed mainly of methane, the lightest hydrocarbon gas', drawn from a number of wells drilled from the seabed down into the reservoir.

It would flow through a pipeline to the 'reception facility' at Ballinaboy. 'There, the gas will be prepared for onward distribution and sale to domestic and commercial users. This method of development is known as a subsea tie-back, where a remote controlled offshore facility is used to supply gas to an onshore facility,' it explained.

It was expected that a total of seven subsea wells would be completed and 'tied back' to a central subsea 'gathering station or manifold', which would in turn link up to the main offshore pipeline. The bulletin continued:

> The pipeline will carry gas from the manifold to a landfall at Broadhaven, County Mayo, and from there, to the onshore facility. The relatively small volumes of water are removed from the gas, enabling it to be exported as dry gas into the Bord Gáis Éireann (BGÉ) national gas grid.
>
> Gas will be exported from the terminal via a £100-million pipeline, supported by the project partners, which will be operated by BGÉ to a tie-in to the national gas grid at Craughwell near Galway. The Corrib gas field will be continually monitored and controlled by a high-tech integrated pipeline monitoring system.
>
> [It will] facilitate the proposed new gas-fired electricity generating plant planned by Rolls-Royce Power Ventures at Bellacorick, County Mayo. The 68-megawatt plant will produce enough power for the whole of County Mayo and will create 130 jobs during construction and up to 60 direct and indirect jobs during operation.

In fact, the Rolls-Royce plant didn't materialize, although it did secure planning permission without any opposition. The bulletin reminded readers that oil and gas companies had spent some £856.6 million on exploration and appraisal in Irish waters since 1970, with Kinsale being the only confirmed result hitherto.

'For those living close to the Ballinaboy site facility, the conditions relating to construction working times and traffic will be especially relevant,' it said. This included designated weekday and weekend working schedules, and prior permission had to be secured from Mayo County Council if such hours were exceeded. 'There will also be strict controls and regulations in regard to all site development and construction activity on-site,' it stated. School opening and closing times were to be 'avoided'.

The bulletin noted that these conditions applied to construction, and separate permission had to be sought for an operating licence from the Environmental Protection Agency (EPA), which would set out the limits of any emissions in relation to air, water or noise. The offshore oil and gas industry 'is probably the most regulated in the whole country with regard to the licensing and planning regime which regulates it. We expect the EPA to be no less strict and rigorous than Mayo County Council when it comes to examining our application.'

Even as the publication was being distributed, work was also in train on a promotional film. Presented by a TV3 newscaster, Alan Cantwell, the *Corrib Gas Energy Report* included shots of the Corrib gas rig at sea, stunning aerial footage of the north Mayo landscape, and references to its 'awe-inspiring rugged beauty', its fragility and the sensitivity of the project. There were shots of the Point of Ayr gas plant in north Wales and an interview with its manager.

There was also a brief interview with a representative of the Irish Business and Employers Confederation (IBEC), Dr Mary Kelly, who assured viewers that the standards set by the EPA for licensing the plant were very 'stringent'. Dr Kelly subsequently became head of the EPA, and was in that post when the Corrib gas developers' application for an integrated pollution prevention control licence (IPPC) was submitted to it. The EPA made it clear then that Dr Kelly would not be involved in any decision on the application, and there was 'no conflict of interest'.

Only one element of the complex Corrib gas project jigsaw required planning approval back in the early noughties: the terminal at Ballinaboy. The existing legislation exempted pipelines from planning, even though this was not the average Bord Gáis transmission or distribution pipeline. A number of other consents and authorizations were required, however, including foreshore licences. As part of the work on licence applications, the developers conducted dye dispersion tests in Broadhaven, using what was described as a very small quantity of a 'harmless biodegradable vegetable dye'. The last of the tests was undertaken in late August. In early September, a coloured slick in the bay resembling a form of 'red tide', or algal bloom, was reported to the Department of the Marine.

The department initiated an investigation into the pollution incident. Enterprise issued a statement responding to 'speculation and comment' on local radio, saying that it had 'no connection whatsoever with the coloured slick' and stressing that its tests had taken place offshore to the 'highest environmental standards'.

The Department of the Marine subsequently said that waste mushroom compost 'may have been responsible for the slick' and it believed it had been illegally dumped.[1] There were several further slightly ambiguous statements, and never any reported outcome of investigations. Residents came to believe that their pollution reports weren't being taken seriously.

In September 2001, Taoiseach Bertie Ahern embarked on a two-day trip of Mayo, with both Achill and Erris on the itinerary. Enterprise invited him to call into the Corrib information centre, which had been established in the village of Bangor Erris, where he could view a model of the gas terminal and receive a briefing from Brian Ó Catháin. He visited Belmullet and the landfall at Glengad near Pollathomas. He was due to visit Ballinaboy but bypassed it – on a back road, which promptly became known locally as the 'Bertie bypass'. Before leaving Glengad, Ahern was heckled by Maura Harrington. Enterprise managing director Brian Ó Catháin signalled to one of his assistants to invite the landowner, Eamon Sweeney, up to talk to Ahern as a distraction. Sweeney demurred initially, then agreed. Someone had forgotten that Mayo was also strong Fine Gael territory.

The fun didn't end there. Marine minister Frank Fahey asked Ó Catháin if he could show their driver the best way to the next stop at Geesala. Ó Catháin knew it wasn't appropriate. 'Who was that awful woman?' were the first words he heard from Ahern in the front of the car, before the Taoiseach grasped that they had a passenger in the back.

'I think they quickly realized they shouldn't be seen with the developer in the car – it was just an accident really – so they dropped me at the next crossroads by John Healy's garage,' Ó Catháin remembered. However, this was in north Mayo, and there was no mobile phone coverage in the area. Fortunately, Fianna Fáil councillor Frank Chambers was en route to Geesala and picked him up. As Ó Catháin recalled, 'He asked me to crouch down in the passenger seat as we arrived at the hotel to avoid being seen by the protesters. Later that night, Maura Harrington was thrown out of the bar for haranguing me in Irish.'

The *Western People* described Ahern's visit as one of the 'first shots in the general election campaign'. It said that Ahern had been let down by his party supporters in north Mayo, with hardly anyone at Bangor when he visited the Enterprise office, apart from school principal Maura Harrington who 'opposes the gas terminal with persistence, passion and conviction'.

The decision to bypass Ballinaboy was 'a spectacular own goal by An Taoiseach on his goodwill tour of Erris', Christy Loftus of the *Western People* reported on 12 September 2001, under the headline 'Bertie bypasses "Battle of Ballinaboy" as goodwill tour bogs down'. The Ballinaboy and Lenamore residents, who had waited for two hours with the tea and sandwiches, were 'not taking it lying down' and 'packed their bodies, their children and their placards into the nearest available means of transport and parked themselves outside' the hotel Ahern was visiting in Geesala.

The Taoiseach met concerned roads people, fishermen, postmasters, jobless, ESB/Bord na Móna workers, leaving the protesters outside. Brian Ó Catháin rejected the suggestion that he might have kept Ahern away from Ballinaboy. The plan had been to show the Taoiseach the landfall site and the route of the pipeline to the terminal. Loftus continued:

The evening before, he had personally invited the residents to come to Bangor to the office to meet the Taoiseach. EEI [Enterprise] were not in the business of deception or causing trouble. They wanted good relations with their neighbours. And still, outside, the protesters kept waiting and kept asking that the Taoiseach come out to meet them. There was an offer to them that he would accept a deputation. They were unsure whether to accept the offer or not. I was in the midst of them myself offering advice – when asked, let it be said – to the effect that to refuse the offer would hand Bertie's alicadoos a stick to beat them with.

There was a lot of heat and little sweetness and light.

The residents finally got their 'slot', after the 'tourism delegation'. They were the final group to see Ahern, with the tea they had made for him long gone cold.

Michael Ring was not a happy man. The poll-topping Fine Gael Mayo TD, formerly an auctioneer, was renowned for his good cheer, his keen wit and his original turn of phrase, which had been put to most effective use on more than one occasion in Dáil Éireann. However, this was no laughing matter. Deputy Ring had tabled a question for Minister for the Marine and Natural Resources Frank Fahey in relation to the Corrib gas project. Ring knew and understood Erris well. He also knew and understood the significance of this scale of investment for the area. And he knew, and understood, the pressure that Mayo County Council was under in its handling of the planning application.

He was one of only two national politicians at the time to make a formal objection to the planning authorities on the terminal.[2] The substance of the deputy's question now related to the terminal application, which had been given approval by the local authority just before a bank holiday weekend two months before, in August 2001. It had been appealed immediately by local residents, who were Ring's constituents. What the deputy wanted to know was

'Had the minister's department contacted Mayo County Council and asked to be informed before a decision was taken on it?'

In a Dáil reply on 15 October 2001, Fahey admitted that his department had made contact, and had asked to be informed before the decision. However, he said that 'Mayo County Council did not make contact with my department in regard to their decision in relation to the application.' A meeting had been hosted by the local authority in late June, at which a department official was present.

'This is not acceptable,' Ring said. 'This is a serious development. It is no wonder people in north Mayo are concerned about the planning process in relation to the terminal building.' Why, he continued, was the Department of the Marine 'interfering in the planning process'?[3]

Several memos obtained by Ring under the Freedom of Information Act gave further clues. Both appeared to indicate an undue level of interest in approvals for the project by the very department that was both developer and regulator. The first, dated 18 May 2001, was addressed to Des Mahon, Mayo County manager, from Michael Daly, principal officer with the Department of the Marine and Natural Resources petroleum affairs division (PAD). It stated that under the Petroleum and Other Minerals Development Act 1960, the minister would be 'granting shortly' a lease to Enterprise, Statoil and Marathon for development of the Corrib gas field. Under the lease terms, the companies would be required to submit a plan of development for the field, and an accompanying environmental impact statement.

The 18 May memo noted that while the minister held these responsibilities he was 'aware' that other bodies were involved in the 'approval process' for certain aspects of the development. These included Mayo County Council, the EPA and the coastal zone administration section of the minister's own department. 'The petroleum affairs division of this department is acting as co-ordinator for approvals to be given with this department for the development,' the memo continued. 'As such it is anxious to ensure that where there is overlapping between other bodies and this department's area/responsibility that the department would be kept informed of the up-to-date position of developments in those areas.'

Significantly, it referred to 'consistency between bodies in decision-making' and said that the department was 'seeking your co-operation and would appreciate if you would inform us in advance of any approval being given for the development'. A note on the memo gave the names of three officials, and was signed by county manager Des Mahon, with a handwritten observation: 'Please ensure that DMNR request is complied with. Des Mahon 22/5/01.'

A second memo on Mayo County Council headed notepaper was dated 12 June 2001, with file reference P01/900. Signed 'T. Stanaway' and addressed to 'SEP', it conveyed the contents of a telephone enquiry from Des Page of ERM – the consultancy hired by the department's petroleum affairs division to examine the environmental impact statement accompanying the Corrib planning application:

> Des Page ERM working on behalf of Dept of Marine rang this
> afternoon re P01/900, to inquire if FI is requested on
> conditions. Please contact him immediately on your return.
> 2 main issues he feels they may have. (1) Project steering group
> to act as a forum for issues on which Minister wishes to be on
> (as discussed with Ray Norton and yourself). Is this possible?
> (2) Does the council have a view on the adequacy of the EIS?
> (Perhaps some conds. re monitoring programmes and surveys
> during construction.) They feel that there are some inexplicable
> gaps in the EIA.

It gave a contact number for a return call, and added, 'Any comments will be sent from ERM to Dept of Marine and hence they wish to speak to you before reporting to Dept of Marine.'

Deputy Ring smelt blood and was determined to pursue the minister within the privilege of the Dáil chamber. Why was a consultancy commissioned by the minister's department to vet an environmental submission by the developers in direct contact with the local authority charged with giving planning approval? On 17 October 2001, two days after the issue came up in the Dáil, Fahey issued a statement on department-headed notepaper in which he 'strongly refuted charges that he or his department had attempted to influence in any way' Mayo County Council's decision on the Ballinaboy terminal planning application. 'My department's sole interest in communicating with Mayo County Council arose from our general responsibility in relation to the regulation of the project under petroleum, foreshore and pipeline legislation,' he said, adding that it was 'entirely legitimate that we be made aware of the timing and nature of any important authorizations by other public bodies as early as possible'. It was in this context that 'letters of 18 May 2001' were addressed to Mayo County Council, the EPA and the foreshore division of his department, Fahey said, and it was 'never the intention of the minister, or his officials, to influence in any way the consideration by the local authority of the planning application'. There was no reference in the minister's statement to the memo of 12 June 2001.

The matter wasn't over there, however: there was a 'heated exchange' between the two deputies in the corridors of the House afterwards. It was witnessed by 'several colleagues', according to *The Irish Times*, which noted that 'Mr Ring has declined to confirm or deny the details of the altercation, but observers say its occurrence was a reflection of the high stakes, and the tensions, involved in the run-up to the general election'.[4] 'It was a very heated row between them in the lift,' Ring told this writer, recalling how his fellow constituency colleague Enda Kenny had to intervene. It would be late November of the following year before Fahey's department would give a more detailed explanation of the matter, following a Channel 4 television report on north Mayo.

*

In November 2001, Enterprise and partners received their petroleum lease from Frank Fahey, who described it as a 'milestone' in Irish exploration history. However, the terms of the Corrib deal were criticized by Des Geraghty, president of the trade union SIPTU. Speaking in Galway where he unveiled a bronze statue by John Behan, which was presented to the city by NUI Galway and the trade union, Geraghty told *The Irish Times* that he believed there 'should always be a "take" for the state. I think that the gas is extremely important as an indigenous energy source for economic development, but I believe from the start the concessions were unbelievable,' Geraghty said. 'There were no jobs in it. There was very little for the Irish economy and we are now suffering the consequences of a very bad policy which former minister Ray Burke has a lot to answer for. I think it goes back a long time, but at this time I think the people of the west are right to demand that we maximize the benefit of that resource for this region.'[5]

Enterprise was still in confident mode, in spite of the planning appeals lodged by a number of groups and residents over permission for the gas terminal – including state heritage agency Dúchas. Enterprise had also appealed some of the conditions. Its second issue of *Corrib Natural Gas News* of November 2001 included a front-page photo of Brian Ó Catháin with Taoiseach Bertie Ahern and Minister for the Marine and Natural Resources Frank Fahey 'on a recent visit to the landfall where Corrib gas will come ashore in 2003'.

It ran a separate article on the 'rigorous selection process' undertaken before decisions were made on the development options for the project, and described the subsea system as being the 'preferred scenario' as 'floating production facilities' were 'not suited to the type of gas production scenario required for a dry gas field such as Corrib'. Such offshore terminals were more suitable where the product could be 'offloaded in a batch', such as oil; it made no reference to Kinsale Head.

There were several more reports, including one on 'significant economic benefits', and landowner compensation. State agencies would be consulted in relation to mitigating harmful environmental impact. Furthermore, a community fund of 'up to £1 million' would be established to support community projects in the barony of Erris.

The bulletin's 'questions answered' panel was very reassuring. Impact on the landscape? 'No significant permanent changes.' Impact on commercial fishing? There would be 'limited exclusion zones over subsea facilities' and compensation for any lost fishing gear – and even extra fuel costs could be covered as a result of avoiding pipeline construction vessels. Fishing grounds would 'not be affected' by a water discharge from the terminal, which would be subject to 'strict mitigation measures'. Any traffic increase onshore would be 'short term'.

Noise would be 'minimized' with 'silencing' and 'screening' of equipment. Atmospheric emissions would be 'very localized' and would 'disperse very quickly'. There would be 'no discernible impact' on the sea and onshore watercourses. Water discharged from the terminal would 'be so clean that it will meet the coveted EU Blue Flag criteria' for clean beaches. There was a brief reference to a planning appeal: 'These permissions and licences involve a high degree of public consultation ensuring that the public has extensive opportunities to have an input into the development process.'

Jacinta Healy had become spokeswoman for residents of sixteen houses in Ballinaboy and Glenamoy that had lodged an appeal to An Bord Pleanála over the gas terminal. 'We called a meeting in one of the houses and asked Fine Gael TD Michael Ring, and he came. He advised us not to say we were against the gas, but to say we objected to the way it was being handled. It was very good advice. After Ring left, I remember how a local man by the name of Francie Murphy called to the house. Francie was very knowledgeable, in his seventies and living in Bangor Erris, and he used to cycle around, sleep out in haysheds sometimes, and he would take tea in many houses. He asked us what we were talking about. "*A stór ó, a stór ó*, coming into the land of the innocents," was his response. I've never forgotten those words of his.'

The hearing was to be held in Ballina in February 2002. Significantly, its brief was limited to the terminal or refinery – not to the entire project, which would include an offshore and onshore pipeline and a discharge pipe. This was 'project splitting' – a concept anathema to the stated European Union view of how planning issues should be conducted. However, from the developers' perspective, the Corrib gas partners were simply acting within the legislation as it stood – even if this was the first project of its type to be considered by state authorities.

Some eight appeals had been lodged, by local residents and by groups including An Taisce, the Friends of the Irish Environment, the Erris Inshore Fishermen's Association, and Dúchas. The hearing's location and duration meant considerable juggling at home and a daily two-hour return journey for residents who struggled to attend in their own time, while the developers and consultants were paid to be there.

Jacinta Healy found it tough going. 'I would leave home at eight a.m., having got the kids ready for school, and my mum and neighbours would help out with the youngest and with the kids returning home afterwards. Some nights I didn't get home till eleven p.m., when the kids were in bed. When I was at home, the phone never stopped ringing, and many's the day the dinner went out the door as I didn't have time to eat it. What was most offputting was that we would meet Iggy Madden haulage company trucks carrying pipes for the project into Erris as we were on the road to Ballina. So we knew it was a done deal, even at that stage.'

The hearing, which ran initially for two weeks, was chaired by Bord Pleanála inspector Kevin Moore, who had retained hydrogeologist David M. Ball to prepare an independent report 'relating to soils, peat, bedrock, groundwater and surface water and their influence on planning issues', as the board put it. Enterprise's team, led by Brian Ó Catháin, persuasively put the case for the project and its benefits to Erris, but at one point an apology had to be issued when a consultant for the developers – describing how the project could transform Erris – tried to paint a contemporary picture of a bleak and poverty-stricken area.

'I was just outside grabbing a cup of coffee,' Healy recalled. 'Someone came running out to get me to come back in. The presentation included a slide of the Spar shop in Barnatra, which the consultant described as "out of character", and a slide of "rubbish" outside a house. That was our turf stack! By the time I had finished with him, he had been made to look like a complete prat, but what broke my heart was that the next morning, as I was leaving home, my husband James was outside trying to make the cover over the turf stack look a bit neater . . .

'Another day, the developers were being questioned about their submission on flora and fauna in the environmental impact statements,' she said. 'We would regularly see badgers and foxes crossing the road, and the inspector asked the Enterprise team members what would the impact be on this wildlife. They said they would "open the gates" of the terminal to let the badgers and foxes through! I'll never forget one of our group turning to me and asking if I thought they might "give the badgers a set of keys".'

The Dúchas objection at the hearing was troubling. Correspondence released under the Freedom of Information Act showed that Jim Moore, Dúchas's deputy regional manager for the west, was not at all happy about the Ballinaboy plans, and a memo of 12 July 2001 outlined concerns relating to effluent disposal, discharge of an 'unacceptable level of contaminants' within the marine special area of conservation in Broadhaven Bay, the capacity of proposed silt ponds and lack of information on site geology.

The Ballinaboy proposal represented a 'significant threat to nearby Special Protection Areas and Special Areas of

Conservation, therefore the National Parks and Wildlife Service of Dúchas must object', the memo said. Mayo County Council issued its approval before the Dúchas objection was lodged, so regional manager Dr Noel Kirby recommended that Dúchas should object on the same issues to Bord Pleanála.

In its appeal, Dúchas expressed concern about the 'inadequate information regarding site geology, which could result in permeation of pollutants from the peat repository to adjacent watercourses'. It said it would have no objection, provided an 'impenetrable layer beneath the repository areas' was imposed as a condition to 'prevent pollution entering the groundwater system'. At the Bord Pleanála oral hearing, Jim Moore outlined the agency's concerns about the impact on three European sites, namely Glenamoy bog complex, Carrowmore lake and Broadhaven Bay, and pointed out that Carrowmore lake was just four kilometres from the northern peat repository area on the Ballinaboy site, and 3.5 kilometres from the main site. Sruwaddacon estuary was even closer, at 1.5 kilometres and two kilometres from both areas respectively.

Many felt that the state agency paid a heavy price for its attempts to conserve the natural environment. After the next general election, it was dismantled in a reorganization of government departments.

Dr Mark Garavan, who attended the hearing, recorded afterwards that it was very important to understand that there was 'no fixed community position'. Certain residents, including workers facing job losses when the Electricity Supply Board's power station at Bellacorick would shut down in three years, welcomed the initiative as a boost to the area.

There were those who were indifferent, or who believed there was little point in opposing something that clearly had government backing. There were those, Garavan noted, who believed that while the proposal might be 'suspect', it was worth making 'every effort' to extract local benefits. And there was a group whose opposition was largely based on the significant health risk that the project posed to the community.

As Garavan noted in a paper published in 2006,[6] the format of this and other oral hearings on planning matters was such that the world of the 'non-expert' – as in a resident who could be most directly affected on environmental, health or other grounds – was silenced within a 'formal public sphere' where 'scientific modes of reasoning' were afforded privilege by decision-making bodies. This was most strikingly reflected, Garavan noted, in a submission by Micheál Ó Seighin on the cultural impacts of the project:

> Culture is not merely some stone remains, or even oaken posts or deer traps under or over the bog, though these kind of artefacts are indeed a part of culture, and we have traditionally given some of them semi-mythical status, for a time of years or maybe centuries. Culture, however, is how we live, what we do, how we mould our environment to cope with the struggle of living and rearing a family; how we and our sometimes harsh environment come to a compromise that enables both to thrive . . . Culture is not a suburban lecture: it is the *dlúth agus inneach* [warp and weft] of our existence . . .
>
> There is a cultural blockage here. According to the stereotype, we, the objectors, should be against the gas – nothing else fits in. But we're not. I hate coal, necessary in space and time but disastrous. Natural gas, once separated from less savoury traces, is as near to a perfect fuel in today's technological milieu as we have access to, apart from wind . . . My problem with Brian [Ó Catháin] and, more so, with the guardians of our democracy – elected or appointed or just doing a job – is a cultural one of incomprehension. The project is said to be good for the 'area' – jobs. If the area doesn't gain much, it is good for the country . . .

Ó Seighin applied humour to good effect in his submission. Two members of his family worked for multinationals and 'haven't been eaten yet', he quipped. The input of jobs and wages represented a 'major, immediate cultural influence', he explained. 'I do not fear the social consequences of five hundred libidinous construction workers descending on the area: Erris ladies are not Sabine or supine, though some of the temporary newcomers could be, like me, lucky enough to be snared by an Erris woman . . . The reality is, however, that construction work on major projects is, and has to be, almost entirely covered by permanent contract staff . . .'

Archaeological work associated with the project had been 'intense, but confined to consideration of that which is academically associated with this area, and so with desktop access', Ó Seighin said. What about the 'hand-cut sod fences in Rossport', he asked, which were of 'a kind that will never be repeated', or the drains, which were a 'testament to the persistence of the human spirit, in its determination to succeed and carve out a viable place, even on the margins'? There were also the visible remains of the open field system based on Rossport House and 'long replaced elsewhere by the enclosures that became modern practice'.

A ballad, 'No Terminal at Ballinaboy', was sung as a submission by Maura Harrington's daughter, Astrid – a first on record at an oral hearing. At one point, hydrogeologist David Ball quizzed the Enterprise team as to their reasons why the pipeline could not be brought into Galway Bay instead of Mayo. The developers were a little vague with their response; Ó Seighin came to their aid with some of his own knowledge of geology.

'Except that I made one mistake,' he recalled afterwards. 'But I kept going. Afterwards, at one of the breaks, David Ball came up to me to point out my little slip. I explained that I was aware, but felt I had to keep going. "Absolutely," was his response. "But wasn't it strange that the developers didn't spot your error?"'

There was integrity, a clear love for their landscape, and quiet confidence about what was local. And there was emotion, as articulated by Jacinta Healy, who warned An Bord Pleanála's inspector Kevin Moore that 'If you grant planning permission, you would be sentencing the people of Ballinaboy to a life of misery and worry.'

As Garavan summarized it, attempts by officialdom, supported by the developers, to downgrade local knowledge on issues like a turf cutter's intimate knowledge of the composition of peat had been met with a community presentation that was a veritable *tour de force* of popular discourse.

In April 2002, in advance of the planning appeal outcome, Frank Fahey published a report by his marine licence vetting committee, which recommended granting full consents to Corrib, but which, as objectors noted, did not investigate the option of a shallow-water offshore processing option. Fahey also approved the plan of development for the field, including consent for construction of a gas export pipeline and subsea structure. The minister was busy with his pen as a general election approached. The next month, May 2002, he signed thirty-four compulsory acquisition orders for access to private land on the pipeline route – the first time that the state had signed over access to private property to a private company. He also granted the key foreshore licence.

Even as all this activity was taking place, and An Bord Pleanála was still deliberating on the appeal, Royal Dutch Shell was poised to buy out Enterprise for a reported €5.7 billion. Enterprise's share price was low, as oil prices were low, but it had great assets in the North Sea, the Gulf of Mexico and Norway. A number of oil companies were interested in purchasing it for this reason, and one company moved to a 'bear hug', which meant that news was leaked to the press. This led to negotiations over the company's sale.

At a certain point, an offer made by Shell was sufficiently higher than the share price, and this prompted the Enterprise board to recommend this offer to its shareholders. Shell had done little work in two areas of the company's portfolio – Ireland and Italy – and had little or no interest in or knowledge of Corrib, according to those involved at the time. It was perceived to be a relatively small part of the reward.

For the residents, the news was something of a hammer blow. This was one of the world's largest oil companies, which, in spite of efforts to restore its image, was still remembered for the Greenpeace protests over the company's efforts to sink the Brent Spar oil rig in deep Atlantic waters in 1995; for its activities in the Ogoni delta in Nigeria, criticized by Ken Saro-Wiwa and his colleagues; and for its controversial oil and gas extraction project on the fragile Sakhalin island environment in Russia, home to the endangered western Pacific grey whale.

In June 2002 An Bord Pleanála sought more information from the developers in a six-page letter that expressed concerns about health and safety, due to the terminal site's proximity to residential areas. The letter said that the company had not demonstrated that Ballinaboy was the best location. The appeals board then re-convened the hearing for later in the year, following receipt of further information from Enterprise.

However, by then Brian Ó Catháin had moved on. In late June 2002, it was confirmed that he would be replaced by Shell executive Andy Pyle, who had been with Shell since 1976 when he worked as field engineer for the North Sea and the Shetland Islands. In 1989 Pyle became engineering manager for Shell's Brent Field operations. His thirty years of experience with the company included working on several joint ventures with Exxon Mobil, and spending time in Houston, Texas. Before moving to Ireland, he had held an information technology role, ensuring software was 'integrated' throughout the Shell group in more than 140 countries. He was approaching retirement. Corrib would be his last Shell posting.

Ó Catháin's departure was by 'mutual agreement', it was said, and Shell's exploration and production regional director Brian Ward paid tribute to his 'highly significant role in the development of the Corrib project'. Ó Catháin explained afterwards that he had worked with Shell for seven years from 1984. 'I did not want to leave Enterprise Oil to go back to Shell as I did not enjoy the Shell corporate culture,' he said. A number of Enterprise staff stayed on for the new owner, including external affairs manager Rosemary Steen, who subsequently departed for a new role with a mobile phone company, Corrib operations manager Mark Carrigy and Corrib environmental, health and safety adviser Agnes McLaverty.

In the summer of 2002, while still awaiting An Bord Pleanála's decision, the Corrib gas developers engaged in excavation work at the proposed pipeline landfall at Glengad beach. It prompted a court action by Brendan Philbin's eighty-four-year-old mother, Mary, of Rossport South, who sought an injunction to stop the work: she argued that a *cillín*, or burial ground for unbaptized children, near Glengad had been removed. Mr Justice O'Sullivan granted a temporary injunction against Enterprise.

'I could not stand by and let this company show such utter contempt to the dead children of this parish,' Mrs Philbin said in her affidavit. 'This is nothing short of desecration. I have never been so upset about anything in my entire life and feel as if my parish has been violated by the most arrogant and uncaring forces.'[7] Peter Sweetman, a planning and environmental consultant who worked for many communities, had found there was no authorization under the Planning and Development Act 2000 for an entrance constructed as part of the work.

Enterprise said that it 'totally and absolutely rejected' claims that the works had interfered with a children's burial ground. Privately, the company was convinced that a pile of builders' rubble had been mistaken for a *cillín*, which was marked on Ordnance Survey maps but not at that location. Work in the 'vicinity' of the mound was being undertaken with department approval and under Dúchas's archaeological supervision, and it regretted that the activity had caused 'concern'.

On 25 June, the developer sought discharge of the High Court orders restraining works at Glengad, denying it had interfered with a *cillín* and that the construction of a temporary access road at Glengad posed a safety risk. It argued that the orders were costing it €85,000 a day, and 'if continued, the order may expose the company to irreparable damage'.[8]

Mr Justice Ó Caoimh was presented with a large number of affidavits from engineers, planning consultants, archaeologists and other experts on behalf of both sides. Affidavits were also submitted from a number of local people in support of Mrs Philbin. Archaeologist William Frazer said he was satisfied there was only one recorded archaeological monument in the field at Glengad, which was linked to the pipeline development by an access road. A second mound in the field, which Mrs Philbin had described as a *cillín*, had no archaeological importance, being a modern 'spoil heap', and this had been removed 'in full compliance with best archaeological practice'.

Mrs Philbin's lawyers argued that it was irrelevant that the *cillín* was not recorded on Ordnance Survey and other maps. Archaeologist Treasa Ní Ghearraigh said in an affidavit that the tracing and identification of a *cillín*, from an archaeological perspective, was difficult.

In another affidavit, Micheál Ó Seighin said the two mounds referred to constituted a sacred place in the tradition of the locality, and should not be interfered with. One had to be 'sensitive to the feelings of people who were aware of their location, largely because of the psychological trauma associated with burying children in unconsecrated ground, usually in secret'.

The developers had applied for consent for 'preparatory works' on the onshore pipeline as part of phase three of the consent programme. Within three weeks, approval had been granted by the Department of the Marine with four conditions. It seemed as if the department did not feel obliged to take into account, or was unaware of the fact, that An Bord Pleanála was still not happy with a key part of the project and was seeking more information.

Frustrated with lack of answers over authorization for the Glengad work, several residents, including Bríd McGarry and her mother Teresa, Maura Harrington and her husband Naoise, her daughter Astrid and partner, went down to the beach where the cliff-face had been opened by the developers. The group wanted to establish their right to walk along a public beach. Bríd Ní Sheighin was about to get married on 5 July, but joined them the day before her wedding.

'Some Shell workers told us that we couldn't stand inside the wire where they wanted to dig, but we just ignored them,' Ní Sheighin recalled. 'It was a small group, there was no trouble.' The gardaí arrived and gave the residents assurances, so they left. Harrington went to school, but decided to check back that afternoon. She spotted a digger working down on the shore, and decided to sit inside its bucket. The driver had to stop. She stayed there for the best part of twenty-four hours, with Bríd McGarry joining her throughout the night and the next day, and neighbours coming down for periods in solidarity. Photographers and television cameras arrived, and there was a delivery of hand-knitted woollen socks from Mary Philbin.

'The driver locked up and went, and we decided we wanted to wait to see the paperwork. It was very cold out,' Bríd McGarry remembered. 'We had blankets and coats and people brought us tea and food, and the gardaí were coming and going during daylight hours but they put no pressure on us. Next day when the sun came out, we actually got sunburned. We were there till four or five p.m., but never got to see the paperwork we were looking for.'

For Harrington, it was her first demonstration of non-violent direct action, and her supporters felt she had right on her side. 'After the gardaí gave us further assurances, Willie Corduff and myself were the last to leave,' she said.

On 21 July, Shell said it was suspending work to undertake an internal review – in fact, work would continue at Glengad all summer, but not offshore as had been planned. Swiss shipping contractor Allseas, which had been booked for summer offshore pipelaying work, filed a €31-million penalty claim. Worried about what was happening at Glengad, Maura Harrington and 'other concerned persons' wrote to Andy Pyle. They referred to an open letter Pyle had sent to the community that same month, and it welcomed as 'commendable' his belief that the venture should proceed only 'with proper engagement and discussion with the local community'.

Would Mr Pyle participate in a conference, at a time of his choosing, where he could meet members of the community 'in an open and transparent manner'? Harrington and her co-signatories asked. Replying on 9 August, Pyle said that the developers were holding a 'series of meetings with representatives of various interest groups in the area', and he felt this was the 'most effective forum' at this stage. He would be happy to meet 'two or three representatives of your group in our Bangor Erris office'. He did not want a crowd.

Pyle referred to the benefits that a successful development of the Corrib field would bring to the community and noted Harrington's protest: 'I am saddened and disappointed that this is the case, and I feel that such activity does not serve well the best interests of the Erris community, the project or the potential benefits that Corrib gas can deliver to the west of Ireland.'

Frank Fahey, who had not been reappointed to cabinet after the 2002 general election, had promised that an environmental management or monitoring group would be set up to monitor developments. The group as constituted wasn't independent, in that the developers were participants, along with a number of state agencies, and the press statements were issued by the department. In July, the department's petroleum affairs division attempted to organize an election of community representatives to the group, at a meeting in Rossport School.

Vincent McGrath, a secondary school teacher and musician, and nephew of Rossport composer and fiddle player John McGrath, had moved back to his native area of Rossport with his wife Maureen and two daughters. He had spent years away in England – working for contractors on the North Sea gas pipeline at one point – and latterly in County Louth. He and several neighbours were disgusted with what they considered to be an attempt to manipulate the vote, and they walked out. He subsequently sent an email to new marine minister Dermot Ahern but got no response.

A department press release of 13 August 2002 confirmed that the environmental management group had paid a 'site visit' to Glengad. It summarized 'activities completed' by Shell/Enterprise as including construction of an access road, the completion of a 'cut' through the cliff at Glengad, excavation of 70 per cent of the pipeline trench from 1.2 kilometres to a full depth of three metres in the nearshore and inter-tidal zone to within three metres of the base of the cliff.

> The main works planned by EEI [Enterprise] for the remainder of this summer include the completion of excavation of the pipeline trench to final level including the rock blasting of some 200 metres of trench in the sub-tidal zone, the back filling of the trench for the winter, the removal of the causeway and the preparation of the cliff and the Rossport compound area for over wintering. Discussions are ongoing with Dúchas and DCMNR [Department of Communications, the Marine and Natural Resources] regarding this work.

It would transpire that planning permission was also needed for the shore works.

However, Dúchas was being dismantled by the new government, and responsibility for heritage was transferred to the Department of Environment and Local Government. The department held the dual mandate of promoting development and protecting heritage and, as one leading academic noted when analysing the impact of this on another significant construction project – the proposed building of a motorway through the high kings of Ireland land-scape, the Hill of Tara in County Meath – 'It is hard to read this as anything other than a move to neutralize the heritage and environment lobby as an obstacle to development.'[9]

In late August, Andy Pyle issued a warning: planning delays could jeopardize the multi-million-euro project. The company was fully committed and would submit a new report on its plan for the Ballinaboy terminal to An Bord Pleanála, Pyle said. However, a 'refusal to sanction the plan was likely to prompt reconsideration of the project'.[10] This was followed by what the *Western People* described as a 'major intervention' by Frank Fahey in its report published on 18 September 2002.

Referring to him as 'former Minister for the Marine and Natural Resources', the newspaper's lead page-one report said he had called for a 'strong positive consensus to ensure that the Corrib project is not lost to the west'. It was the 'most important strategic economic development for this region in the history of the State', Fahey was quoted as saying. He warned that the project would 'come to an end if the planning application before An Bord Pleanála failed'. Speaking 'as a west of Ireland TD', he said, 'I believe this would be a most detrimental blow to the economic development of the north-west and Co. Mayo in particular.' It was an 'opportune time' for 'those with an interest in the economic development of the west to recognize how great an opportunity would be lost if the project failed . . .

'Recriminations will be too late in the event of a decision by Shell to abandon the project as a result of a Bord Pleanála planning refusal,' Fahey continued. He 'recognized the rights' of the appeals board to make its decision and the 'rights of the objectors to high-light their case', but said that he 'questioned the motivation' behind 'some of the most vehement and vocal objections. All politicians, business leaders, community and local authority leaders who support the project should now come together to highlight the absolute necessity that it should proceed,' Fahey said, and 'revealed' that he had already made contact himself with people in positions of influence. Among those he said he had contacted were 'senior' Mayo TDs Enda Kenny, Beverly Flynn, together with Mayo county manager Des Mahon, and the Western Development Commission, a state body charged with promoting development in seven western counties, including Mayo. The junior minister also confirmed that he had 'written' to Des Geraghty, president of the largest trade union, SIPTU, asking for 'clarification' on the union's 'support for the project'.

> Making it clear that a negative Bord Pleanála decision would end the Corrib gas project for 'this and a number of generations', Fahey said the choice facing the west was 'stark'. 'Now is the time to be clear on what our priorities are. Do we want the west of Ireland for the first time in our history to become self-sufficient in our energy needs with the ability to generate electricity economically, or do we want to accept the argument that this project is of benefit primary to private enterprise exploiting our natural resources?'

Michael Ring was incensed. 'The minister should stay out of the planning process,' he declared, in a statement issued on 18 September 2002, the day that the *Western People* report was published. He said he was 'amazed' that the junior minister could make such an intervention at a time when An Bord Pleanála was due to announce its decision on the terminal. 'I wonder would Mr Fahey make such an intervention if this was in the middle of a court hearing. That, in effect, is what he has done,' Ring continued, adding that he had met with Andy Pyle, managing director of Shell E&P Ireland/Enterprise Energy, and 'made it clear to him that he could not support the terminal project, as it was trampling over the rights of the people of Ballinaboy'.

'I support bringing the gas ashore in Erris, but not at the expense of the people there,' Ring said. 'Their rights and concerns have not been adequately addressed by Shell/Enterprise Energy and until that happens I am standing by the people of Ballinaboy.' Fahey's appeal had shown 'scant regard for the independence of the planning board, and was typical of the Fianna Fáil approach to authority', Ring declared. 'As far as Fianna Fáil are concerned, rules are there to be broken.'

In October, the Department of Communications, the Marine and Natural Resources – as the reconstituted department under minister Dermot Ahern was named after the 2002 election – had confirmed to *The Irish Times* that the European Commission's environment directorate was seeking 'clarification' on assessment procedures in relation to a number of aspects of approvals for Corrib, including safeguards for protected habitats, reasons for approving a discharge location from the proposed terminal, treatment and discharges of waste water, and the impact of seismic surveys, exploratory drilling and blasting in Broadhaven Bay on whale and dolphin populations.[11]

Ballinaboy was the 'only viable option' for a gas terminal, Pyle said publicly, in November 2002, shortly before the reconvened Bord Pleanála hearing opened. An offshore plant was not an economic proposition, he said. 'It is our hope that we will receive a favourable response from An Bord Pleanála in the near future and that we will be able to commence work on the onshore terminal in the early part of next year [2003].' 'Every effort' would be made to have the 'first gas ashore in 2004', he added.[12]

6

DANCING ON THE BEACH

The days were growing shorter and Christmas was approaching as Jacinta Healy made the daily trek to the hearing and home again. Bríd McGarry, Mary Corduff, Monica Muller, Maura Harrington, Edward and Imelda Moran from Belmullet, Uinsionn MacGraith and Micheál Ó Seighin did likewise, in their own time and at their own expense with their own legal advice. Other objecting groups included the Erris Inshore Fishermen's Association and An Taisce.

The resumed session had just opened when Channel 4 television broadcast a documentary on north Mayo, which included an interview with the now former marine minister Frank Fahey and referred to Michael Ring's allegations of interference in the planning procedure in 2001. Reporter Tanya Sillem drew reaction from Fine Gael leader and Mayo TD Enda Kenny, who said that this would have amounted to undue political interference.

The documentary asked why Mayo County Council did not conduct its own independent environmental audit of the area, rather than relying solely on the audit submitted by Enterprise/Royal Dutch Shell, which then had to be withdrawn and resubmitted when the first audit included non-existent villages. Sillem said that undue pressure had been placed on locals to sign away access to their property for the onshore pipeline, and security men had filmed locals. And she explored the legislative changes overseen by Frank Fahey to permit compulsory acquisition of private land.

Sillem questioned the route of the pipeline on safety grounds, and explored political lobbying and claims of political donations to Fianna Fáil in 1998. She could find no evidence of any donations to Fahey. Previewing the documentary in the *Irish Independent*, senior journalist Sam Smyth said:

> The report suggests that the Department [of the Marine] has acted like an implementation arm of Royal Dutch Shell, using the blanket explanation that it is 'in the national interest', rather than undertaking a critical audit of the project.
>
> The government, *Channel 4 News* say, has given away natural resources to Royal Dutch Shell with none of the gas revenues going back to Irish taxpayers – a less advantageous deal for the Irish than the same oil company negotiated with the Nigerian government. Beyond a few hundred short-term construction jobs, Mayo gets nothing from the deal, not even the natural gas, because it is not commercially viable for Bord Gáis to supply Erris. Beyond Royal Dutch Shell's project, there are no concrete proposals for investment in the area, although Rolls-Royce say they will build a natural gas power station in Bellacorick to supply the national grid when the pipeline is operating. However, even that will only provide a limited number of jobs for locals.[1]

Channel 4 had circulated copies of the programme to politicians. On the day of the broadcast, 26 November 2002, Green Party TD and marine and natural resources spokesman Eamon Ryan called on the government to 'outline the precise details of the deal it has struck with the Royal Dutch Shell Company, Enterprise Energy Ireland to bring gas ashore from the Corrib field off the coast of Mayo'.

'A number of disturbing facts seem to be emerging regarding the Corrib field,' the south Dublin TD said in his statement, and he called on the government to give 'precise answers. For example, why did former Minister for the Marine and Natural Resources Frank Fahey change the law to give himself the power to issue compulsory acquisition orders? Why did the Department of the Marine exert pressure on Mayo County Council's planning committee? Why did the Department of the Marine refuse to disclose how much Royal Dutch Shell paid for the land acquired for the inland terminal site, even though it was purchased from the state-run agency?

'The real question which must be asked about this gas field is this. Is the Royal Dutch Shell deal a good deal for Mayo and a good deal for the Irish taxpayer? Reports suggest that not only is it not a good deal for Mayo, but it is an even less advantageous deal for Ireland than that negotiated between Royal Dutch Shell and the Nigerian government. Apart from the obvious economic questions that need answering, it is quite clear that Enterprise Energy Ireland's [EEI] proposed route for the pipeline contravenes two EU habitat directives, as it cuts through a Special Area of Conservation.'

Ryan, a tour-company operator and founder of the Dublin Cycling Campaign, had been elected to represent the constituents of Dublin South on the other side of the island, but he had a particular interest in what was happening in Erris beyond his opposition brief. He had brought cycling groups along the southern shore of the Sruwaddacon estuary, through Pollathomas, and they had picnicked regularly on Glengad beach. He was concerned about the plan to run a high-pressure pipe through beautiful but unstable bogland.

Sinn Féin's natural resources spokesman, Martin Ferris, also commented on the Channel 4 report. He said that the state's handling confirmed the worst suspicions of those who opposed it; he would raise the issue in the Dáil and he would demand an 'immediate debate surrounding the role of former minister Frank Fahey'. Two days after the Channel 4 programme was broadcast, a lengthy statement from the Department of the Marine and Natural Resources defended its role and that of the former minister.

The department said it was responding to 'recent allegations in both print and television media' about its 'role in the Corrib gas field project'. Laying out the 'factual position', it 'absolutely' refuted 'the allegation that "huge pressure" was placed on Mayo County Council's planning committee [sic]' in relation to approval for the Ballinaboy terminal, and said that its petroleum affairs division had written to a number of bodies, including the local authority, arising from its 'sole interest in communicating with it', due to its 'general responsibility in relation to the regulation of the project under petroleum, foreshore and pipeline legislation'.

This was 'entirely legitimate' and it was 'never the intention of the minister, or his officials, to influence in any way the consideration by the local authority of the planning application', the press statement said. It also dealt with a number of points raised in the Channel 4 programme and other reports, including the department's failure to inform people attending the Geesala public information meeting in July 2001 that they were being filmed; the purchase of land from the state forestry agency, Coillte, for the terminal; the issue of discharges; and changes in legislation to facilitate the onshore pipeline.

It also referred to the 'suggestions' that the minister wished to become part of the project steering group on planning issues, and said that this 'appears to be a total misunderstanding'. The 'allegation' appeared to be 'confused with the setting up of the environmental monitoring group'. In fact, the minister had established an environmental management group, which included representatives from his department, the developers, Mayo County Council, Dúchas, the North Western Regional Fisheries Board, local fishing interests and residents.

*

As at the earlier Bord Pleanála hearing, Enterprise/Shell was represented at the resumed hearing by senior counsel Eamon Galligan, along with at least a dozen staff members and some twenty-five consultants. Mayo County Council retained a senior counsel, its solicitor, and was represented by its senior planner, Breda Gannon, senior engineer Ray Norton, chief fire officer Seamus Murphy and several officials.

On day four, 28 November 2002, Jacinta Healy reiterated her concerns about the impact on the immediate environment, and the fire hazard. The nearest main fire station was an hour away from Ballinaboy in Castlebar, and the sub-station was in Belmullet, fifteen minutes away. Belmullet was a part-time 'retained' service. An Bord Pleanála had asked the Health and Safety Authority to investigate the safety of the terminal site under the EU Seveso II Directive on control of hazardous substances. The authority's assessment stated that the risk of fatality arising from the proposed terminal's operation would be one death in a million years – and this plant was due to operate for twenty years, or so it was believed.

During questioning, the Health and Safety Authority's representative acknowledged that residents' fears could not be taken into account in a risk assessment since the organization didn't 'have a mechanism that factors that in' and it 'is not within our capacity to relieve anxiety'.

Why was there only one reference in the report from the Health and Safety Authority to the danger of a terrorist attack? Jacinta Healy persisted. This was, after all, only a year after the 11 September attacks in North America. 'We consider a terrorism attack should have been considered, considering current world events, and Shell, the parent company to EEI, doesn't exactly have a good reputation in relation to terrorist attacks either,' she said.

Joan Geraghty, reporting for *The Irish Times*, recorded that 'experts could still offer no certainties that a gas explosion would not occur as a result of operations at the plant', and noted that 'representatives of the Health and Safety Authority could only reiterate that it was "extremely unlikely" a major accident would occur'.

'Assurances given at this hearing don't make the residents feel any safer or more confident about the terminal being built,' Healy said. 'We live in a small village. If there is a fire, help will have to come from Castlebar. I'm sure Castlebar fire station has no knowledge of dealing with a fire at a gas terminal. The fact is we will be going to bed thinking, If there is a fire there or some major explosion, am I going to be there in the morning and if I am, will I be able to live my life the same as every day now?

'We will never have peace of mind if that terminal is built. You can have all the experts you want, but common sense would tell you, why put a fire hazard in the middle of a forest plantation, for the love of God?'[2]

The oral hearing had been the longest to date in the state's history, but history had also been made at another level, Liamy MacNally wrote in his De Facto column in the *Mayo News*. 'It showed that lived experience of local people is as important as the views of "experts". The inspectors asked local people to speak about their experience of working with turf,' he noted. He quoted the summary given at the hearing by Jim Moore of Dúchas – 'We do not inherit the land from our ancestors, we borrow it for our children.'

What irked MacNally most was the 'speculation' that Shell was already engaged in buying out properties close to Ballinaboy. This 'buying out' would not reduce opposition, and was a form of 'ethnic cleansing', he wrote. What was worse, Fine Gael party leader and Mayo TD Enda Kenny supported the concept, although Mr Kenny denied this later. 'If this is the best Fine Gael, as a party, can offer the Corrib gas project, then it should do the honourable thing and disband.'[3]

There had been precedents for buy-outs, and for state compensation for offshore islanders who agreed to move to the mainland, but it wasn't always successful. Back in the 1960s, the west Cork

island of Whiddy, a US air base during the First World War, had
been earmarked as a site for an oil terminal. Islanders dependent on
farming and fishing were offered compensation that seemed good at
the time, according to ferryman Tim O'Leary, whose father had
sold more than a hundred acres. Some of the families moved into
Bantry on the mainland, and dozens of houses were bulldozed. It
'broke the community', O'Leary remembered. Many who sold
never got over the disconnection, and the fact that they missed out
on some good fishing years in the 1970s and farming in the 1980s.

In January 1979 a French oil tanker, *Betelgeuse*, blew up while
unloading a cargo at the Gulf Oil terminal, killing fifty people
including the French crew of the ship and seven local men from the
Bantry area. O'Leary's father had tied up the tanker that night,
hours before the explosion on board. It was his last shift before he
went on leave. He never quite got over the shock. A population of
up to eighty dwindled to twenty. The terminal, latterly owned by
Conoco Phillips, has its own pier and has little contact with what's
left of the community. Staff travel in from elsewhere. The disconnection is palpable, according to O'Leary. 'Your place is your
place.'

The north Mayo residents' 'sustained onslaught on scientific
reality' at the Bord Pleanála hearing on the terminal 'frustrated
Shell', Mark Garavan noted. The company's closing submission in
December 2002 was 'filled with references to the unprecedented
length and level of detail of the hearing, the extraordinary scrutiny
to which they [Shell] had been subjected, and the failure of the
appellants to accept the logic of their scientifically based
conclusions'.

Shell had argued that it had a group of 'highly experienced and
competent experts – the best in their field', but this had been 'an
unusual and unsettling oral hearing' from its perspective, Garavan
wrote. 'The rules of engagement were clearly not as they had anticipated,' he said. 'Rather than being locked into an adversarial
struggle on scientific or conventional environmental issues, they

were confronted with a complex register of voices', and 'their obvious discomfort in addressing this subtler situation undermined the aura of power and competence often attributed to multinational corporations in such settings'.[4]

Jacinta Healy was sitting in a waiting room with her daughter Sinéad at Mayo General Hospital, Castlebar, when the news came on the last day of the month, 30 April 2003. 'I jumped up and laughed, and the nurses came running out to know what was wrong. "Fair play to ye," they said, when we told them that An Bord Pleanála had refused permission for the terminal. It was a great feeling at the time, like we had taken on the Almighty and won. We had a big bonfire at our house, and a party in Glenamoy a few days later – although by then I had read the full decision from An Bord Pleanála and knew the developers had been given a way out.'

Out at Ballinaboy, Maura Harrington, enveloped in a brown cloak, held up her right hand in a victory sign for photographer Keith Heneghan. She was 'delighted', she said. The company's 'disdain for the communities' who lived on and beside the bog at Ballinaboy had 'led to its downfall', she said.[5]

An Bord Pleanála's decision would prompt dancing on beaches, but would cause consternation among business interests in the west, and in board rooms and government departments back east. Shell's subsidiary Enterprise had issued a statement several weeks before the ruling, denying a newspaper report that it might quit the Corrib project for the Gulf of Mexico unless the appeals board's decision was 'hassle free'. Trade union SIPTU's offshore commit-tee said in turn that it hoped the original newspaper report had not emanated from Enterprise Oil, and called on Minister for Communications, the Marine and Natural Resources Dermot Ahern to issue a statement supporting the independence of An Bord Pleanála.

On 23 April, Shell/Enterprise reported that it was 'pleased' with a hydrocarbon find on the Dooish well off Donegal and intended to resume drilling there on 1 May. The Shell subsidiary also denied an assertion by SIPTU's offshore committee that it was putting 'implicit pressure' on the appeals board over the Corrib gas decision.[6]

Bord Pleanála inspector Kevin Moore gave little quarter in his 377-page report, which was based on the findings of the February and November–December 2002 oral hearings. His conclusions and recommendation ran to almost seven pages in which he focused on his request of June 2002 for further information from the developers. 'Fundamentally, the applicant failed to respond to the board's request,' Moore wrote, in relation to his specific request for information on alternatives to the proposed 'subsea tie-back', involving processing gas inshore.

The current proposal would make this subsea tie-back 'the second longest in the world' if built, he noted. Projects similar to Corrib in other parts of the world were tied back to offshore processing platforms, not to land-based terminals, he said. Where terminals were sited on land, they were in coastal, rather than remote inland rural locations.

'The developer has not proposed that the alternative option is unviable,' Moore said. Significantly, 'the value of the Corrib gas resource has not been quantified in any of the information now before the board. Thus, the considerations on the cost of the proposed development compared with the alternative cannot be determined,' he added.

Nor was there any obvious socio-economic benefit to the Erris region, and 'The claim by the applicant that Corrib gas would provide in excess of 50 per cent of the country's gas requirements in its key production years must seriously be called into question as a consequence of the control and method of gas sales by the three individual partners who are developing the Corrib gas field.'

The plan to build a terminal on a blanket bog threw up more problems, and Moore was particularly unhappy with the proposal to develop repositories for peat removed. There was also 'uncertainty' with regard to 'handling large quantities of fire water in the event of a serious accident at the terminal', and there would be problems associated with substantial truck traffic during the construction. The development would have 'significant environmental costs', which were 'compounded by the new submissions of the applicant on visual impact' of the terminal.

The industrial complex would be 'highly obtrusive from a regionally valuable area of special scenic importance, Carrowmore lake area of high amenity, which is 2.5 km from the site, and the adverse impact cannot be mitigated', he said. The 'material and significant adverse environmental cost should not be borne by the local community as there is effectively nothing gained in return by the community for the injury caused'.

On health and safety, Moore was scathing of the National Authority for Occupational Safety and Health, which had failed to provide appropriate 'technical advice' to the appeals board on the issue of hazardous substances at the terminal. There was 'no detailed risk assessment relating to areas of public use, as was requested by the board', he said. The authority had not considered a 'wide range of credible scenarios arising from accident hazards'.

Nor was there independent advice available to the board on how trees surrounding the terminal 'area' should be controlled to reduce the risk arising from forest fires. No public water supply was available in the area and there was an 'unacceptable risk being imposed on the local community'. He described Coillte's sale of public lands as 'regrettable and short-sighted'. The sale 'appears to be a purely financial consideration that fails to give due regard to the wider socio-economic and public amenity values of this land', he said, and went on:

It is my opinion that the sale of these lands for the industrial purposes intended does not rest well with [Coillte's] principal objectives.

I must stress that it should not be regarded as acceptable to permit a development such as that which is proposed with its seriously damaging environmental impacts on the pretence that there may be greater economic benefits to the region. The possibilities, potential or otherwise from the gas field (with a limited 15–20-year lifespan) should not pre-determine that a large processing terminal, located in a remote, infrastructurally deficient rural area, with consequent serious adverse environmental impacts, should override the proper planning and sustainable development of the Erris region . . . It is my submission that the proposed development defies any rational understanding of the term 'sustainability'.

It is my submission to the board that, from a strategic planning perspective, this is the wrong site; from the perspective of government policy which seeks to foster balanced regional development, this is the wrong site; from the perspective of minimizing environmental impact, this is the wrong site; and consequently, from the perspective of sustainable development, this is the wrong site.

He recommended that permission be refused for the development for three reasons. The first related to conflict with planning policies, due to imposition of a large industrial development in a remote inland rural location seriously deficient in public infrastructure: its visual obtrusiveness; irreversible alteration of the landscape; degradation of fragile ecology; environmental and public safety implications derived from construction; and significant increase in traffic and heavy goods vehicle movements at construction stage on a 'substandard' road network.

The second reason related to the 'likely instability' of proposed bunds, or liquid containment facilities, and storage of waste peat in repositories. The developers had failed to design an effective surface-water drainage system that would ensure their integrity on a slope above a regional road, and in close proximity to watercourses draining to the Glenamoy river and Sruwaddacon Bay. This would pose a consequent safety risk to the local community and a pollution risk, he said.

Moore's third reason for refusal related to the provision of the EU's Seveso II Directive on transmission of hazardous substances. The proposed development would 'give rise to an unacceptable risk to members of the public due to the proximity of the terminal site to residential properties and areas of public use to which the directive applies'.[7]

The appeals board, which had final say over the inspector's report and recommendations, overruled Moore's first and third reasons for refusing permission. However, it accepted the second, which had been informed by expert examination of the peat stability issue by David Ball, the hydrogeologist employed for the appeal. Defending its rejection of Moore's first and third reasons, the board said that the visual impact argument would not warrant a refusal, given the 'strategic nature' of the development. It also argued that the project would not necessarily be incompatible with the goals of the Mayo County Development Plan and the strategies of the Border, Midlands and Western region.

Curiously, the board did not accept the inspector's reason for refusal on health and safety grounds. This was because the National Authority for Occupational Safety and Health had not advised against the granting of planning permission, subject to a number of conditions. Yet Moore had been scathing in his criticism of that organization for failing to provide adequate information to the board.

The board's very selective acceptance of their own inspector's damning report was somewhat overlooked in the public response. And there was little doubt that Moore's report represented a 'devastating critique', James Laffey, then journalist with the *Mayo News*, reported on 7 May 2003:

> His mammoth 377-page report should become compulsory reading for every current and future member of Mayo County

Council, because this sort of planning fiasco must never again be allowed to occur in the county.

It is difficult to countenance how a group of law-abiding citizens in a picturesque village in north Mayo could be let down so badly by their own local authority but that is clearly the verdict of Mr Moore. He doesn't just uphold the overall stance of the Ballinaboy residents, he comes down on their side on every single issue of importance pertaining to this multi-million-euro development . . . There are simply no hiding places in this report for the promoters of the Corrib gas project.

There isn't really a whole lot more that can be said about this sorry saga. Kevin Moore has said it all in his epic report. The great tragedy in all of this is that the Corrib gas field had the potential to become the flagship project for Mayo in the 21st century. That this has failed to happen is not the fault of the Ballinaboy protesters; it is the fault of government politicians who have once again let down this county by attempting to foist an entirely inappropriate development on an unsuspecting rural community. Sadly, the majority of local politicians and the senior officials of Mayo Co Council also stand indicted. At one stage, the council actually suggested to the residents of Ballinaboy that the value of their properties would be increased by virtue of the fact that there was a gas terminal on their doorsteps. The sheer arrogance of this sort of statement takes one's breath away . . .

Unusually, the government issued a statement about the decision, as conveyed by the minister responsible, Dermot Ahern. The government 'regrets' that the Corrib field's development was 'further delayed', Ahern said. 'While I fully respect the decision today of An Bord Pleanála, it is to be hoped that the ruling does not end efforts to tap into this rich vein of energy off our coast and the jobs, investment and infrastructure which it has the potential to deliver,' he continued. His predecessor in the marine and natural resources post, Frank Fahey, who was now a junior minister with responsibility for labour affairs, also issued a statement: if the project didn't go ahead, it would have a 'severe, negative impact' on the west of Ireland.

The decision was welcomed by the appellants in north Mayo, by Green Party MEP Patricia McKenna, who had filed a complaint with the European Commission over the project in 2001 after planning permission had been initially granted for the terminal by Mayo County Council, by An Taisce, by Friends of the Irish Environment, also appellants, and by SIPTU's offshore committee. New independent TD for Mayo, Dr Jerry Cowley, who had campaigned during the 2002 general election on health issues but also on inadequate state benefits from offshore oil exploration, had already called for a review of the government's original deal with Enterprise in the light of the Royal Dutch Shell €4.3-billion valuation of the Corrib field. The general practitioner welcomed the board's concern for health and safety. However, he said that 'Mayo and the western region need this gas, and we can only hope that it can be brought ashore in an environmentally responsible manner.'

In early May, standing orders were suspended at a Mayo County Council meeting. Michael Duffy of the *Mayo News* reported that the majority of the council members 'took to their feet to voice their "disgust" and "disbelief" at the Bord Pleanála decision'. Bord Pleanála officials were accused by councillors of being 'whizz kids'. County manager Des Mahon said he had contacted Shell managing director Andy Pyle to let him know that the local authority was available for 'any pre-planning concerns which may arise from a new application', and he referred to the 'hurtful criticism of the council in the local media'. Fine Gael TD Michael Ring was a lone voice in his criticism of the conduct of the developers.

Junior minister Frank Fahey wasn't going to let the matter rest. He invited west of Ireland deputies and senators to a meeting in the Leinster House audio-visual centre on 7 May, which would be addressed by Shell E&P Ireland managing director Andy Pyle. 'Further to the recent Bord Pleanála decision on the Corrib gas field, there is now major concern that the project will be delayed indefinitely,' Fahey wrote. The purpose of the meeting, he said, was

> to get a political consensus in support of the project and to urge Enterprise Energy Ireland/Royal Dutch Shell to proceed with the new application for planning permission, which will comply with the requirements of the planning authority and An Bord Pleanála.
>
> I have asked Andy Pyle, managing director, Energy Ireland [*sic*], to meet with West of Ireland Oireachtas members who wish to see the project proceed so that we can demonstrate the wide-spread support that exist [*sic*] in the west of Ireland for Corrib Gas. Given that the planning application was refused by An Bord Pleanála on one issue only, that of the storage of peat, I believe it is important that we show a cross party political consensus for the continuation of the project.

'Trusting that you will be able to attend,' he concluded, signing the letter of 2 May 2003 as Frank Fahey TD, Minister for Labour Affairs, and giving contact details for RSVPs.

Dr Jerry Cowley TD, who had been invited to the meeting, said that it was 'inappropriate' to invite Andy Pyle. In spite of the clear reference to him in the letter, Fahey said that Cowley had misunderstood him; he had requested a meeting with Pyle, which would follow from the discussion with fellow politicians. Cowley said he would attend, but would be calling on the developers to build an alternative platform offshore in shallow water. It was 'crucial' that the development went ahead, Cowley said, but the 'subservient' approach being adopted by Fahey was 'a bit of an insult to the western deputies'.[8]

Fahey's description of the Bord Pleanála ruling as a 'technicality' was picked up by Michael Ring, who said that this was 'incorrect'.

Speaking on the day of the Fahey meeting, Ring said he would support construction of an offshore platform as the safest option and one that would generate more jobs. 'I am all for this gas coming ashore,' he said, but not by the method planned. He attributed the impasse to the 'bullying tactics' of Enterprise and Fahey. 'If they hadn't tried to bully the people of Mayo over the gas terminal, we wouldn't be at this stage,' Ring said.[9]

However, one western deputy was incensed. Labour TD Michael D. Higgins, a former minister in the coalition government, refused to attend the meeting, describing it as 'inappropriate'; he said it could set a 'very bad precedent in terms of the relationship between public representatives and the planning process', and added that he fully supported the gas development. The meeting also raised questions about the identification of public representatives with a particular corporate body, as in the developers of the gas field, Higgins pointed out.[10]

A regional lobby group, the Council for the West, said it was very disappointed by the Bord Pleanála decision, and hoped a new planning application for the terminal would be lodged, but pointed to another dimension of the saga. 'It would be helpful if the government and Bord Gáis Éireann recognized that their foot-dragging on the issue of a natural gas service strategically throughout Connacht and Donegal has not helped to create an environment for acceptance of the Ballinaboy project,' its chairman, Killala-based businessman Seán Hannick, said. His group wanted a 'cast-iron commitment' that towns in Connacht and Donegal would have access to gas.

The 'Nuremberg rally', as it came to be known locally, and remembered by Micheál Ó Seighin, Maura Harrington and others, took place in early June, just a month after the Ballinaboy ruling. More than eight hundred people turned up for the meeting in Belmullet, called by Erris Chamber of Commerce, Erris Tourism and the Erris Pro-Gas Group. Denis Daly of the *Western People* reported on 4 June that politicians present included Minister for

Community, Rural and Gaeltacht Affairs Éamon Ó Cuív, who had recently articulated his concerns about the growing east-west economic divide at a time when the economy had taken off, and junior minister Frank Fahey, along with Fine Gael TD Michael Ring.

Ó Cuív told the meeting that it was 'ninety-nine per cent likely that Shell will go back to the Ballinaboy site, because they know, as a result of the Bord Pleanála decision, what they have to deal with to overcome the problems'. He stressed that it was important that the local area and Mayo benefited from the gas and said he was giving his commitment that, if gas did come ashore, he would do all in his power to ensure a pipeline was built from Ballinaboy to Belmullet.

Fahey indicated that he had heard from Andy Pyle of Shell that the company would 'go ahead' with the project, and the 'problems' could be dealt with. 'And that,' he said, 'is the best news in a long time for this area and for Ireland.' There was a call for a show of hands, and the majority indicated that they wanted the project to go ahead. The methodology was the real issue – but there was no vote on that. Michael Ring recalled afterwards that the atmosphere was extremely hostile to objectors, and he believed that the meeting had been called to try and damage him. 'It was politically motivated to do me down. But I pointed out that if this development was taking place outside anyone else's doorstep, and they weren't happy, the first person they would come to would be me. That silenced the critics, but Fine Gael councillor Gerry Coyle was the only politician who stood by me that night.'

And then, just over three months later, the landscape changed – literally. On the evening of 19 September 2003, heavy rain began falling over Dooncarton mountain (also known as Glengad Hill) and the communities of Glengad, Pollathomas and Barnacuillew on the southern side of Sruwaddacon estuary, and the mountain itself, began to move.

7

THE SLIDE

Martin Moran, eighty-two years of age and living in Barnacuillew, spoke of hail and rain hitting his roof 'like bullets' on the night of 19 September 2003. Teenagers Siobhán McGrath and Anthony McGuire were driving home from a pub quiz along the coast road on the southern shore of the Sruwaddacon estuary when their car struck what they described later as a wall of water and mud. 'We thought we were going to die,' Siobhán told Mayo journalist Tom Shiel. 'We sat in the car, hugging each other. We thought that was it – that we would never see each other again.'

Siobhán's mother, Teresa, and her partner Patrick Flannery were following in the car behind. 'Their car was like a toy bathtub,' Teresa said. It 'kept turning and turning in the swirl. We thought it would go into the sea.' Somehow, Anthony McGuire's car came to a halt at the edge of a drop. They had one torch between the four, and with it they managed to find their way to safety in pitch dark – the electricity lines were down.

It was a night that they and their neighbours would not forget, as three separate mudflows among more than forty individual landslides on Dooncarton enveloped houses, sweeping away part of Pollathomas graveyard, smashing tombstones, and covering the area in rocks, bushes and fifteen-foot chunks of soil.[1]

Belmullet-based Garda Superintendent Tony McNamara was attending a colleague's wedding in Athlone, halfway across the country. When he arrived back west, he described what he saw as like a scene from the film *Apocalypse Now*. It was, he said, a 'miracle' that no lives had been lost – the pub quiz might have saved many. Three miles of coast were affected in all, with bridges destroyed, one vacant house completely demolished, roads torn up, sheep killed and farmland and fencing washed away. The Ballinaboy river flooded, and debris was swept down to the landfall for the proposed pipeline. Some forty-two families were evacuated from their homes initially and given temporary accommodation, until Mayo County Council could assess the situation with geological advice.

Local postman John Joe Barrett was one of the hundred people affected, along with his wife, Kathleen, and his two younger children. His forty-mile daily circuit on a bicycle from his home in Barnacuillew became something of an ordeal over the coming weeks, while the local authority worked to repair roads and bridges, and the Irish Red Cross distributed humanitarian aid. His home was categorized as 'medium-risk': walls were cracked, his sewerage system had sunk and new fissures appeared daily for up to a month afterwards.[2]

At mass that Sunday, Aughoose priest Father Declan Caulfield tried to comfort residents as he confirmed that the contents of four graves, including one of a family of five, from Pollathomas cemetery had been swept to sea. The *Western People* reported that there were 'poignant scenes' as 'family members searched, sometimes in vain, for their loved ones' final resting place'. Martin Mills, his brother Seamus and friend Declan Healy were among the many who began work on the graveyard. 'People have been ringing from England and America, trying to find out if everyone is okay first and then they are desperately seeking news on the damage done to the graveyard,' Martin told the newspaper.[3] Minister for Community, Rural and Gaeltacht Affairs Éamon Ó Cuív had been in London but visited the area the following Tuesday and described it as a 'natural disaster that could not have been foreseen'.[4]

Fine Gael TD Michael Ring, who had received one of the first phone calls about the emergency, and Independent TD Dr Jerry Cowley were among politicians who called for a compensation package. The government said the Irish Red Cross would distribute emergency aid. The local authority's geological advice was that further landslides could occur. Met Éireann confirmed that some 89.33 millimetres of rain had fallen on that area on 19 September, most of it within a four-hour period between 8.30 p.m. and 12.30 a.m., compared to a total monthly mean of 108 millimetres for Belmullet some fifteen miles away.[5]

There were further geological evaluations, with six occupied properties, including McGrath's pub in Pollathomas, classified as 'high risk'. The wife and family of Publican Paddy McGrath's son Tony watched the mud slide past their house; Tony was on the road delivering shellfish and couldn't reach them for hours. The pub had to close down for the first time in a hundred years, but reopened just over a fortnight later. Following a visit by the new Bishop of Killala, Dr John Fleming, it was confirmed that counselling was being offered to residents. The damage to the old cemetery had had a 'huge effect' on the local population and on people living all over the world who had relatives buried there, he said.[6] It was a reflection of people's attachment to, and sense of, place.

Taoiseach Bertie Ahern flew up a month later, following criticism of his failure to do so earlier. At this stage, the west had experienced a further geological event, but it was well south of Dooncarton, in the Slieve Aughty mountains of south Galway. On 16 October 2003, a 100-metre-wide cascade of blanket bog slid more than a kilometre as contractors were preparing the area for a 60-megawatt wind farm, comprising seventy wind turbines, developed by Hibernian

Wind Energy, a subsidiary of the ESB. Fortunately, there were no reported injuries, but there was extensive pollution and at least fifty thousand wild trout were killed after the landmass spilled into a gorge attached to the Lough Cutra river system. Some sixty-two farmers were affected, with at least 550 sheep lost, machinery and farm buildings damaged and fodder destroyed.

Unlike north Mayo, south Galway had not experienced torrential rainfall, but there had been considerable tree felling and construction work on the wind farm site; and, as it transpired later, shortcomings in its environmental impact statement that would be highlighted in a subsequent court case. This would in turn result in a significant European ruling.[7]

So, was Dooncarton's movement a 'natural disaster that could not have been foreseen', as Minister Éamon Ó Cuív had asserted? Some residents felt that erection of an aircraft radar dome on the mountain's summit some years before might have been a contributory factor. A local had witnessed lightning hitting the dome, which could, residents felt, account for the fissures running out from its foundation down the mountainside. Others suspected a link with work carried out by Enterprise in the Glengad area as part of investigations into pipeline routing and the landfall. The issue of landslide frequency had been raised at the terminal's Bord Pleanála oral hearing.

The Irish Aviation Authority confirmed that the building and equipment for the radar had not been affected, and back-up generators had continued to provide power to the facility, which provided air traffic control information to Shannon airport. Shell/Enterprise issued a statement saying that it had been 'in contact with the relevant agencies since Saturday [20 September 2003]' and it was 'being kept up to date on developments'. It would 'consider any advice we receive from these agencies as we continue in our assessment of the future viability of the Corrib gas project'.

The landslide risk had been identified by the company during its environmental impact assessment of Pollathomas. As former

Enterprise managing director Brian Ó Catháin recalled later, it had been dropped from an early draft of the document at the request of a senior Mayo County Council official, due to concerns about a possible impact on land values.

Professor John V. Luce, Trinity College Dublin orator and classicist, offered his opinion on Ó Cuív's 'natural disaster' theory in a letter to *The Irish Times* on 30 September. 'No, Minister, this was not like a volcanic eruption,' Professor Luce wrote. 'This was a man-induced disaster, which was in principle predicted for the area by Bord Pleanála when it gave reasons against the siting of the proposed Atlantic gas terminal at Belmullet [*sic*].'

Overgrazing by sheep could have also been a factor, Professor Luce indicated. Years ago, 'far-sighted environmentalists', like the late Professor Frank Mitchell, had warned against the damage caused to upland bog cover by overgrazing – encouraged in turn, many said, by EU policies. Michael Boland, executive secretary of the Institute of Geologists of Ireland, had another angle. His professional group had recommended in 2002 that the geological section of environmental impact statements should be signed off by a professional geologist. Currently there were no professional geologists employed by county councils anywhere, including Galway and Mayo. 'It is necessary to employ experts who understand the natural environment,' he stated.[8]

On the day that Taoiseach Bertie Ahern flew to north Mayo in October 2003 – opening a hotel in Castlebar and attending a local Fianna Fáil party meeting in Ballinrobe – the Geological Survey of Ireland published a preliminary report for Mayo County Council on Dooncarton. It ruled out the impact of the aircraft radar dome, and said it could not find any link with overgrazing, or with Enterprise's 'gas preparations' in the Glengad area the previous year. It attributed the landslides to the exceptionally high rainfall on the night of 19 September, which the 'thin' peat cover and the schist and sandstone rock of low permeability had been unable to absorb.

The preliminary report underpinned a theory held by Professor

Mike Williams of NUI Galway, who believed the previous long, hot and dry summer had caused the peat to perish and its roots to shrink and shrivel, leaving no firm hold on the underlying bedrock. He had charted previous landslides, including four in the County Galway area from 1740 to 1890, and discounted overgrazing as a possible cause. Although it was a 'major problem' in the west, this was only in a previous twenty-year period, and new rural environmental protection schemes (REPS) had been introduced to tackle this.[9]

A local landslide action committee was established in Pollathomas, and a further report on design of engineering solutions was commissioned by the local authority. This report recommended erecting 'kinetic energy-absorbing rock-anchored barriers' as part of a project that would also involve restoring the graveyard.

Even as people began to get their lives back together in the weeks that followed, it emerged that the Taoiseach had been involved in very significant discussions in Dublin on the night of the landslide, and those discussions related to north Mayo.

As Ahern confirmed in the Dáil on 14 October and again on 19 November 2003, he had met Tom Botts, managing director of Shell E&P Europe, on 19 September as part of a delegation involving the Corrib gas developers. The ministers for the environment and natural resources also attended, as did department officials. The purpose of the meeting, Ahern said, 'was to hear the company's concerns about progress on the Corrib gas project'. The government was at that time considering fast-tracking legislation that would allow approvals for significant infrastructural projects to be expedited – perhaps bypassing local authorities and placing a question mark over the role and future of An Bord Pleanála.

Ahern rejected Sinn Féin charges in the Dáil that the Shell meeting interfered with the planning process, pointing out that it had not taken place until six months after An Bord Pleanála had issued its ruling. He told Sinn Féin TD Caoimhghín Ó Caoláin that it 'would be highly inappropriate for me not to meet with people wishing to legitimately invest their resources in generating projects and activities in this country'.

Ahern explained that Shell had 'wanted to know if the government's proposed National Infrastructure Board would be in place in time to deal with a renewed application, should that arise', and that 'it was unlikely the new Bill would be in place before next summer when it was hoped they could commence work. I recommended they stay with the normal planning structure,' Ahern told the Dáil, and the company had said it 'would take that information into account'.[10] When the issue arose again in the Dáil, the following month, Ahern insisted there was nothing unethical about his discussion in September with the senior Shell executives. There were 'no deals or arrangements' with Shell, he insisted, adding that 'other countries have ways and means of treating large companies, which I do not agree with. I have had a fair few meetings over the years that might border on the unethical, but I am not guilty of it in this case.'

Green Party leader Trevor Sargent said it was 'tempting to ask about the other meetings which the Taoiseach thinks were ethically borderline, but we will leave that for another day'.

'If this resource works effectively, as some of the early studies suggest, we could be exporting gas to the UK,' Ahern added. 'Shell does a great deal of business and has invested heavily in this country.'[11]

Four years later, in November 2007, the RoyalDutchShellplc.com website run by Alfred and John Donovan – long-time critics of the multinational – published details of minutes of a meeting of Shell group managing directors on 22 and 23 July 2002. Planning refusal for the Ballinaboy gas terminal in north Mayo was discussed, according to the website, which quoted from the minutes: 'The committee queried whether the group had sufficiently well placed contacts with the Irish government and regulators. Paul Skinner undertook to explore this issue further in consultation with the country chairman in Ireland.'

Shell had already been working on a new planning application, and it was lodged with Mayo County Council on 17 December 2003. The new plan involved overcoming the peat storage issue by trucking some 450,000 cubic metres of peat away to a neighbouring blanket bog run by semi-state company Bord na Móna at Srahmore, some eleven kilometres south of Ballinaboy. Independent TD Dr Jerry Cowley called on opponents in Mayo to look carefully at the new planning application before objecting to it. 'Given that north Mayo is one of the most deprived areas of the state, there is no point in a knee-jerk reaction to a revised planning application,' he said. The area's deprivation and high unemployment was something that he had raised in the context of compensation for the victims of the Pollathomas landslide only months before.

SIPTU offshore committee spokesman Padhraig Campbell accused Cowley of doing a 'volte-face', given that he had expressed support for an offshore terminal. 'No area should be jeopardized for corporate profit.'

Cowley acknowledged that he had been a critic of the €800-million gas field project, and had called for a review of state tax terms and royalties for offshore exploration and development. 'However, the company has made it clear that an offshore terminal is not feasible, so I think we need to look very carefully at this new plan, which will be subject to an environmental impact statement. And I think the government can help by making sure that the people of Pollathomas get more than the €300,000 earmarked for them as compensation for the recent landslide.'[12] A scheme for 'hardship cases' affected by the landslide had been established by government after much pressure.

Concerns about the new Shell plan were raised by a Galway-based structural engineer with north Mayo roots, Brian Coyle. In a submission to the local authority, he highlighted the unstable nature of the proposed terminal site at Ballinaboy and referred to the

Dooncarton landslide and the generally fragile nature of blanket bog. The terminal would be the only inland one of its type, with the longest pipeline link of its type, and would constitute a 'Seveso II high-risk event', given the proximity of housing.

Coyle contended that Shell had failed to provide 'in-depth, site-specific information' on alternative terminal sites[13] – a point made forcefully by Bord Pleanála inspector Kevin Moore in his report of April 2003. In a later comment, Coyle also called on the Health and Safety Authority to commission an independent safety audit of the gas terminal project.

Responding to Coyle's concerns, Shell said that it would examine all submissions through the planning process, rather than 'comment publicly', but it was 'happy that the terminal was in the correct location and that all aspects of the project were environmentally and technically safe'.[14]

Former Department of Finance secretary and highly respected economist Dr T. K. Whitaker was also uneasy. He had a home in Glencullen, north Mayo, which was formerly a nineteenth-century school building. It was on the route proposed for transporting peat to Srahmore. Dr Whitaker confirmed to *The Irish Times* that he had contacted the company, and had submitted his 'reservations' about the plan to the local authority in a private capacity. His concern focused on the impact of the intensive truck traffic on the foundations of his house, and the transporting of peat could cause sliding and silting, with serious consequences for the local river systems linked to Carrowmore lake and their wild salmon stocks.[15]

Marine tour operator Anthony Irwin was seriously concerned. He had been commissioned to provide his boat for a study carried out by University College Cork's Coastal and Marine Resources Centre for the Corrib gas developers. When it was completed, the study of marine mammals in Broadhaven Bay painted a very significant picture. It quoted more than 220 sightings of seven cetacean and other species, and described Broadhaven Bay as an 'important area for marine mammals and other species' with 'few, if any, comparable examples of a relatively small, discrete bay in Ireland containing all five Annex II marine mammal species (bottlenose dolphin, harbour porpoise, grey seal, common seal and European otter) with such frequency'.

It noted that mitigation measures to protect cetaceans from the effects of proposed drilling and blasting works within Broadhaven Bay as part of the project would be 'difficult to implement effectively'. Curiously, the study, initiated in September 2002 and completed in March 2003, was conducted in co-operation with Dúchas, the state heritage agency, but was not forwarded to the Department of the Marine and Natural Resources at that time. Yet the department was the lead agency in approving much of the Corrib project – and it had appointed a marine licence vetting committee to conduct an environmental assessment of the foreshore licence applications and the plan of development.

Neither was the whale and dolphin study submitted as part of the developer's foreshore licence application. By the time it was released to the press in February 2004, as a result of Irwin's persistence, the marine licence vetting committee had already sanctioned the project, subject to certain conditions. Irwin believed that the committee should review the foreshore licence in view of this important information. He also noted that the position of the outfall pipe – a matter of such concern to fishermen – had been determined before this information had been made available to the vetting committee.

In response to queries from *The Irish Times*, the Department of the Marine said that the study was 'not a condition of any of the consents, permits or licences approved for the proposed development of Corrib', and most of the work had been carried out after the awarding of a foreshore licence by Minister Frank Fahey. On 2 February 2004, Shell told *The Irish Times* that, in parallel to the study, it had carried out detailed site investigations in Broadhaven Bay's inner reaches and it was 'currently thought that it will not be necessary to use explosives in the pipeline construction'. Shell was 'fully committed to protecting fish species and marine mammals from any negative impact arising out of construction activities', and would be consulting with marine mammal scientists.

Two days later, on 4 February 2004, Shell told the newspaper that Dúchas had received a copy of the report in November 2003. Curiously, it confirmed that the Department of Communications, Marine and Natural Resources had not. The department had 'requested a copy of the report and will receive one this week'.

Mayo County Council approved Shell's new application in April 2004, subject to seventy-five conditions, and the decision was appealed once more to An Bord Pleanála. At this point the developer's parent company was dominating international news. On 19 April 2004, Bloomberg financial news service reported that the Royal Dutch Shell group had lost its AAA credit rating with Standard & Poor's over a shares scandal, involving overstatement of oil and gas reserves between 1998 and 2003. In August 2004, British and US financial regulators imposed fines of £17 million sterling and $70 million, and group chairman Sir Philip Watts was ousted.[16] Statoil was recovering from its own problems – a 20-million Norwegian kroner fine and top management resignations – after it was found to have secured contracts in Iran in 2002–3 through bribery.

Just over a year after the Pollathomas landslide was marked with a memorial mass and blessing of the restored cemetery, An Bord Pleanála announced approval of planning for Ballinaboy. The final permission, issued with forty-two conditions, included payment of €6.5 million by the developers towards provision of roads and water, the Mayo fire service and recreational and amenity projects, and establishment of a monitoring committee that would have local representation.

The approval was based on the developer's plan to deal with the peat issue – one of inspector Kevin Moore's three reasons for refusal the previous year but the only one accepted by his board. Opponents noted that the decision was made at the time when the proposed new critical infrastructure legislation was still hanging over the appeals board.[17] For Andy Pyle of Shell, it was 'all systems go'. Ballina Chamber of Commerce joined business groups, and former marine minister Frank Fahey, in welcoming the decision, but warned that the region would not fully benefit unless the government ensured that Ballina and north Mayo were connected to the national gas grid.[18]

Applications for a judicial review of An Bord Pleanála's decision on the Corrib gas terminal were lodged in the High Court by Martin Harrington, an electrician and brother of Maura, and environmental consultant Peter Sweetman. An Bord Pleanála's responsibilities in relation to the EU Seveso II Directive was cited – as was the board's failure to seek an environmental impact assessment for the entire project.

Micheál Ó Seighin and Brendan Philbin, two of the notice parties to the judicial review, would read about Justice Fidelma Macken's decision to reject the legal challenge while serving in-definite time in Dublin's Cloverhill prison. In her reserved judgment, published on 26 July 2005, Mrs Justice Macken said that she found Martin Harrington had not established to the court's satisfaction that there were substantial grounds for permitting him to commence judicial review proceedings of An Bord Pleanála's planning permission.

8

'WHY NOT JUST HAVE ALL OF THEM COMMITTED?'

Gale-force winds were blowing in Erris when there was one of several altercations over attempts by Shell representatives to gain access to land in early 2005. As Willie Corduff recalled, a dog wouldn't put his nose out of the door on 10 and 11 January, when Shell representatives arrived to check out land on the onshore pipeline route. Winds of 97 m.p.h. were recorded at Belmullet, and Maura Harrington remembered that the weather was so bad local schools had to close. The company wanted to undertake survey work, however. It had its consents – the pipeline route was exempted from planning permission – but seven landowners who had not signed up to their compensation offer owned 50 per cent of the property on that route.

Project engineer Paul Gallagher claimed that he was 'opposed and obstructed', while the residents, who expressed serious concern on health and safety grounds, said he had not sought permission and felt that he had been 'unnecessarily provocative'. Gardaí were called, and Superintendent Tony McNamara said afterwards that there was no trouble, and differences between the company and landowners were a 'civil matter'.

Vincent McGrath was at home – his house was the closest to the pipeline route – when he saw the Garda squad car arriving, and 'a lot of people in close pursuit'. He hadn't thought Shell staff would be out on such an 'atrocious day' when they kept 'telling us that health and safety was their main concern'.[1]

The landowners questioned the legality of compulsory acquisition orders issued, given that they had applied to Enterprise Oil rather than Royal Dutch Shell. Shell said it had legal entitlement as it had taken Enterprise over in 2002. 'They did not produce the relevant paperwork to demonstrate that they had a valid consent,' Bríd McGarry remembered.

Monica Muller told *The Irish Times* that Shell staff had a video-camera with them, and she said that they 'hid' it when she questioned why she was being filmed on her own property. She had also received an unsolicited visit from a company representative on a week night. Denny Larson, a North American environmental lobbyist, witnessed some of the activity – Maura Harrington had contacted him. He was a director of Global Community Monitor, a non-governmental organization, and the co-ordinator of an alternative Shell corporate social responsibility study known as *The Other Shell Report*.

Shell made another attempt to gain access to lands in February. Willie's father had died – 'I remember we got a letter informing us, and I don't remember half of the funeral as a result,' Mary Corduff recalled. 'We were in an awful state – people warned us they might come when we were at the funeral.'

Shell said that it might seek a court injunction to gain access to the property.[2] Indeed, a few days later, the company issued the group with an ultimatum by post, including a copy of the certification of compulsory acquisition rights granted by former minister Frank Fahey. Eugene F. Collins, Solicitors asked the landowners to give an 'unconditional undertaking' to Shell E&P Ireland 'in writing' within seven days of receipt of the letter. That undertaking involved agreement to 'immediately cease and desist from all efforts and actions', which were 'designed and intended to obstruct and/or frustrate' Shell's 'efforts to exercise its lawful rights under said orders'. Otherwise, proceedings in the High Court would be issued.[3]

It would be another few weeks before the company moved on its court option, by which time questions were already being asked in the Dáil of Noel Dempsey, who had been appointed as marine and natural resources minister following a cabinet reshuffle.

'The Corrib onshore pipeline has certain design considerations, which are unusual and unique both within Ireland and also within Europe, and for this reason there is no direct precedent,' Dempsey said in his reply of 24 February 2005 to Michael Ring TD.

As a result, the Irish design standard applied to all Bord Gáis Éireann transmission and distribution pipelines was 'not optimized for wall thickness and less practical to construct', he said – in other words, it was unsuitable. A British design code, BS8010, was being applied, with the Bord Gáis design code of IS328 used as a 'supplement' where 'it was considered beneficial', he said.

Independent TD Dr Jerry Cowley sought an independent quantified risk assessment of the pipeline. The company had told him it could provide him with a 'non-technical' summary of its own analysis as, 'if viewed in isolation, the report may be misunderstood and could be selectively used, misrepresented and potentially taken out of context'. Cowley believed the whole document should be released as a gesture of goodwill. 'Shell are treating the people of Erris as imbeciles and with mistrust,' he said.[4]

Gardaí were called again to Rossport on 1 March when ten Shell representatives tried to gain access to land. Monica Muller told *The Irish Times* that she believed the visit had been choreographed and that they were trying to provoke a row. She and her neighbours had sought copies of a map outlining the co-ordinates of the way-leave, but this was 'refused', she said. Bríd McGarry also refused the company entry at Gortacragher.

'They were once again told that we had not been furnished with the relevant completed legal paperwork,' Bríd recalls. Vincent McGrath remembered that 'All their moves seemed to be well rehearsed . . . they would surround you and try to intimidate you . . . the dark glasses and the whole paraphernalia seemed to be the equivalent of war paint.' Both TG4 Television and MidWest Radio recorded one of the stand-offs.

'We were talking about the pipeline design code that they were using,' McGrath recalled later. 'I was questioning them about the fact that they were using a code – that's BS8010 – that was no longer in operation. They asked me was I an engineer. I remarked that Shell would have intelligent pigs [PIG, as in pipeline integrated gauge] going through the pipeline and that we, the residents, would be the guinea pigs living alongside. One of their engineers burst out laughing.

'He came over to me later and asked me was I Vincent McGrath who plays the accordion. He was an accordion player himself from near Aberdeen and not too far from where the great fiddle player and composer Scot Skinner was born. We had a great chat about music, but unfortunately we both had to go back to our respective sides of the divide and wait for the gardaí to arrive. I said to him, "We won't mention the war." That was the human side . . .'[5]

The company reiterated that it had 'all necessary permissions for the way-leave'.[6] A few days later, five landowners, Bríd, Monica, Philip McGrath, Brendan Philbin and Willie Corduff, and environmentalist Peter Sweetman were served with a plenary summons to appear in the High Court on 14 March. 'Brendan Philbin and I decided to represent ourselves as defendants in the High Court,' Bríd recalled. 'We had practically no other option open to us at that stage. The "equality of arms" principle was not in evidence as far as we were concerned.'

The company applied for an order seeking an interlocutory

injunction, restraining the six named from preventing access by the company on to their own lands. High Court president Mr Justice Joseph Finnegan deliberated for four days, and granted it on 4 April 2005. In its affidavit, the company claimed that its engineer was 'pushed' by one of the landowners during a second attempt to gain access to lands on 1 March.

It claimed it would incur costs of €25,000 a day if the contract work, already signed for, couldn't begin by 1 June that year, as a result of delays in preliminary work on the routing of the pipe to Ballinaboy. It would incur a remobilization fee of about €2.5 million if construction was delayed till the following year. It told *The Irish Times* that it regretted this course of action, but 'all other means have been exhausted' and it would 'continue to engage with all relevant stakeholders'.

Maura Harrington, speaking for the residents to the press, accused the company of engaging in 'scare tactics; and said that the cited delay costs were minimal for a multinational which had recently filed record profits, equivalent to stg£1 million an hour'.[7]

Residents held a silent protest march on Good Friday, 2005, stopping at a compound at Rossport that Shell had established. It was reported locally and on Indymedia, the Irish-based media collective, which was part of a global network. The original independent media centre was established to provide grassroots coverage of World Trade Organization protests in Seattle in November 1999. One of its Irish correspondents, Terry Dunne, had taken an interest in what was happening with Corrib when studying at NUI Galway.

Concerns in Erris were further exacerbated when marine minister Noel Dempsey confirmed that he had ordered a new review of part of the project, following the company's reactivation of its application for consent under 'phase three' of the offshore pipeline. Dempsey said his review would be carried out by Andrew Johnson, a consultant who was also the author of a quantified risk assessment for the department in 2002 when the original application had been made. Independent TD Dr Jerry Cowley questioned its independence, and accused Dempsey's department of working 'hand in glove' with the developer.[8]

Dempsey released information about the review in late May, when he indicated that he would approve the phase three works as part of the phased consent procedure for the project. Cowley said he would seek a meeting with the Norwegian ambassador, as Statoil was a partner in Corrib. 'Norwegian safety standards would be quite high . . . and we need to be sure the Norwegian government can stand over this,' he said.[9]

Several days later, however, to the minister's embarrassment, it emerged that consultants British Pipeline Agency (BPA), hired by his department to review the quantified risk assessment for the pipeline, were part-owned by Shell. Shell said that it had actually alerted Dempsey's department to the potential conflict, after it was contacted by BPA with a series of technical questions. The consultants had been appointed by tender and had not raised the issue of any potential conflict of interest, Dempsey noted, when confirming that he had instructed his officials to order a further review.[10]

However, Cowley and his constituency colleague, Fine Gael TD Michael Ring, were incensed, and called on the government to insist on the company building a shallow-water offshore terminal. Shell said it was 'regrettable' that public representatives should 'seek to undermine public confidence further' in the Corrib gas field project, which was designed, it stressed, to 'world class standard'.[11]

Deputy Ring was adamant. 'The people of Erris who oppose this high-pressure pipeline are not unreasonable, but they are frightened, and now it has been shown once and for all that they cannot trust anybody. It is time that the government and Mayo County Council ask Shell to do what is being done everywhere else in the world, and not allow this terminal or this pipeline onshore to put people's lives at risk. This should have been made a condition of planning approval at the very outset, but life was made too easy for the exploration companies by the government and Mayo County Council.'[12]

One of the difficulties faced by residents from the outset was the fact that Ireland had not translated the Aarhus Convention on access to information, public participation in decision-making and access to justice in environmental matters into Irish law – one of the only EU member states not to do so, as highlighted by the Green Party, then in opposition. In late March, Rossport resident Monica Muller made one of her many attempts to find out more. She sought eighteen pieces of information from the Department of Communications, the Marine and Natural Resources under an EU directive 2003/4. The response, on 24 April, informed her that this particular directive had not been transposed into national legis-lation. Her request was turned down as some of her questions did not 'relate to environmental matters' and some, including, curiously, the environmental management plan, were 'still in the course of completion'.[13]

On 30 May, the day after the two TDs made their call for an off-shore terminal, Shell staff had to halt peat excavation at the terminal site at Ballinaboy due to pollution of a local river, and a warning from the North Western Regional Fisheries Board that it would consider legal action if discharges of silt into Ballinaboy river continued. It was what engineer Brian Coyle had warned of in his objection to the new terminal planning application. Several weeks before the pollution was noticed, several trucks carrying peat to the transfer site at Srahmore had accidentally left the road, depositing their loads, and a hairline fracture had also been detected in a grotto close to Srahmore and just outside Bangor Erris. The company agreed to rebuild the grotto, while stating that it was not taking responsibility for the damage.[14]

As spring turned to summer and there seemed to be no resolution, Mayo was a fraught subject for Fianna Fáil generally, and for its marine minister Noel Dempsey in particular. He and party colleagues were in the midst of talks to form new branches and identify new electoral candidates in the constituency, following a virtual rebellion by supporters of the TD Beverly Flynn who had been expelled from the party in May 2004.[15]

A loose group was coming together in support of the Erris residents, and that group required a campaign name. It was devised during an informal discussion in the Breaffy House Hotel, Castlebar, involving Maura Harrington and Lelia Doolan and the late author John O'Donohue of the Burren Action Group. The action group had won a protracted legal campaign to stop an interpretative centre being built in the sensitive Burren landscape of County Clare. Doolan, film-maker and producer with a stellar career spanning RTÉ, the Abbey Theatre and the Irish Film Board, had a particular affinity for Erris, as she had worked as research officer for the Combat Poverty Agency in the area in 1979. And so 'Shell to Sea' was born, and it soon had its own website – ironically representing an area with little or no access to broadband – thanks to the efforts of a Donegal-based community group keen to use new technology to support people whose voices were not being heard.

A Shell to Sea gathering took place on the June bank holiday weekend of 2005 in a field belonging to Ray Corduff, a distant relative of Willie, in Rossport. Some of those who travelled up were students at NUI Galway and the University of Limerick, who had heard talks given there by Maura Harrington and Padhraig Campbell. 'We didn't know what to expect or how many people would turn up but I felt that there should be at least some little bit of information available. Hence one of my earliest tentative efforts to produce an information brochure,' Vincent McGrath recalled later.

Speakers included Werner Blau, physics professor at Trinity College, Dublin, who had bought the McGrath family home in Rossport as a part-time residence, Maura Harrington, Willie Corduff, and Mary Hurley from the Cork Harbour Alliance for a

Safe Environment, which had been formed in 2001 to campaign for clean industry and oppose a hazardous waste incinerator in Cork harbour. The gathering was reportedly observed by the gardaí.

On 16 June 2005, Shell suspended further attempts to gain access to lands for the onshore pipeline, following objections by the landowners. Bríd McGarry was sure of her ground. She claimed the company had breached the terms of a ministerial order, as it had conducted extensive works when it was only entitled to survey, set out and fence off lands.

The company said that protests at Rossport and nearby Gortacragher represented the 'third occasion' on which it had been 'met with obstruction from this group', and the landowners were 'in contempt of a High Court ruling prohibiting them from disrupting Shell E&P Ireland's preparatory work', following the order it had secured in April. The company also said it had been subjected to a road 'blockade'.

Vincent McGrath remembered it differently. He had parked his car along the side of the road to highlight the lack of a traffic management plan on the small roads around Rossport. 'It was not a blockade, because ordinary traffic could easily pass. I had photos taken to illustrate this point.' The gardaí were called and took names. Frustrated by what it felt was lack of sufficient action by the gardaí to remove protesters, the company said it was considering the 'options open; including possible further recourse to the courts', and noted that access and compensatory arrangements had already been agreed with a 'majority' of landowners on the route.

Independent TD Dr Jerry Cowley said he was 'with' the protesting landowners, and Shell should be asked to 'pack up and go', if it was not prepared to pay for an offshore terminal. 'I would urge the minister to be courageous and take a stand,' Cowley said. 'The penny is dropping now in Mayo that benefits from this whole project are going to be very limited. A social service would employ more people than this multinational is promising in this area, and there is no guarantee that any town in Mayo is going to see any of this gas.'[16]

On 18 June, Bríd McGarry drove to Belmullet Garda Station and made a statement claiming that Shell was carrying out drilling work on the pipeline route opposite her family farm at Gortacragher. Ten days later, Shell confirmed that it was seeking a committal order in the High Court, which could see some of the landowners committed to prison for contempt of court. It was assumed that this would be against those who had refused to sign up to access, and who had been named in the April injunction. In fact, as details of memos published in *The Irish Times* on 5 July 2005 recorded, Shell's Andy Pyle felt that all non-compliant landowners should be jailed if necessary.

The memos for meetings dated 7, 8 and 10 June 2005 gave an insight into the company's thinking.[17] In discussions that Andy Pyle and his senior management held with lawyers on 7 June, concern was expressed about the 'latitude' afforded to two of the land-owners, Bríd McGarry and Brendan Philbin, by Mr Justice Joseph Finnegan, president of the High Court. Both had 'insisted on introducing references to many issues that were not the subject of these proceedings', the memo recorded. 'Unless the trial judge (which is unlikely to be Finnegan P.) is prepared to be much stricter in his control of the lay litigant defendants, then there exists the distinct possibility that the case will endure for much longer than the three-week estimate [for the trial].'

Five members of the Corrib Onshore Pipeline Steering Committee, including Andy Pyle, attended a separate one-hour meeting with a lawyer on 10 June. The lawyer stated that a 'major issue was the lack of progress with the preparatory works on the lands of the non-consenting landowners in light of the urgency which we pleaded before the court during the injunction proceed-ings' of the previous April.

Shell identified two main reasons for the delay – the time taken in obtaining necessary consents for pipeline installation, and the fact that 'It was important to make a conscious effort to meet and talk with the non-consenting parties.' However, the lawyer said he believed, 'It would be preferable to make an attempt now to enter the lands of the non-consenting landowners and decide whether we want to take matters a further step, i.e. to seek to have the land-owners who continue to prevent access to their lands held in contempt.'

Andy Pyle asked him to outline the 'procedures involved in enforcing our right'. He was told that the company would have to 'firstly attempt to get on the lands and start the preparatory works, and when/if they were obstructed from doing so in contravention of the injunction then an affidavit would be sworn by the party . . . who made the attempt'.

A 'committal' was unlikely on the first attempt, as the defendants 'would probably just be advised to stop obstructing the plaintiff', Pyle was told. As the procedure could take 'weeks', it was crucial to attempt to enter lands 'as soon as possible'. While the company should try to gain access to all of the land, it was not necessary to seek to have all of the defendants jailed.

'AP [Andy Pyle] asked why not just have all of them committed?' the memo recorded. 'RS [Rosemary Steen] pointed out that from a public relations point of view this was not the best course of action.'

The meeting agreed that the attempt to enter lands would take place on Wednesday or Thursday of the week ending 18 June, in advance of the next court hearing on injunction proceedings on 29 June, and landowners would be 'notified informally by Shell before the next attempt is made'.

The 10 June memo also referred to a conversation between Shell and Michael Daly of the department's petroleum affairs division, in which Daly said that the new review of the quantified risk assess-ment was expected the next week and he 'would hope to have issued consent by late that week or early the following week'. Shell told their lawyer that the department's 'PAD [petroleum affairs division] did not want the original quantified risk assessment [of the pipeline] made public' until the 'consent on foot of the updated quantified risk assessment has issued'.

'Tactics going forward' were also discussed, and Shell's lawyer advised that it would be 'important to impress upon the trial judge to keep a handle on the witnesses for the defence to prevent the trial being used as the forum for all of the defendants' grievances which may have nothing to do with the issue before the court'. There was discussion of Shell's witnesses, and the company noted that 'training similar to that which their witnesses received in advance of the oral planning hearing would be useful'. There was also discussion on the implications of any failure to obtain a permanent injunction and 'could we then sue the state for the costs incurred?'. A subsequent report on the memos in *The Irish Times* on 5 July bore the headline, 'Memos show Shell talked of suing the State'.

Notices of High Court proceedings were served on three landowners, Willie Corduff, Brendan Philbin and Philip McGrath, on Vincent McGrath, whose house was beside the pipeline route, and on retired schoolteacher Micheál Ó Seighin of Carrowteigue, who had had a triple heart bypass in 2001 and was therefore not an ideal candidate for prison – but who was known by the developers to have a strong influence on the objecting landowners.

Dr Jerry Cowley convened a meeting of fellow TDs on 28 June, the night before the date for the High Court hearing, in Leinster House to try to work out a compromise. 'Shell hasn't even got all the ministerial consents to install or commission the pipeline, and yet it is being given total control of this area of north Mayo by the government, and people are going to be sent to jail,' he said.[18]

Up in Mayo, jail was the last thing on the minds of people who had gathered to watch what was described by the *Mayo Echo* as a 'strange event'. Shell's compound at Rossport had become the focus of anxiety. Locals believed it included a septic tank, for which there appeared to be no evidence of planning permission. Roadbridge, construction contractors for Shell, denied this, as did Mayo County Council initially.

Following a series of complaints to the local authority, and a late-night 'inspection' of the compound by several locals, Paddy Mahon came out to inspect for Mayo County Council on 28 June. Shell and the contractors argued that it was a holding tank, overflow pipes had been blocked, and sewage was removed by tanker. The *Mayo Echo* and MidWest Radio, which had been invited by locals to witness the inspection, noted that Mahon 'listened to both sides' and left the site, promising to return.

Gardaí were called in the interim and took statements when residents refused to leave. Mahon returned, ordered an excavation, and for several hours it appeared as if there was nothing to find. 'As the last scoop-full of soil and rubble was being removed, we got a glimpse of something yellow in amongst the material, something that looked out of place,' the newspaper recorded. 'All eyes turned to the spot, and as the digger removed more material, a large yellow sewage pipe began to be uncovered. Gasps were heard from the assembled audience . . . The residents had been vindicated, Shell had been caught with their pants down . . .'[19] Mayo County Council issued the company with warning notices on 29 and 30 June, but there was no further action.

The morning after, Willie Corduff, Philip and Vincent McGrath, Micheál Ó Seighin and Brendan Philbin travelled to the High Court in Dublin. They represented themselves: solicitor Daniel Coleman, retained by two of the men, did not turn up. Philip McGrath said that he just wanted to protect his property. His brother, Vincent, a resident but not a landowner, said the route was seventy metres from his house, and his main concern was that no state body was taking responsibility for his safety. Willie Corduff described how he wasn't sleeping at night, and begged the court for 'justice' as Shell was 'using the courts to bully us'.

Brendan Philbin said the court had heard a 'one-sided version of events' from Shell. He had reason to believe a quantified risk assess-ment of the pipeline had not been carried out independently, there were no valid ministerial consents, and Shell was 'seeking to make criminals of us'. He added that was a 'poor state of affairs if the judiciary was working hand in hand to facilitate a private company over the rights of Irish men and women'.

Micheál Ó Seighin said that the proposed pipeline route was technically and materially wrong, in breach of EU regulations and Shell's normal standards. It was not under the control of the energy regulator, and the state's Health and Safety Authority had said it was beyond its remit. A safety review of the onshore pipeline had been ordered by Minister for Communications, the Marine and Natural Resources Noel Dempsey the month before (May 2005), but it had emerged that the consultants hired were part-owned by Shell. The minister had gone back to the drawing board, and ordered another 'independent' evaluation when this had become public, and had suspended final approval to install and commission the pipeline till this was carried out. But there would be no need for such an evaluation, no need for a pipeline, no need for compulsory acquisition orders signed by the government for access to private land, if the gas terminal was built offshore.

Ó Seighin had prepared a six-page A4 letter for the court hearing before leaving north Mayo. It described how he, as a sixty-five-year-old resident of Ceathrú Thaidhg, originally from Gleann Bruachán in County Limerick, had enjoyed a 'rather miserable legal record' in the forty years he had spent in the area, working for most of that time as a teacher. He had recently retired after a coronary bypass. He recounted his previous 'conflicts with the law'. In 1965, he said, he had been summoned to court and fined two pounds for not having a parking light on his car outside Healy's hall in Glenamoy. 'The then sergeant of the day . . . was a native of my home parish and had been taught by my father.'

In 1968, he said, he had been fined ten shillings for 'speeding' through Charleville (Rath Lúirc) on 18 March that year. 'The gardaí in Ballina paid the fine as it was not worth their while collecting it,' he said. And then during the national Christmas 'round up' in the late 1990s, which had focused on detecting driving under the influence of alcohol, the then sergeant had arrested him for 'being under the influence while being in control of a vehicle'. As it transpired, when the blood sample was returned, his blood alcohol level was all of 50 micrograms. 'No further legal action was then attempted,' he recalled. All in all, it seemed 'a poor apprentice-ship for my pending exalted criminal status!

'I believe and know that I have no choice but to oppose the imple-mentation of this injunction and also that I have a lawful excuse for so doing, to prevent a greater harm than the injunction facilitates,' Ó Seighin continued. He outlined his fears about the impact of a high-pressure pipeline on the landscape, which would become 'unfit for agriculture' if dug up with heavy machinery. He cited associated cultural issues. The land system, specifically its highly skilled drainage, represented 'human archaeology at its most basic'. Such painstaking work would never be repeated, he said.

Monetary compensation could not repair the damage done to such work, or restore the function of a 'basic economic asset', he said, given the minimal size of the landholdings. His second reason for opposing 'the encroachment' by the partners involved in the Corrib gas project related to the 'major issue' of 'safety of residents . . . of farmers and of visitors or of children playing or people gathering cockles or, indeed, cars driving along the road in close proximity, or at one spot, driving over this rare bird of a pipeline . . .'

The proposed pipeline would run 'hardly twenty-five metres' from where one resident 'turns off the road into his yard', Ó Seighin wrote. He quoted recent pipeline failures, such as a Shell-operated pipe near Brussels that had ruptured, engulfing a wide area and killing twenty-one people in July 2004. There were 'dozens' of other recorded pipeline failures, he said. He acknow-ledged that pipelines were not designed to fail but the reality was that they still fractured or leaked. He noted how a 'learned judge' at an earlier court hearing had expressed the opinion that 'we walk over or around such Bord Gáis pipelines every day'.

'*Mo léann, nach siúlann*,' Ó Seighin noted – as in, 'I'm sorry, we don't walk'. This particular pipeline planned for Erris was not a distribution or transmission channel. It was 'exceptional' in terms of pressure. Were it otherwise, 'we five would not be today occupying the High Court in this less than productive way'.

The five were sent to prison indefinitely. MidWest Radio senior reporter Liamy MacNally had a lump in his throat as he watched them being led away. A garda assured Micheál Ó Seighin that they would not be handcuffed or shackled if they would just follow him, and Ó Seighin readily agreed. Outside in the bright sunshine, Mary Corduff collapsed on the pavement. MacNally, who went to her assistance, found himself talking to a familiar senior garda who was now assigned to the Four Courts and wanted to know what all the commotion was about.

MacNally filled him in. The senior garda assigned a colleague to escort Mary Corduff inside. Once off the street, she was shown into a room, where her husband was brought to see her for a few brief minutes before she was led away. MacNally would remember it as one of those random acts of kindness and singular humanity from an officer of the state.

Fine Gael's Michael Ring said that Ireland was now living in 'a dictatorship within a democracy' while Dr Jerry Cowley and fellow independent and ex-Fianna Fáil TD Beverly Flynn called on the minister to intervene. The jailings were 'disastrous news', Flynn, a supporter of the gas project, said, adding with prescience that 'This is only going to galvanize opposition to the project.'[20] For the *Western People* of 5 July, it was 'a national disgrace' – this was its front page headline in the immediate aftermath of the jailing. 'This country is either being democratically governed by Bertie Ahern or it is being dictatorially ruled by Andy Pyle and Shell,' its editorial said.

The issue was raised in the Dáil the following day, when the minister, Noel Dempsey, asserted that Shell had consent for preparatory work but did not yet have consent to lay the pipeline. Solicitor Greg Casey, acting for several of the five jailed, told the High Court the following week he would use the minister's statement to seek a dismissal of Shell's April injunction, on the basis that there were no valid ministerial consents for actual installation of the pipeline. However, Shell said it intended to press ahead.

From within Cloverhill prison on 30 June, the five men issued a statement through relatives calling for a halt to 'all illegal development at Rossport'; a withdrawal of 'all threats of imprisonment and financial ruin' hanging over the people of Mayo; a renegotiation by the government of oil and gas exploration deals agreed under the 1992 Finance Act to ensure a better return to the state; and a call on Shell to refine the Corrib field gas at sea.

As Beverly Flynn had predicted, opposition did 'galvanize', with even the landowners who had signed up to compulsory acquisition having serious second thoughts. Chris Tallott, a leading member of the Pro-Erris Gas Group and the North West Mayo Action Group, which involved many staff who had worked with the Electricity Supply Board (ESB) or Bord na Móna at the Bellacorick power plant, was stunned. 'I know these five men. I regret very much what has happened, and perhaps it has wakened us up to reality,' he said. 'Those men are not going to be released now unless Shell can give a firm assurance about safety, and that means looking at other options which will take this pipeline away from people's houses and lands,' he said.[21]

Engineer Brian Coyle, who had objected to the second terminal planning application, said that the minister and Shell were technically correct when they said that the pipeline was designed to the highest standards, as 'no engineer wants to design a pipeline that fails'. The reality was that 'pipelines do fail, and have done so all over the world with fatal consequences'. One only had to scan the columns of the weekly maritime industry publication, *Lloyd's List*.

When they entered prison on 29 June, the five men were 'almost entirely unknown outside a cluster of small villages and communities', Dr Mark Garavan, Galway-Mayo Institute of Technology lecturer, recalled in 2006.[22] 'Five quiet and private individuals had become the "Rossport Five". They were men who had brought the €900-million Shell project in Mayo to a complete halt.'

9

'LIKE CATCHING A FOX . . .'

The steel cages in the van from the Bridewell in Dublin to prison were like 'dog boxes' to the five men who had 'kissed goodbye to the daylight', as one of them described it. They were stunned, but also exhausted, having risen shortly after 4 a.m. that morning to travel east from Mayo. As Vincent McGrath recalled, Justice Finnegan 'gave us ten minutes to reflect on the consequences after we had refused to purge our contempt . . .'[1]

At Cloverhill prison in west Dublin, they were weighed and photographed, they were handed towels, instructed to strip and directed to walk through a security screen before being given uniforms and allocated cells. For Willie Corduff, so used to life on the land with his animals, digging for lugworms on the shore, the confinement was almost unbearable. It was 'like catching a fox and putting him into a dog's cage, or catching young wild ducks', he said later. There were nights where 'we cried ourselves to sleep'.

For Philip McGrath, younger brother of Vincent and the youngest of the five, it was a shock that turned to depression. He was offered a job in the kitchen at €3.25 a day. 'I can't wait,' he recalled himself saying. One day, he was assigned to work alongside a Russian who was in jail for murder. Philip was a little alarmed, for they had access to kitchen knives. Eventually, he plucked up the courage to ask the man how he had managed to get marked for kitchen duty. 'It's okay, my friend,' the Russian replied, as if reading his mind. 'I killed him with my bare hands!'

One of the first pieces of advice that had been given to them by the prison authorities was to 'stick together' when in the exercise yard, given that they were 'from the country'. As it transpired, the 'farmers', as they became known, received assurances from fellow inmates that they would be safe. Not only that but they drew admiration when it became clear that they weren't going to return to court to say sorry. Ó Seighin and Vincent McGrath shared a cell thirteen foot long by nine foot wide, with an open toilet and washbasin beside the door.

Philip McGrath, Willie Corduff and Brendan Philbin shared a second cell, with the most challenging period being the long confinement from early evening to the following morning. There was television, but no radio. They watched reports of the London bombings of that July, the impact of Hurricane Katrina in New Orleans . . . and reports from their own home in Mayo. The post began arriving in bags, and then sacks. 'I remember I got a letter in Irish from a woman in Spiddal with a cheque in it to go into the account, which I couldn't do until I came out,' Ó Seighin remembered. 'I got two letters from two kids in a secondary school in Dublin. This was later on. They were just back to school in September. They were in transition year. The whole class had discussed the Rossport Five immediately they came back to school. I knew then that this is different.'[2]

Outside, there were farms and families to mind. Mary Corduff, Aggie Philbin, the two Maureens married to brothers Philip and Vincent McGrath and Caitlín Uí Sheighin had to plan visits to Cloverhill, and to court. The daily six-minute phone calls permitted by the prison were never easy. Dr Jerry Cowley undertook to use his status as a political representative to visit Cloverhill several times a week, driving east especially for this purpose during the Dáil summer recess.

Bríd McGarry was upset and frustrated. Five of her neighbours were behind bars on a point of principle. She had been with them in the High Court and had told the judge that the pipeline posed an unacceptable risk to the community. By rights, she should be in prison also, as owner with her mother, Teresa, of 20 per cent of the land on the onshore pipeline route, and one of the named landowners in Shell's original injunction. In her view, the memo of Shell's meetings with lawyers in early June, enclosed in legal documentation forwarded to the defendants, explained it all. It would not be good from a PR point of view to have all of the objecting landowners jailed – particularly if one of them was a woman. If there was one consolation, she recalled, it was that the community's case had been taken up on national television and radio.

Bord na Móna workers hired to help Shell's transfer of peat from Ballinaboy to Srahmore walked off site after the men's jailing, and there was a series of protests in Mayo, Galway and Dublin in the following weeks, when the five were brought back to court several times, but refused to purge their contempt. The Shell to Sea website, maintained by a team of volunteers, was averaging 600 viewers a day and up to 2,000 at peak periods. The fact that Google and other Internet search engines ranked the website for 'Corrib gas' searches higher than Shell's own website was 'helpful', according to one of the website's founders, who observed that it 'presented the opportunity for the global public to view all sides of the debate'.

At a press conference in Dublin, Micheál's daughter, Bríd Ní Sheighin, called on the public to boycott Shell and Statoil petrol stations. Separately, Green Party TD Eamon Ryan called on Shell to waive the court order and said it was 'remarkable' that Shell had 'pursued this course even when the Minister for Communications, Marine and Natural Resources Noel Dempsey has not yet given full and final consent for the contentious pipeline project'.[3]

Shell, which made it clear that it had no intention of stopping its work in north Mayo, said it had never claimed to have overall consent for the project at this stage, as the process involved a series of 'phased consents'. It was concerned about the 'level of misinformation and myths' about the pipeline, even as the men's imprisonment dominated the airwaves and print media. The pipeline was 'designed to world class standards, would never run at 345 bar pressure, and would normally operate at a pressure of 120 bar'. The 345 bar option was for 'fault conditions' only, if the subsea modules failed.

The 'bar', or unit of pressure, was confusing to a public that knew little of the difference between production, transmission and distribution pipelines. Several business commentators confused the issue further by stating that pipelines ran under motorways and housing in the capital and there was no big deal. In fact, these transmission and distribution pipelines carried treated gas at much lower pressures than that designed for the sea to inshore link in north Mayo. The minister had acknowledged in the Dáil that it was 'unique'.

Labour Party president Michael D. Higgins added his voice to those calling on the minister to act. It was 'unthinkable' that the five men could be allowed to remain in jail for an indefinite period, and recourse to the courts was 'not the appropriate way to deal with a complex dispute like this'. A cross-party delegation of politicians made a visit to the men in Cloverhill prison, and a senior officer in the Irish Farmers' Association offered his services as mediator.

The day after the men were jailed, Glenamoy farmer P. J. Moran hitched up a converted horsebox, which he used as a trailer, and drove it to the gates of the terminal at Ballinaboy. A picket was planned and the weather was hot. Refreshments would be required, and his sister, Mary Moran Horan, had volunteered to provide them.

The first in a series of major rallies was held in Castlebar in the first weekend of July, when there were demands for both the minister, Noel Dempsey, and his predecessor, Dermot Ahern, to resign. Speaking on the platform in the car park of the Travellers' Friend hostelry, Mary Corduff accused Shell of trying to 'divide and conquer and split neighbours and friends', while Brendan Philbin's son, Chris, said that responsibility lay not primarily with Shell but with the government and with 'Mayo County Council, where it all started'. Dr Werner Blau, professor of physics at Trinity College, Dublin, appealed to Shell's shareholders to remind the company of its corporate responsibilities.

Dr Blau said he had examined the technical data, and the proximity of the high-pressure pipeline to houses would not even meet US standards, which were 'pretty lax'. He noted that the US Office of Pipeline Safety had recorded 1,586 pipeline rupture incidents, including 61 fatalities, 235 injuries and over US$408 million in damages from 1986 to 2004.

Two days after the Castlebar rally, there was a protest at Ballinaboy and attempts were made to halt lorries involved in the peat transfer to Srahmore. The gardaí were called, and Inspector Michael Murray spoke to those present. His tone was sympathetic, for he understood the depth of feeling – and the fact that people had the right to protest. However, they must remain within the law, he said. It was a conciliatory moment that objectors would not forget when, just over a year later, the Garda approach was to change.

In Galway, writer Jennifer Johnston and Lelia Doolan attended a forty-strong picket outside the constituency office of Minister of State for Labour Affairs Frank Fahey. Several days later, ten Erris priests issued a statement calling for the 'restoration' of the five men to their families, and expressing the hope that all genuine worries about health, safety and the environment could be addressed. One of the ten, Father Michael Nallen, said he intended to stay on a picket line mounted at Ballinaboy.

Another, Father Kevin Hegarty of Kilmore, was held in high

regard at national level. He had been removed as editor of the Catholic bishops' magazine, *Intercom*, after he raised the issue of clerical child sex abuse in 1993 and 1994. He had written about the Corrib gas project in *Céide* magazine, April/May 2001, calling then for an independent environmental, health and social impact study of the entire Corrib gas project. As he told *The Irish Times* four years later, it was the only option for the government now, given the 'barrage' of scientific evidence on 'both sides'.[4]

Within a week of the men's jailing, a group calling itself Rossport Solidarity established a camp on Philip McGrath's land, led by Bob Kavanagh and Terry Dunne among others. The families of the five men welcomed the support from visitors, several of whom had already visited the area through their affiliation to student societies in Galway, Cork, Dublin and Limerick. Others had travelled from parts of Europe, having read about the situation on the Indymedia and Shell to Sea websites, or viewed footage posted on YouTube. They offered to help out on the farms and with saving turf, when they were not holding pickets at the Shell compound at Rossport where, as Maura Harrington remembered, they were 'eaten alive by midges'.

By mid-July, it had become a 'massive influx of people', according to one of the camp founders, and it was not without some difficulties. 'Everything happens up here without any collective organization – and, yes, we would like to change this,' a memo of 13 July recorded. 'Everything is individualism and personal relationships.' A number of 'incidents' had occurred, including some communication breakdowns at local level and an unwarranted visit from 'some pissheads from Ulster', leading to moves to establish ground rules at the camp.

'This is not camping out for the weekend, this is a serious political action that involves hard work and discipline,' the memo stated. There would be no alcohol and no drugs, as 'this is what the media/State/corporations want – not us'. Decision-making would be 'collective'. A compost toilet was created, and waste was to be separated and sent for recycling in Ballina.

The long, hot summer was difficult for the large number of consenting landowners, who had signed up to a pipeline through their land in return for compensation. The money had been welcome: farming was tough, and tougher in such an isolated area. They were reluctant to be interviewed as the press arrived into Erris, but some admitted privately that they had not realized it was a high-pressure pipe – one that they would not be able to work near.

On 6 July, *The Irish Times* reported that a review of safety aspects of the pipeline commissioned by the department of the minister for Communication, the Marine and Natural Resources had been written by a company that listed Shell among its clients. This was the second time that a company hired to review Shell's own quantified risk assessment had been found to have links to the developer. AEA Technology had, like the British Pipeline Agency before it, indicated 'no conflict of interest', the department told the newspaper. It meant that there would be a further delay to the minister's decision on consenting to the laying of the pipeline.

Dr Cowley called for a return of the Dáil, while Erris Fianna Fáil councillor Tim Quinn said he thought the gas should be processed at sea, as people were genuinely worried and the issue had 'caused a lot of pain'.[5] By the end of the first week, when the men returned to court and solicitor Greg Casey highlighted the minister's Dáil statement about lack of consents to lay the pipeline, the five were told by Mr Justice Joseph Finnegan that their fate was in their own hands.

Outside the court, Shell appealed for 'calm' and for 'reasoned and constructive dialogue', as it confirmed that it was suspending work at Ballinaboy due to continued protests by locals. A man on one of the pickets had been injured by a Land Rover. Adding to the tensions, several days later there was an incident at Ballyglass, when a Scottish contractor working for the developers on a sixty-foot barge hired for inshore pipeline work damaged four local fishing boats. The man was arrested and charged with having excess alcohol taken while in charge of a vessel.[6]

Two weeks into the jailings, the minister ordered a new and 'comprehensive' safety review of the pipeline, and Shell said it had given Dempsey an undertaking that no work would take place on the pipeline 'for the present'. Shell to Sea handed in a letter to the Norwegian embassy in Dublin, calling on the Norwegian government to intervene, given Statoil's role as partner in the project.[7]

Examining in jail the minister's terms of reference for the safety review, the five men accused the minister of 'deliberately excluding the risk of rupture and explosion'. They would have been unaware that Dempsey had talked to senior Shell management about the jailing. The Shell team that met him comprised Dutch, English, Scottish and Irish company executives. Dempsey tried to explain the local sensitivities. Eventually, exasperated when they didn't seem to 'get it', he was heard to exclaim that he could forgive a Dutch or Scot or Englishman for failing to appreciate the significance of land ownership in the county of Land League founder Michael Davitt, but he could 'never forgive an Irishman' for failing to understand the same. Their statement was issued through Irish Farmers' Association vice-president and voluntary mediator, Raymond O'Malley.

There was another headache for the minister as the men entered their third week in jail. Both the *Western People* and *The Irish Times* carried photographs by Peter Wilcock of a gas pipeline above ground and under construction in Erris, just across from Bríd McGarry's farm. Shell and the department told the newspaper that as 'matters were the subject of legal proceedings, it would be inappropriate to comment further'. The company did not have consent for this – and it had given an undertaking to the minister the previous week that it would suspend preparatory work.

Bríd McGarry told the newspaper that this formed part of the works she had reported a full month earlier, before the men's jailing, to the gardaí at Belmullet on 18 June.[8] The minister moved fast.

He sent petroleum affairs division staff up to inspect the welded pipeline, he published details of all consents so far granted to the developers for Corrib gas, and he established a new high-level body, known as the technical advisory group, within his department to monitor the project. He informed the company on 23 July that a '*prima facie* breach' of consents was 'very serious'. Just under three kilometres of onshore pipeline had been strung together and welded between a crossing at Sruwadaccon Bay and the terminal site at Ballinaboy. The company reiterated that it had all 'necessary' consents, and a spokesman told MidWest Radio that the 700-metre welded section visible from the road had arrived like that. Listeners recalled thinking it must have been a 'very long truck'.

Shell publicly blamed 'ongoing protests' on a decision to lay off thirty-five sub-contractor staff at the landfall site.[9] Meanwhile, there were unattributed press reports of 'intimidation' of Shell contract staff in Mayo. Chief Superintendent Tony McNamara, formerly based at Belmullet and now head of the Mayo Garda Division, said his staff had received no such reports, apart from one logged at Belmullet station involving a verbal exchange in a pub in Belmullet several weeks before.[10] However, in a small rural area, intimidation, whether it be refusing to talk to someone or following them home, was almost impossible to report.

One man whose position on the whole issue was not clear was Fine Gael leader Enda Kenny, a constituency colleague of Michael Ring in Mayo. So far, Ring had made all the running on this issue. Back in 2000, when details of the gas find were first in the public domain, Kenny had called on the government to ensure that all of Mayo benefited from it, and said he wanted taxes levied on the company over the twenty-year lifespan of the field to be put into a 'rolling fund' for that purpose. Kenny, then a TD, said he had met with developers Enterprise and had learned that a 'small number of jobs' would be created by the project.[11]

Both he and his brother Henry, chairman of Mayo County

Council in 2005, had visited the men when they were imprisoned, but Henry Kenny did not win any Brownie points with the five by his lack of support for a motion calling for an offshore terminal for the project at a Mayo County Council meeting in mid-July. As Vincent McGrath and Micheál Ó Seighin recalled, the Kennys and so many other politicians who made the same trip didn't understand the point. 'The first thing they would ask was "How do we get you out?"' McGrath remembered. 'But what they had in mind was some face-saving formula that would make us look good when we climbed down and purged our contempt . . .'[12]

The men were further upset by a remark made by Enda Kenny on one visit. 'We were talking about the pressure we were under and how Fine Gael wasn't doing anything for us, and Kenny replied that life was worth very little today, and that people could be "taken out" for as little as five hundred or a thousand euros,' Philip McGrath recalled. 'He tried to backtrack then, but I was so upset that I wrote to him about it.' Mr Kenny said that his several visits to Cloverhill were 'to visit constituents who were in prison on a charge of contempt and to inform them of the best legal opinion that I could get from our party as to how this might be dealt with to bring about a resolution to their being able to be released back to their families and communities'.

'The reference to professional hit men was part of a different conversation entirely, and nothing to do with the previous issue. Any attempt to do so is completely misleading and wrong,' he told this writer.

The Fine Gael leader was not best pleased when Dr Jerry Cowley highlighted his absence at a protest rally in support of the men in Ballina in mid-July. Kenny was on holiday. 'The fact that he [Cowley] repeatedly attacks me and the Fine Gael party, rather than the government who are completely responsible for this mess, speaks for itself,' Kenny said. 'The current licensing regime was agreed by the Fianna Fáil government and a Fianna Fáil minister, and it was Fianna Fáil who controlled Mayo County Council, which granted the initial planning permission for the terminal. It was Minister Frank Fahey on behalf of a Fianna Fáil government who authorized the pipeline from sea to terminal without planning permission or compulsory orders being required.' The 'only initiative taken in the current impasse' had been the minister's new safety review and this had been 'at my instigation', Kenny said.

Kenny stopped short of supporting the offshore terminal demanded by his party and constituency colleague, Michael Ring. Cowley said it was 'legitimate' that the Fine Gael leader should 'make his position clear'. The spat between Kenny and Cowley continued, with the party leader calling for an apology from the independent TD when his staff had been, he said, 'abused' during an eight-hour picket of his constituency office in Castlebar.

Kenny would later draw the ire of the five men in prison, when he claimed on RTÉ that the Cowley picket had been organized by Sinn Féin.[13]

Dr Mark Garavan, Galway–Mayo Institute of Technology (GMIT) lecturer, had become the men's spokesman while they were in Cloverhill. Speaking after another court hearing on 25 July, when they again stood firm, Garavan explained that the issue was not simply one of purging contempt in court: 'This does not just involve an apology. It involves giving an undertaking not to protest against Shell's activities on the ground in Mayo and this is something that the men cannot do, given the risk posed to them, their families and neighbours by this high-pressure pipeline ... One would have thought that when the liberty of five citizens was at stake, the minister or his representatives would seek to clarify the information before the court.'

Significantly, the company's unauthorized welding of just under three kilometres of pipeline onshore, as identified by the minister, had called into question the nature of the injunction secured by Shell, he explained.[14] At this stage, the developers had acknowledged a 'technical breach'.

Dr Garavan's wife Pauline was due to complete her final year in art at GMIT. As her husband became heavily involved in efforts to resolve the impasse, she wrote daily letters to the men, enclosing newspaper cuttings in the envelopes. She also began dissecting articles from the *Western People*, for eventual release on to a 12-by-18-inch piece of gesso board.

Nouns such as 'Rossport', 'gas', 'pipeline', 'health', along with verbs like 'commit', 'resolve' and 'jail', and phrases like 'national interest' and 'security of supply' formed water drops, hundreds of them. They filled an outline of an Atlantic wave sketched by the artist, from a photograph taken off Achill Island in south Mayo one summer evening. One of two such pieces from her 'sea of words' was subsequently shown in the Royal Hibernian Academy in Dublin.

As if tensions weren't high enough, a new row broke out in Mayo over claims that the company was about to start laying the offshore section of the pipeline – as per its plan of summer 2002. The Erris Inshore Fishermen's Association (EIFA) said it had been told by the company on 28 July that the world's largest pipelaying vessel, the 285-metre (935-foot) *Solitaire*, would arrive within the next fortnight in Broadhaven Bay. The department confirmed that no consent had yet been issued by the minister for the offshore work. The association chairman Eddie Diver said he was 'warned by a Shell official on the phone that the law would be enforced if the vessel's activity was impeded in any way by the fishermen'.

Diver told *The Irish Times* that he felt 'intimidated' by this; the fishermen's association had informed the company earlier that month that its members would no longer co-operate with Shell on the project while the five men were in jail. 'I advised the company that, given the tensions onshore, and if they were intent on proceeding, they would be better advised to start at the wellhead,' Diver told the newspaper. 'However, I was informed that they would be starting in Broadhaven Bay.'

Garavan noted that if the company went ahead with the work off-shore, it would be an effective rejection of an offshore processing option and would 'prejudge' the minister's safety review. Green Party spokesman Eamon Ryan, who was due that day to visit the five men in prison with his party colleague John Gormley, called on the company to suspend all work, pending publication of the minister's pipeline safety review.[15]

Shell's 'technical breach', in welding just under three kilometres of pipeline without ministerial approval, was described as a 'departure from the terms of the consents issued' in a letter from Shell's Andy Pyle to Noel Dempsey in late July. He added that the company had 'been meticulous in complying with all of the conditions attached to the various consents issued to the project to date'. The minister ordered the company to dismantle the pipeline section, and said it would be subject to closer monitoring. The minister also asked Shell to confirm what its understanding of a 'consent' actually was.[16]

A day later, the company would 'neither confirm nor deny' that it had applied for the consent for the offshore section of the pipeline, even as the EIFA said that such work would 'inflame tensions' in north Mayo. Shell's 'often-stated policy of wanting to be a "good neighbour"' to the people of Erris was disingenuous and for public relations purposes and was now 'well and truly shattered', the association said.[17]

Dempsey was advised that he had no option but to approve the application for phase four of the seven-stage consent procedure, under the plan of development approved by former marine minister Frank Fahey in 2002. The decision, issued when the minister was still waiting for the company's response to his order to dismantle the three-kilometre onshore section, and when the five men were still in jail, was described as 'insane' by SIPTU's offshore committee spokesman, Padhraig Campbell. 'Totally spineless,' said Dr Jerry Cowley.

Mayo Fianna Fáil councillor Tim Quinn criticized the minister's timing, saying he could not see any pipeline being laid on or offshore in north Mayo due to tensions over the imprisonment, and Fianna Fáil councillors in Mayo were 'united' in calling for an offshore processing terminal. The councillors had supported the Ballinaboy planning application but the entire project, including the onshore pipeline, had never been laid before them. The minister's department made it clear that if the company went ahead with its offshore work in advance of the safety review of the onshore pipe, it was at its own risk.[18]

Eddie Diver was almost lost for words, but not quite. 'It is devastating that the minister should have given this sanction before he has investigated the safety concerns of residents in relation to the onshore pipeline,' he said. Shell welcomed the minister's sanction, stating that it provided an 'important reassurance of certainty in the regulatory process'.[19]

That first week of August, the wives of the five imprisoned men held a six-hour sit-in at Mayo County Council offices, supported by members of the Shell to Sea campaign in Mayo, in which they demanded that the council meet to discuss the situation. Another attempt by the minister to break the impasse failed when he proposed the appointment of a formal mediator for the men in prison, and was rebuffed. Meanwhile, Shell agreed to defer laying the offshore pipeline, saying a 'temporary suspension' would 'not materially affect the project's completion schedule' and would 'allow for a period of discussion and dialogue'. Amid sighs of temporary relief all round, Garavan urged the company to 'use this pause to reconfigure the entire project, make it safer and better with an offshore terminal, and listen to the people of Mayo'.[20]

Several days later, the company also agreed to dismantle the section.[21]

Dempsey published the details of the scope of his new safety review, with a tight deadline, and promised that the Health and Safety Authority, An Bord Pleanála and the Environmental

Protection Agency would be involved in a final safety audit before issuing phase seven of the series of consents. And he also named the high-level technical group that would monitor the project.

Dempsey outlined his various actions in a letter to Micheál Ó Seighin, sent by courier to him in prison on 12 August 2005, in which the minister stressed his commitment to ensuring that the 'entire project operates to the highest international safety standards'. Ó Seighin believed the efforts to be genuine. 'I very much regret that anyone should be in jail because of their concerns regarding this project,' Dempsey said in his letter.

Ó Seighin also received a request for a visit from a 'friend, Charles Feeney'. It was none other than Chuck Feeney, the Irish-American billionaire who had donated millions to educational and other projects in Ireland through his foundation Atlantic Philanthropies. Ó Seighin was very impressed by him. 'He said very little but obviously wanted to hear and get a grasp of the issues that put us in jail,' he recalled. 'I recognized this and just talked. He told me he was going to meet the then Taoiseach [Bertie Ahern] and tell him what he thought of our jailing. As he left, he smiled, brandished his fist and said, "Don't give in to them, we're all behind you!"'

In mid-August, several days after Labour Party leader Michael D. Higgins called for an offshore terminal at a Galway rally, Shell initiated a new public relations campaign. It promised a 'programme of dialogue' with 'local stakeholders'. Separately, in an open letter from prison, the five men offered to hold direct talks with Shell and its 'government partners' if the injunction against them was lifted. The company described it as 'significant', but rejected the move. It was conscious that it had to 'preserve' its legal position on construction of the onshore pipeline, and protect its staff from 'unlawful interference', Andy Pyle explained.

If August had been a wicked month, the following one was fraught with tension. In early September, Taoiseach Bertie Ahern was 'doorstepped' by reporters in Ennis, County Clare. Ahern felt that the five men had made their point 'many weeks ago', and 'for their families' sake, they are not actually making any further point by staying in jail'. His minister Noel Dempsey had 'literally stood on his head all summer' to assure the men and their families that their safety concerns about the high-pressure pipeline would be taken into account. Dr Jerry Cowley noted that the Taoiseach did not seem to realize that it was up to Shell E&P Ireland to lift the injunction before such a move could take place.

Up in Erris, Shell representatives said they were unable to inspect a water-purification plant at Ballinaboy due to protests, and on several occasions that summer gardaí were called. Much of the time there was a 'carnival' atmosphere during the pickets, with children playing football on the road, tea and scones for protesters and gardaí in Mary Moran Horan's 'silver trailer' outside the gate, and a song about the Rossport Five penned by Michael J. Togher blaring over speaker systems. At this stage, the trailer had been equipped with a generator and boiler for continuous cups of tea. When Paul Claffey of MidWest Radio dedicated Big Tom's 'You're Going Out The Same Way You Came In' to the people of Rossport and Ballinaboy, it became a local hit.

Cowley and members of the five families, including Caitlín Uí Sheighin and John Monaghan, wife and son-in-law of Micheál, Brendan Philbin's son Chris, and Rossport resident Anthony Irwin, prepared for a visit to Norway to highlight Statoil's role. The trip was financed using personal funds and monies raised by the Shell to Sea campaign. One of Shell's senior management described it as a 'junket' for Dr Cowley, but issued an apology afterwards.[22]

After their meeting with Statoil, the Norwegian company's senior vice-president Helge Hatlestad said that it was 'fully confident in the safety of the pipeline and the project'. However, a member of the new Norwegian government, Labour MP Tore Nordtun, promised to alert the Oslo administration to the urgency of the situation in Ireland. 'We cannot have such a big company having an argument with our friends in Ireland,' Nordtun told *The Irish Times*.[23]

In Dublin, the Green Party called for a full oral hearing into the Corrib gas pipeline, and its energy spokesman Eamon Ryan said that the state's failure to hold one before compulsory acquisition orders for way-leave through private land had been signed by Frank Fahey in 2002 could be subject to legal challenge. A full hearing would deal with many questions relating to the pipeline's safety, but also the manner in which, Ryan said, 'Fianna Fáil rushed this aspect of the project through. The failure at any stage to allow local people to raise wider objections to the project is one of the reasons why we are now at an impasse on this issue. Real questions have to be answered about how Mayo County Council, the department and Shell have acted. Instead of just debating yet another desktop review of the pipeline design, there needs to be a public forum where questions can be raised about the overall project plan and the approval process that has been followed.'[24]

Both Fine Gael leader Enda Kenny and Labour Party leader Pat Rabbitte seized on legal precedents that would allow the company to withdraw its interlocutory injunction and facilitate the men's release. Former Labour Party leader Ruairí Quinn noted the success of the Labour Party in the recent Norwegian general election and said he had contacted his colleagues there: 'The Norwegian Labour Party has made it clear that it wants no involvement by a Norwegian state company in such a scandal.'[25]

By late September, the minister was able to inform an Oireachtas committee that he expected Shell to begin dismantling the illegally welded section of the onshore pipeline the following week. The company had claimed it had been unable to gain access for the work because of protests, but residents promised there would be no attempt to block it.

However, the minister also used the occasion to question the motives of Shell to Sea, and to reiterate that, despite the safety review he had ordered, he was 'personally satisfied' that the pipeline was being built to the 'highest international standards', and he would continue to hold this opinion unless the new review advised to the contrary.

As a prelude to the review, a two-day hearing was held in Mayo by senior counsel John Gallagher. Consultancy Advantica had won the review contract; it had also carried out work for Shell in the past, and was a sister company to Transco, which had been fined €22 million just the previous month over health and safety breaches that had claimed four lives in a gas blast in Scotland in December 1999. On their first fact-finding visit to Erris, the consultants were given a Garda escort – a move criticized by the Shell to Sea campaign.

There was one essential element missing, however. The Dáil was due to return after its summer recess; the minister needed to secure the men's release from jail. On 30 September, the company suspended its injunction, and that night the five men found themselves in a south Dublin hotel with their families. Instead of watching it in prison, they would be appearing on one of RTÉ's most popular weekly programmes, *The Late Late Show*.

10

'SPACE SHUTTLE SYNDROME'

Bonfires had been lit west of the Shannon, and it was around two in the morning when the five men and their families reached Ballinaboy that weekend, where several hundred people had been waiting for them. 'They were gathered right in front of the refinery gate,' Vincent McGrath recalled. 'The location they chose was interesting. They were making their own statement. I can't put into words what it was like. You could see how the mood had changed and that people now realized that it wasn't just about a few people in Rossport. It concerned them too. There was hugging. There were tears. There was pride.

'One woman said to me, "They gave ye back to us." I thought that phrase was very, very emotive. I felt she was saying, "You're one of ours. You belong here. What they do to you they do to us." It was as if we had been taken away, wrenched away from the community, the extended family. There were echoes in that remark of snatch squads in other parts of the world or a folk memory of things that happened in our own history . . .'[1]

The two cells they had left in Cloverhill seemed like a lifetime away, and yet their time there had left an indelible impression on them, as they had recounted the previous night to host Pat Kenny on *The Late Late Show*. Prolonged applause, wolf whistles and cheers heralded their arrival on the show. 'How does it feel, guys, to be out after ninety-four days inside?' Kenny asked.

Micheál Ó Seighin, wearing a red cravat, elicited some laughter when he replied that he hadn't found it 'particularly difficult'. It was, he told Kenny, 'probably the first time in penal history that the entire staff and inmates felt the same way about a particular issue', and so they had received tremendous support. Kenny, who had interviewed the men's wives some weeks earlier, noted that there were people who had committed far more serious crimes who had been in prison for far less time. Vincent McGrath reminded him that it wasn't quite over for them: the judge had wanted them to give an undertaking that they would not protest. They were due back in court on 25 October. 'We're not too sure what will happen then . . .'

Before they had left for the west the following day, there was a victory rally outside Leinster House and more interviews. 'It was the first time many of them were listened to in their lives,' Vincent McGrath said, of his conversations with fellow prisoners. 'They would tell you more and more about themselves and their back-ground and the mess they had got themselves into. Drugs and women seemed to be the source of their problems in many cases.

'You could see how they had no way out really when they were released. They'd go back to the same environment and they'd be back in. Even within the three months that we're talking about, we saw the revolving door syndrome . . . At visiting hours, their babies would be brought in. This was very sad to see, and sadder to know that nothing was going to change for them. I often thought of what we could do with all the money from our oil and gas that was given away for nothing . . .'[2]

Several thousand people attended the city centre rally, which was addressed by the five men, by Dr Owens Wiwa, brother of the late Ken Saro-Wiwa, by the Irish Congress of Trade Unions' general secretary David Begg, by the SIPTU president Jack O'Connor, by politicians and other key supporters. The issue was not over, however. Shell had made clear it still intended to pursue its case for a permanent injunction to secure access to private land for the pipeline.

The men were invited to participate in mediation with the company, and the minister set 12 and 13 October as dates for a public consultation by consultants Advantica as part of their pipeline safety review. 'It's a bizarre scenario,' the men's spokesman, Dr Mark Garavan, said. 'Here we will have three processes running along parallel tracks when, in fact, the simplest thing to do would be to go back to basics and conduct an entire review of the project. It is already running two years behind schedule, so why not just start again and get it right?'

Micheál Ó Seighin echoed these sentiments. 'We're not telling a commercial company how to do their job, we're just asking that it be done properly, and if so we will co-operate.' Shell, meanwhile, was described in the press as 'coy' about its decision to drop its temporary injunction, which had led to the men's release, and why it had taken so long to do so. A 'key' factor was the timing of the minister's safety review, a spokesman told *The Irish Times*.

'We have already suspended all onshore work while this review is being carried out, and once it became clear that the review wouldn't be ready till late October, we knew that the temporary injunction wouldn't be necessary. We are into a period when no work is being carried out anyway due to weather conditions.'[3]

The impact of the Norwegian visit organized by Dr Jerry Cowley and the fact that key Statoil executives had visited Ireland shortly before the release was played down by the lead company. Corrib partner meetings were a 'regular event', Shell said. Cowley was convinced that the Norway trip had been a turning point. Fine Gael TD Michael Ring said there were 'serious questions' to be answered about the government's role in the 'whole débâcle'. Corrib was on the agenda for the following Tuesday's Dáil session, but Ring was seeking a full debate and not just 'statements'.[4]

During comments on the minister's statement on 4 October 2005, Enda Kenny said that 'the pipeline from the landfall to the terminal should have been part of the transparent, accountable planning process in the first instance', as he believed this would have 'dealt with issues of design, safety and health, and would not have resulted in ordinary people being put in fear of their lives' because of 'legitimate' concerns over these issues.

The five men were convinced that the government should participate in mediation, and also made it clear they would not attend the two-day oral hearing into the Advantica review, due to its terms of reference, which were confined to the safety of the pipeline rather than to the safety of the entire project on sea and land. The following month, they stated this formally when Micheál Ó Seighin attended the hearing briefly on 12 October in Teach Iorrais, Geesala, and Dr Mark Garavan read out a statement.

The men acknowledged the principle of consultation and public engagement, Garavan said, but believed this should have occurred before approvals were granted. The review's terms of reference didn't leave room for alternatives, and the entire project required a 'thorough health and safety review'.

There was an informal and welcoming atmosphere as senior counsel John Gallagher and representatives of Advantica opened the hearing with a presentation from the Department of Communications, Marine and Natural Resources technical advisory group. Eddie Diver, chair of the Erris Inshore Fishermen's Association, said he was confused about the consultation's purpose; former Statoil director Mike Cunningham spoke about the safety implications of splitting the project into parts for separate approvals, and argued for reconfiguration of the entire venture; Maura Breathnach of Ballinrobe in south Mayo, who described herself as a 'student of the universe', asked Advantica and the government to bear in mind that 'safety locally is safety globally'.

Aughoose resident Gerard McDonnell, community representative, chair of the government's environmental management group, and chair of the Dooncarton Landslide Committee, had nineteen points to make, including a recommendation that the onshore pipeline be moved and that an analysis be conducted of onshore and offshore processing options. He also called for major investment in the area, as he said local people felt they were being 'used and abused'. An existing Corrib gas fund should be increased to €20 million upfront, with a further €1 million for every year of the project's lifespan, he said. Summarizing the atmosphere of the two-day hearing for the *Western People* on 18 October 2005, Orla Hearns said that the overall mood was one of frustration at the limited terms of reference and a 'mindset difference' between officials, developers, consultants and the local community:

> Throughout the entire history of this development very few of the experts involved on behalf of the developer or indeed the government appear to have been briefed or suited to dealing with the public and, more importantly, dealing with the people of Erris. The defenders of this project were not prepared for the immense pride and devotion these people feel for their homeplace and their environment . . . They regard themselves as guardians of the land, charged with caring for it and passing it on to the next generation . . .[5]

Mediation talks had still not begun when there were further delays in Shell's promised dismantling of the three kilometres of pipeline welded without consent. The minister had been contacted by the Pro-Erris Gas Group, which suggested that the dismantling be deferred until the safety review of the onshore pipeline had been published. The group, founded by Micheál Ó hEalaithe, principal of Coláiste Chomáin in Rossport among others, suggested that Shell could instead be asked to pay a minimum of €250,000 towards community projects in Kilcommon parish, rather than 'wasting money'.

The minister took the proposal seriously, and said that Shell had sought legal advice. Garavan said any such wavering by Dempsey would place a serious question mark over his own regulatory regime. That same week, an advertisement for a 'slightly rusty experimental gas pipeline unwanted by local people' in County Mayo fetched all of eight bids on eBay, the highest at twenty-one euro, before it was declared 'invalid' and withdrawn.[6]

The minister opted to proceed with the pipeline dismantling, acknowledging that any deferral in return for payment of money would not help. Several days before, Micheál Ó Seighin had noted publicly that all but three of the thirty-four landowners bound by compulsory acquisition orders were among sixteen thousand signatories to a petition in support of Shell to Sea's campaign.[7]

That November, Chris Philbin, Brendan's son, and John Monaghan painted a mural of Ken Saro-Wiwa, which was unveiled by a visiting family from Ogoniland at the gates of Ballinaboy. Every African in Ireland had been invited to north Mayo for the day, and African food was prepared for a 'monster party' in Healy's Hall, Glenamoy. Various African church groups, the Irish Nigerian Association and Integrate Ireland pledged their support. Nine crosses, bearing the names of Saro-Wiwa and his executed comrades, were placed across the road. The mural was subsequently erected on a shed facing the Shell compound at Rossport at Brendan Philbin's mother's house.

As Dempsey was working on the appointment of a mediator there was a further hiccup. Shell confirmed that it was conducting 'ongoing research' as part of its programme drawn up by its new communications team. The team included former RTÉ and BBC journalist John Egan, who had reported for the BBC from Nigeria in the aftermath of the Ogoni hangings, and was now replacing Rosemary Steen, who had left the company. The 'research' was being conducted by Field Work Future, on behalf of Irish International, on behalf of Shell. Selected individuals were contacted by phone and offered fifty euro to participate in 'focus group meetings' in Belmullet. Several of those who had participated told *The Irish Times* afterwards that they had been 'misled', with one participant told she was questioned for a 'food survey'. She and others were also concerned that the discussions about the Corrib project had been recorded.

Garavan questioned how the data was to be used, adding that 'If this is a genuine attempt to sound out opinions in the area, at this stage of the controversy, it is also a remarkable indictment of Shell's approach.'[8]

On a frosty morning in late November 2005, a group involving a key member of the judiciary gathered in a hotel in the north Mayo town of Ballina. Mr Justice Feargus Flood was chair of a private body established earlier that year to investigate matters of public importance in Irish political, public and corporate life. The Centre for Public Inquiry (CPI) was modelled on the Centre for Public Integrity in Washington DC and was funded by Atlantic Philanthropies, established by Chuck Feeney – the Irish-American businessman who had given so much of the fortune he had made from his duty-free retail chain back to worthy projects in Ireland, and who had visited Micheál Ó Seighin in Cloverhill prison.

The CPI, headed by journalist Frank Connolly, had already published one report of public interest since its formation, on a controversial hotel development in County Meath. This second publication, *The Great Corrib Gas Controversy*, examined the history of oil and gas licensing in the state, the role of politicians, and the involvement of Sir Anthony O'Reilly, head of the Independent News and Media Group, in offshore exploration.

Focusing on Corrib, the authors, Frank Connolly and Ronan Lynch, said that within a week of senior executives of the Corrib consortium – Shell, Statoil and Marathon – meeting the Taoiseach in September 2003, the developers were given 'unusual access' to An Bord Pleanála to express their concerns over planning delays for Ballinaboy.

The centre had also asked a US consultant, Richard Kuprewicz of Accufacts Inc., to examine the safety of the onshore pipeline. Kuprewicz, an acknowledged expert, had tendered, unsuccessfully, for the minister's safety review, which was awarded to Advantica. He judged that the pipeline had a uniquely large rupture impact zone with potential for high fatalities, due to its 'unacceptable close proximity' to people and dwellings. The government and developers were caught up in a 'space shuttle syndrome', driving the entire Corrib gas project forward, Kuprewicz remarked.

The Kuprewicz analysis found that the piping specified for use was 'not invincible' to leak or corrosion. Shell's difficulties with building an offshore processing plant were overstated, in his view, and routing analyses for the onshore pipeline system were seriously deficient. He criticized the government's reliance on a quantified risk assessment to assess the Corrib project, when the pipeline was the first of its kind in Ireland. He quoted previous pipeline ruptures, such as that in Carlsbad, New Mexico, in August 2000, which had claimed twelve lives, as a compelling reason for a 'reality reference check', given that the proposed pipeline ran within seventy metres of houses in Rossport, County Mayo. The minimum safe distance from houses for this pipeline, which he described as 'unique' – a word once used by the minister also – had to be between two hundred and four hundred metres.

The government said that its high-level advisory group on Corrib would refer the CPI study to its consultants, Advantica. Dr Jerry Cowley said the minister must now halt the project, while the Green Party's marine spokesman Eamon Ryan said the minister must direct Shell to plan a safer means of bringing the gas ashore: the report's criticism of the government's handling of planning, legislation and execution of the project meant that it was now incumbent on it to 'give a comprehensive, honest account of the entire project for once and for all'.[9] Shell E&P Ireland said it was also 'studying' the report. It pointed out that the Corrib project had been through an exhaustive public consultation and regulatory process, and the final project design had been fully endorsed by all relevant authorities.

However, several days later, both the company and An Bord Pleanála took issue with what they regarded as an inference by the CPI authors that the company's new application for the terminal at Ballinaboy had been 'fast-tracked'. Bord Pleanála chairman John O'Connor confirmed that the board had met senior executives of Shell, Statoil and Marathon as part of an Irish Offshore Operators' Association delegation on 23 September 2003 – four days after the Shell meeting with the Taoiseach. It was 'wrong', he said, to infer that this meeting 'entailed special treatment for this particular application', which was finally approved in October 2004. Minutes of the meeting and related correspondence were published on the board's website on 23 November 2005.[10]

In the first week of December 2005, the CPI faced closure after a decision by Atlantic Philanthropies to withdraw its funding.[11] Its executive director Frank Connolly, a leading investigative journalist, was brother of one of three Irishmen arrested and sentenced to imprisonment in Colombia in August 2001 on suspicion of teaching Farc rebels how to use explosives. Minister for Justice Michael McDowell had been passed a Garda file which showed that Connolly had been interviewed by the force in 2002 on allegations of travelling to Colombia with a false passport; in 2003, the Director of Public Prosecutions had decided not to bring any charges.

Just weeks before Christmas, Minister for Communications, Marine and Natural Resources Noel Dempsey issued Advantica's draft safety review of the Corrib onshore pipeline, in which it said it had been 'designed to meet or exceed appropriate standards' and accorded with 'best international practice', but also that Shell had taken 'no account of societal risk to the local population as a whole' in its risk assessment. It recommended the pipeline's pressure should be limited to a maximum of 144 bar (down from the 345 bar maximum it was originally designed for) to enable it to be reclassified as a 'class two suburban pipeline'.

Advantica also raised the possible emergency of the contaminant hydrogen sulphide or 'sour gas' in the Corrib system. The contaminant is caused by biodegradation or thermal decomposition of organic matter, which can affect cellular respiration in a manner similar to carbon monoxide or hydrogen cyanide, and can cause corrosion in pipelines if driven at high pressure. It recommended provision of a formal integrity management plan and an independent audit and inspection procedure. It called for a review of the consent and permission system by the minister's department, and said Ireland should adopt a 'risk-based framework' for assessing the safety of major infrastructural projects.

Shell's detailed response to the draft, which came several days after publication, accepted the 'principle' of limiting gas pressure, but took issue with a number of points, including the reference to hydrogen sulphide. It was 'very unlikely' that the pipeline would be exposed to hydrogen sulphide concentrations that would exceed design limits, it said, as no traces of the compound were measured in any of the hydrocarbon samples taken at the Corrib reservoir. However, it did propose a 'monitoring procedure' for the Health and Safety Authority as a safeguard. Shell sought an assurance from the department that the draft would be 'removed from the public domain once the final report is issued'.

The report was 'illogical', according to Dr Mark Garavan, spokesman now for both the Rossport Five and Shell to Sea. The 144-bar pressure could not be guaranteed, he said. The only guarantee was distance from homes, but this was 'excluded from the consultants' terms of reference', which had focused on the pipeline design. Green Party marine spokesman Eamon Ryan called for a major redesign of the pipeline, based on the findings. Chris Tallott of the North West Mayo Action Group said he was disappointed that there was not enough clarity in the report, and it had not addressed concerns of those living close to the pipeline and terminal.

Ballinaboy resident Jacinta Healy said that the report had

vindicated the Rossport Five, and the study should have been carried out years ago: 'The government let us down badly.'[12]

The following Sunday, just after 6 a.m., there was a series of explosions at Buncefield oil storage depot, owned by Total in a joint venture with Chevron in the Hertfordshire town of Hemel Hempstead, forty kilometres north-west of London. Forty-three people were injured, houses had to be evacuated, and buildings in the nearby industrial estate and some homes as far as three miles away suffered severe structural damage. The blaze burned for three days and was described by the BBC as the largest such fire in Europe since the end of the Second World War. At one point, satellite images showed black smoke covering much of south-east England.

Total UK, Hertfordshire Oil Storage, the British Pipeline Agency and others were subsequently prosecuted by the British Health and Safety Executive and the Environment Agency. Total admitted exposing staff and members of the public to risk, and to allowing water below the depot to become polluted following the blast. It was a different type of facility, but the images of the inferno on television and reports of the vapour cloud caused by the rich fuel/air mix sent a shiver down spines in north Mayo.

The Rossport Solidarity Camp, which was suspended for the winter, would return in 2006 to a new marquee close to the dunes at Glengad. Shell to Sea protests in Dublin and throughout the country continued to highlight the issue in Erris, focusing on Shell and Statoil petrol stations. However, Shell was already in advance negotiations to dispose of its retail and commercial fuels business to Topaz Energy Group, and the stations would eventually be rebranded, as would those owned by Statoil.

The Shell to Sea website issued a pre-Christmas appeal for 'gifts' such as night-vision goggles and bolt cutters that could be used to breach heavy-duty steel fencing. Gift vouchers similar to those issued at Christmas by charities were available for this purpose, with options for purchase including waterproof clothing, a wind turbine, a digital camcorder, the aforementioned goggles and bolt cutters, and a sack of lentils.

'Since the government supports the scheme, it is feared that gardaí and defence forces will be used to support Shell's contractors in their attempts to force the high-pressure pipeline through the scenic area. The camp will act as a focal point for those who wish to help defend the area against the controversial scheme,' the website appeal said.[13]

The camp was located on a stretch of beach surrounded by *machair* (grass) overlooking the 'currently beautiful Broadhaven Bay', Terry Dunne recorded on Indymedia. He went on:

Due to the *machair* being a sensitive ecosystem we are taking steps to ensure the camp will not have a negative effect . . . Willie Corduff helped to move materials to the campsite using his tractor and trailer and several other locals including Terence Conway, Ray Corduff and Sean Harrington to name a few joined with campers in the construction. Delicious vegan food was prepared by the Bitchin' Kitchen radical cookin' collective.

Much of the equipment and material used to build the camp was donated and collected by various groups and individuals including Cork Shell to Sea and Greenpeace. The camp is being built to facilitate up to 50 people and it is hoped that people from all over Ireland and further afield will visit, help in whatever way they can, find out more and show their solidarity with the people of Erris.

Formal mediation eventually began in the new year, led by former Irish Congress of Trade Unions secretary general Peter Cassells. That same week of late January, former Dutch prime minister Wim Kok led a delegation from Royal Dutch Shell's Corporate and Social Responsibility Committee to Ireland. He had been appointed non-executive director to the committee; its remit included reviewing 'standards, policies and conduct of the company' relating to the 'safe condition and environmentally responsible operation of the company's facilities and assets'.

However, the committee's trip to Erris became a bit of a public relations fiasco when it emerged that written invitations for meetings had been issued to 'consenting' landowners only on the pipeline route. Two opponents who asked to meet the delegation in a personal capacity in Belmullet were ushered out of the hotel by security staff. Shell said that 'verbal invitations' had been issued to the Rossport Five and to 'non-consenting landowners' through an 'independent third party'.[14] Before he left Ireland, Kok said his delegation understood 'the hurt' caused the previous year in north Mayo over the project, and said that a 'partnership with the local community is the only way forward'. He said that it was his delegation's intention to 'bring the clear messages we've heard over the past two days to the highest levels of management in Shell'.[15]

Within days of mediation opening, it stumbled. It wasn't helped by the fact that Cassells was accompanied by one of Rossport's consenting landowners on several visits to the homes of objectors. The five men suspended their participation, claiming 'interference' by the minister, as he had sought briefings from Cassells. The minister had also indicated in a Dáil reply on 25 January and an interview on MidWest Radio on 1 February that he had never intended mediation to be confined to the Rossport Five and Shell, in spite of a clear indication that this was the case in an RTÉ Radio *Morning Ireland* interview on 30 September, hours before the men's release, and in a speech in the Dáil on 4 October, several days after.

Cassells sought to clarify the situation, saying that his talks involved a number of elements, including formal mediation between Shell and the Rossport Five, mediation with several other non-consenting landowners, including Bríd McGarry and Monica Muller, and consultation with other landowners and the wider community. Lack of natural gas provision for Mayo had come up in the discussions, and this was what he had referred to the minister, he said.[16]

Even as the talks remained suspended, Shell appointed a second journalist to its new communications team. *Western People* reporter Christy Loftus, who was also a former president of the National Union of Journalists, had written extensively on the Corrib gas project in its early years and had been highly critical at times of the way it was handled. He had been appointed the company's new external affairs adviser.

In a statement on 13 February 2006, Shell referred to the fact that Loftus had been a member of the Mayo 2000 group, which had campaigned in the late 1990s to ensure that the Corrib gas find was brought ashore to Mayo. 'Shell recognizes that it had a huge amount of work to do to rebuild the trust and confidence of the people of Rossport and the wider community in Erris. I hope I can play some role in helping to restore trust, rebuild relationships and support Shell's goal to work in partnership with the local community,' Loftus said, in a statement.

The company later confirmed that former Mayo County Council secretary Padraig Hughes, and former Garda Chief Superintendent John Carey, who had retired from his post as head of the Mayo Garda Division in early 2005, had also been engaged on a consultancy basis. 'Carey joins Shell's "dream team",' read the headline of 14 March 2006 in the *Western People*. 'As a former chief superintendent, a former Mayo County footballer and team captain and GAA all-star, John Carey knows Erris and County Mayo intimately,' Shell's Mayo area manager, Mark Carrigy, said in the newspaper report. 'He is a respected figure throughout the length and breadth of the county.' Carey was quoted as stating that the Corrib project would bring benefits to Erris and Mayo, the development was essential to Ireland, and 'We hope that by engaging openly and honestly with local groups and individuals, including those who are currently strongly opposed to our proposals, we can reach a consensus which will allow this project to go forward.'

The new team's plan was a little upset by a report by Mayo journalist Áine Ryan in *The Sunday Times* of 9 April 2006. 'Shell's Rossport spin doctor calls for alternative pipeline,' read the head-line to Ryan's report, which said, Ryan wrote, that Loftus had predicted that the onshore pipeline would 'never run through Rossport. Christy Loftus, recently appointed as external adviser by Shell, has advised his employers to start looking for alternative routes.'

'I've told Shell that I believe this pipeline will never be laid along that route, and I think they are listening to me,' Loftus said. But the former *Western People* journalist, who said he had been 'un-compromising and confrontational' during his job interview with Shell, said that the company would never process the gas offshore because of the 'economic implications' and also said he would not defend the company's mistakes and 'wouldn't be their apologist or spin doctor in north Mayo'. His conscience was clear in that he had supported the project from the outset. 'However, if I were in the shoes of any one of the Rossport Five and I had safety concerns about a gas pipeline running close to my house, I would have acted in the same way,' he told Ryan.

Ryan sought the reaction of Mary Corduff, who said it was 'typical of Shell's cynicism to use Mayo natives in an attempt to drive a wedge among the people. The only people Shell feel sorry for is themselves.' Ryan also quoted Maura Harrington as saying that morale was 'high' among 'anti-Shell protesters', and the Rossport Solidarity Camp had reopened at the beginning of the month, 'after many of its members spent the winter on lecture tours around England and Europe espousing the cause'.

'All we need now is our first camp baby,' Harrington told Ryan.

Non-violent direct action is synonymous with Martin Luther King and Gandhi, with anti-nuclear and anti-war activists, and with environmentalists such as those seeking to protect sensitive landscapes from motorway developments, like Twyford Down in Britain and, more recently, the Hill of Tara landscape in County Meath. In 1998, local farmer Michael Kunz had placed himself repeatedly in the path of ash trees being felled for the N11 road development at the Glen of Downs in County Wicklow, as gardaí controlled traffic through the valley. Yet, perhaps after several decades of conflict in Northern Ireland, non-violent direct action as a form of civil disobedience was not something that middle Ireland was too comfortable with.

In February 2003, up to a hundred thousand people, including many families with young children, had attended an anti-war march in Dublin over the invasion of Iraq. A month later, when a group known as the Grassroots Network Against War (GNAW) confirmed plans to pull down a perimeter fence at Shannon airport to protest over the airport's use by the US military, two leading anti-war organizations opted not to attend but to organize a prayer vigil in Dublin instead. 'Direct action isn't new to Ireland, but non-violent direct action is pretty new,' Ciaron O'Reilly, Catholic Worker activist, said in a subsequent interview on US radio.[17]

Erris residents had never envisaged they would be engaged in any such activity, let alone sit-downs or lock-ons, when the first concerns were raised about the Corrib gas project. With loyalties divided between Fine Gael and Fianna Fáil, residents of parishes like Kilcommon, encompassing Rossport, were as conservative as their farming and fishing counterparts in neighbouring counties. Token support for other parties, such as Labour, tended to be based on personalities running under such mantles at local level.

With each step, however, whether it was Maura Harrington and Bríd McGarry's decision to sit in the bucket of a digger for twenty-four hours on Glengad beach, or the decision by five men to go to jail rather than purge their contempt of a court order, certain residents were being carried on a particular path. On 17 February 2006, the concept of 'direct action' was confirmed as an option by Tadhg McGrath, co-ordinator of a Shell to Sea protest in Dublin, when he spoke of how funds had been raised internationally over the winter and members had prepared for a 'Greenham Common'-style protest.

Later that spring of 2006, former Fianna Fáil councillor Paraic Cosgrove responded to Shell's public invitation to visit the water-treatment facilities at Ballinaboy – still the subject of disquiet over aluminium levels. He and fellow Bangor resident John Carolan attended a mandatory safety course, were kitted out in safety gear, and were shown around. He took photographs. When it was time to leave, he said he was 'blocked' from leaving the site by protesters. 'I had an option of leaving by the side gate, gate two, but I wanted to go out the way I came in, by gate one,' he said later. 'We were held up for thirty-two minutes. I was in the process of ringing the Department of Foreign Affairs to say that we were being held hostage in our own country, with no passports and no identification, when things changed slightly and we were able to go. I was very angry about that, as I was only trying to get information that would reassure the public.'

Green Party leader and TD Trevor Sargent and other party members were among those photographed with Shell to Sea campaigners bearing placards and a Rossport Solidarity Camp banner outside the offices of Shell E&P Ireland. A number of bicycles had been chained across the office entrance; the protest was good-humoured, the emphasis being on a photo opportunity to continue highlighting the aims of the Shell to Sea campaign.

There was another peaceful protest, this time with a long section of plastic piping, outside Statoil's offices several months later. Gardaí were called to halt the protesters' attempts to negotiate a revolving door with their 'pipe' delivery. On the same day, Shell and the Rossport Five had something to agree about. Both welcomed the decision by the High Court president, Mr Justice Finnegan, not to punish the five men further for their contempt of court.

*

In early April, an employee of an English contract firm engaged by Shell contacted *The Irish Times*, claiming that the Corrib gas project pipes stored in Killybegs, County Donegal, had been treated for corrosion but the contractor hired had been less than thorough. In a statement of 27 April 2006 to the newspaper, Shell confirmed that the pipe material had been stored in Killybegs for the past four years, and had been cleaned at the beginning of 2005 as a precursor to planned installation of the offshore section.

The pipes had been inspected on 'regular occasions' since the decision to defer laying the offshore pipe in July 2005, the company said. They had 'oxidized', which was a 'natural process', but were subject to an ongoing inspection regime, which ensured that their 'intrinsic quality remains excellent'. They would be further inspected and cleaned before being loaded on to barges for laying on the seabed.

The minister published the final Advantica report on the onshore pipeline in early May 2006, along with a series of safety measures as recommended by his technical advisory group. He conceded that monitoring of pipeline safety would no longer be undertaken by his department, and legislation would be prepared to allow this function to be taken over by the Commission for Energy Regulation. Further consents would be subject to Shell accepting a series of extra safety measures, and pressure would be limited to 144 bar, which was still twice the pressure of an average gas transmission pipeline. The report also said that a landfall isolation valve at Glengad would have to be redesigned to ensure this pressure limit. Significantly, there would be no change to the pipeline route.

The Advantica authors noted some common ground between their study and the report by US pipeline expert Richard Kuprewicz for the now disbanded CPI. On the 'critical issues' identified by Accufacts, Advantica's consultants agreed with half of the points raised by Kuprewicz, but could not comment on six other issues as they were outside their terms of reference. Richard Kuprewicz, speaking from the US, urged interested parties to read the two reports in tandem.[18]

Shell and its Corrib partners described the Advantica report as 'very thorough and detailed'. Dr Mark Garavan, for the Rossport Five, expressed 'deep sadness' and warned of a 'countdown to conflict' unless the entire project was reconfigured. The minister's safety recommendations were 'irrelevant' to the resolution of the dispute, Garavan said, as 'we are still left with a production pipeline traversing a village and within seventy metres of some homes . . . Nothing has changed.'

Fuelling his concern was a BBC Radio interview with the minister on the day of Advantica's publication. The minister told BBC journalist Diarmaid Fleming that if a pipeline failed three metres from a person that person would be safe. Fleming asked him if he might like to correct this. Dempsey declined, pointing out that a seventy-metre proximity to the pipe gave an extra sixty-seven-metre safety margin over the three-metre safety zone.

On 3 May 2006 Green Party energy spokesman Eamon Ryan said: 'The Minister for Communications Noel Dempsey and Shell will put a positive spin on the [Advantica] report but the reality is that it provides a further indictment of the way that they have both managed this project.'[19] Essentially, Ryan welcomed the report, stating that it vindicated many of the residents' concerns. Several of the report's criticisms, he said, 'show up what has been a bungling approach by the government from the start', and he called on Shell to 'go back to the drawing board'.

The day after the Advantica release, Shell managing director Andy Pyle held out what appeared to be an olive branch. He said that the company was looking at alternative routes for the pipeline. *The Irish Times* led its front page with his offer on 5 May 2006. Pyle wouldn't be drawn on specifics, but said that the company had looked anew in recent weeks at locations considered in the original environmental impact statement, and was 'willing to talk' to north

Mayo residents about 'all options'. He acknowledged there was a 'widely held view' that it 'would be very difficult to put the pipeline through Rossport. If people at the end of the day say, "It's prison or nothing," then we have to find a solution,' Pyle said. 'We can't go forward with people going to prison because of their rightly strongly held beliefs . . .We didn't recognize how strongly they held these concerns and now we need to find a way around it.' He also expressed 'regret' for the previous year's imprisonment of the Rossport Five. 'Mistakes have been made,' he told the newspaper. 'We regret the part that we played in the jailing of the five men last summer. For the hurt that this caused the local community, I am sorry.'

Questioned by the newspaper on his reported support for the 'committal' option at a meeting with lawyers on 10 June 2005, Pyle denied he had favoured jailing the five men. The comments attributed to him were a 'misinterpretation', he said. Micheál Ó Seighin was contacted by the newspaper and by RTÉ for his response on behalf of the men. He said that any apology would be accepted, but Pyle also needed to be 'realistic'. Mediation was 'back on track', contrary to comments made by the company, Ó Seighin added. The five were simply waiting to hear back from mediator Peter Cassells.[20]

It seemed there had been a major breakthrough: the company was, for the first time, willing to discuss 'all options'. A similar message was transmitted to and published in other newspapers, including the *Irish Independent* on 5 May, under a headline reading 'hopes of breakthrough as Shell says sorry . . .' Jason O'Brien of the *Irish Independent* quoted Micheál Ó Seighin as welcoming Pyle's apology, while also cautioning that 'We will have to wait and see if this is more than just the PR requirements of Shell.' Pyle had confirmed to the newspaper that the 'options of moving the pipeline or developing the gas at sea would be looked at', but added that 'The selected option was selected for very strong reasons and I still believe that this is the only viable one.'

Within twenty-four hours, Pyle had rowed back on his stance. At a media briefing in Castlebar on 5 May – the day his earlier comments were published in the national press – he told reporters that an offshore platform was 'not' viable. 'We have to be realistic,' he said, almost taking his cue from Micheál Ó Seighin's turn of phrase the day before. 'We did not believe an offshore platform was viable before. We do not believe it is viable now.'[21]

The shift was not lost on one of several business journalists who had been lined up the same week for interviews with Pyle as part of a post–Advantica publicity campaign. Richard Delevan noted in the *Sunday Tribune* that Pyle was 'fairly definitive' on the onshore option on safety grounds. 'To put a platform on the Atlantic, with people housed on it, with helicopter flights, it's a less safe option. And that's one of the reasons we would say going to sea is not an option,' Pyle told him.

Delevan noted in the article, published on 7 May, that 'Both Thursday evening's RTÉ *Six One News* and Friday's lead story in *The Irish Times* gave the impression that Shell is far more flexible on the issues of the pipeline route and the site of the terminal than it may actually be.' The journalist concluded that 'both sides' were 'trying to reframe the outstanding questions of the dispute in the media'.

Micheál Ó Seighin was asked to participate in a Newstalk radio discussion on the issue that weekend. A Chambers of Commerce of Ireland spokesman said that gas pipelines of over 144 bar ran all around Dublin, and even over the M50 motorway. Ó Seighin could hardly believe it as he tried to clarify the difference between production, transmission and distribution pipelines and their various legal pressures. He had a petroleum affairs division handbook with him when he quoted the stipulation that pressure in the national gas grid must be less than 85 bar.

To some residents on the pipeline route, it was as if Pyle's apology had been snatched right back. Had his initial offer received the full backing of his Corrib partners? Or was he using the media, shifting his position when he had secured Ó Seighin's acceptance of his 'regret'? Within several days, his own press spokesmen were describing *The Irish Times* report of alternative pipeline routes as 'speculative', in spite of Pyle's on-record comments published on 5 May.[22]

Ó Seighin was not surprised, but came under attack from within his own community for accepting the apology in the first place. 'In reality, there was nothing else to do except to react to what was, in my opinion then as now, an effort to wrong-foot us on the basis of Cassells's advice that we were very rigid in our stance,' Ó Seighin said later. 'I did not believe that Shell operatives were sorry in the sense of "I'll never do it again, Father," but to turn a public apology, well choreographed, into the subject instead of the tactic it was would have been severely counter-productive for our cause.'

11

NO 'MARTYRS'

After the intriguing dance step that followed publication of the Advantica safety review in early May 2006, it was very much back to business for Shell in north Mayo. In his series of interviews with the business press, co-ordinated by his communications team in the week of the Advantica report's publication, Shell's Andy Pyle had emphasized that the majority of landowners in north Mayo supported the onshore pipeline crossing their land. The 'vast majority' of people in the county backed the project, and he referred to the '700' construction jobs, with the prospect of fifty permanent staff when the infrastructure was built.

Asked by Emmet Oliver of *The Irish Times* about the 'kind of things' he had 'endured' from a 'small number of protesters' since taking up his Irish post in 2002, Pyle referred to 'a lot of . . . pretty aggressive verbal abuse' and 'indirect intimidation' such as 'people being followed in cars, followed home at night'. He had had 'occasions when people phoned me' and 'a number of us have experienced phone calls, of what I would call a threatening type'. He said he had once found himself blocked in by cars after visiting someone in the area. Despite this, he said Shell was committed to further test drilling in Ireland and had started off the Donegal coast.[1]

Shell was still publicly committed to mediation while hoping to secure a permanent court injunction against the six named defendants – Bríd McGarry and Brendan Philbin, Willie Corduff and Philip McGrath, Monica Muller and Peter Sweetman. The government appeared to regard these two apparently conflicting positions as separate matters. Yet this 'project splitting' approach, which the European Commission was known to disapprove of, had contributed significantly to the impasse.

'By the time the legal action reaches a full hearing, we hope to have reached a negotiated settlement,' Shell said, on 7 May 2006. However, the six defendants in the action had been given permission by the High Court to include in their defence claims that there was no valid legal consent for the pipeline. And McGarry and Philbin, who had opted to be represented by a separate legal team, had successfully applied for the minister for the marine to be named alongside Shell in the action.[2]

The Pro-Erris Gas Group, which said it represented residents and business interests in the area who were concerned about any delay in the project, came up with a suggestion within a week of the Advantica safety report's publication. Why not run the pipeline down the middle of the Sruwaddacon estuary? As Micheál Ó hÉalaithe, Coláiste Chomáin school principal explained, this would keep everyone happy. The first part of the crossing from the land-fall would run through the property of consenting landowners, and would then turn right into the estuary to avoid commonage and the property of dissenting landowners. Former Corrib owners Enterprise Oil had ruled the estuary's route out primarily on environmental grounds, as it was protected under the EU Habitats Directive, but there were also technical and cost factors. Shell said it had no specific comment to make, other than that it was prepared to discuss 'all options'.

Mediator Peter Cassells, who was still trying to get direct talks on track, said that he would ask the company to clarify its position. Dr Owens Wiwa, brother of the late Ken Saro-Wiwa, believed he already knew the answer. The Canadian-based medical consultant had flown into Ireland on the invitation of justice and peace non-governmental organization Afri, which had asked him to lead a commemorative 'famine walk' held annually in Doolough, south Mayo. Andy Pyle's statement of 'regret' to the Rossport Five in early May was a 'public relations' move, timed to mollify share-holders before the parent company's annual general meeting (AGM) in The Hague, Wiwa told *The Irish Times*. He went on: 'I have seen this pattern many times before. The company makes conciliatory statements before its AGM, and then it resorts to "Shellspeak" when the AGM is over.' This had happened with other controversial projects, Wiwa said, and he was adamant that Shell could well afford to process gas offshore. Royal Dutch Shell had recently banked declared profits equivalent to £1.5 million sterling an hour for the first quarter of that year, he pointed out.[3]

In the Dáil, Minister for the Marine and Natural Resources Noel Dempsey promised a review of licensing terms for oil and gas exploration, following the defeat of a private members' motion backed by Dr Jerry Cowley on the issue. The Irish Offshore Operators' Association said it hoped that any recommendations for change would not be retrospective, as 'constitutional' issues might arise. Association spokesman Fergus Cahill said it would await the review 'with interest', adding that 'We'd far prefer to be in a high tax regime and finding oil and gas than in a low tax regime, finding nothing.'[4]

If this promised government review of 'state take' seemed like a concession to Shell to Sea, along with the Advantica safety review, there was another potential addition to the objecting residents' 'victory' list in early June 2006. Shore work previously carried out by Shell at Glengad – which had led to protests – was declared invalid by An Bord Pleanála. This contradicted Mayo County Council's assertion that the works, including a road from the foreshore up to the main road, and a beach valve facility for the pipeline, were not subject to planning. An Taisce, the heritage and environmental group, had submitted a series of questions about authorization for the work.

The company described the works as 'minor', but An Bord Pleanála's view was that they were sufficiently major to require planning permission. 'This ruling raised a more serious concern,' Ian Lumley of An Taisce said. 'It means that a major element of the Corrib gas development in County Mayo is unauthorized, and is proceeding without planning permission.' Shell said it would review the decision and 'consider its implications'.[5]

Several days later, it said that the 'planning problems' could be resolved easily by narrowing a road access route, and it accused An Taisce and Shell to Sea of 'overplaying' the significance of the appeal board's ruling. The company also sought to play down press reports that there was potential for further oil or gas finds off the west coast and that Ballinaboy could expand, with many pipelines feeding into one terminal or refinery. There had already been a reference to this at An Bord Pleanála's oral hearing on the terminal. Shell to Sea supporters seized on this as evidence that there were complex reasons for the company's attachment to Ballinaboy.

On 29 June 2006, the five men who had been jailed a year before gathered at Mayo County Council offices in Castlebar to mark the anniversary, then visited the grave of Michael Davitt to lay a wreath. Journalists anticipated a head count – there had been speculation in the press in previous days of a 'split' between the five. The fact that Brendan Philbin had withdrawn from mediation talks with Peter Cassells, and that he and Bríd McGarry had opted for a separate legal team in the permanent injunction action sought by Shell had fuelled the rumours. Yet Philbin was present in Castlebar, a broad smile on his face.

So was Willie Corduff, keen to play down any rumours. 'Bertie Ahern and his backbenchers are split ... Mary Harney and

Michael McDowell are split . . .' Corduff quipped, referring to members of the cabinet. 'If they think there's a split [here], let them come on to the land with their pipe again.'

Shell had appointed a new deputy chief executive to its Irish division, Terry Nolan from County Carlow. At a briefing breakfast in Dublin for Chambers Ireland, the umbrella organization for chambers of commerce, he outlined how the company was committed to the existing onshore pipeline route.

There were also unconfirmed reports that the company was not happy with the progress of mediation, and believed that Peter Cassells had gone 'native', but it denied this when asked to comment.[6] Two weeks later, Peter Cassells confirmed that mediation had failed. Shortly beforehand he had contacted the Rossport Five. As Micheál Ó Seighin recalled, 'He told us that he had told Shell he could not come back to us again unless the company made some offer or movement, but it did not and Cassells's next meeting with us was to announce the end of mediation.'

'No agreement is likely in the foreseeable future,' Cassells said, following seven months of 'intensive discussions' and 'detailed consultation with the local community'. Shell accused the 'principal objectors' of 'refusing to engage in face to face dialogue' and 'presenting the company with an unrealistic ultimatum'. Dr Mark Garavan, for the five men, said the accusation indicated the company had never been serious about mediation. Cassells said he would produce a report in a couple of weeks.[7]

That report, published on 28 July 2006, recommended modifying the route of the onshore pipeline to avoid housing at Rossport, and it advised that the semi-state company Bord Gáis become involved in the project to 'provide added assurance that local concerns would be adequately dealt with'. Cassells referred to genuine fears held by residents about the impact of the project on the local drinking water supply at Carrowmore lake, south of Ballinaboy. Water run-off and discharges from the terminal must be 'closely monitored' and local people should be kept informed, he said.

Cassells said it was clear that a majority of people in Rossport, the wider Erris area and County Mayo were in favour of the project – a statement that local supporters of Shell to Sea would not have faulted, as they had always emphasized that they were in favour of the project, if done safely. He described three categories of residents: those who were 'in favour of the project from day one and have not changed their minds'; those who were in favour of the project but 'still have genuine concerns'; and those who were 'opposed to any development in the area'. He said that during his consultation, concerns were expressed about 'intimidation', and he had asked that such complaints be referred to the Garda Síochána. Some consenting landowners told him that Shell had treated them 'fairly' and the company had 'visited their homes on a regular basis to brief them on the project'.

Cassells advised that local involvement in environmental monitoring be reviewed, and he recommended that a 'proactive, transparent system for dealing with local concerns and complaints be established'. Significantly, he also advised that financial incentives be introduced, including a review of compensation to landowners on the pipeline route and establishment of an investment fund for north Mayo.

This latter recommendation was the report's strength, or its Achilles heel, depending on perspective. It did not address the fundamental issue of ensuring that both the state and local area benefited equally, as per the Norwegian and Shetland models. It implied that serious concerns about health and safety and the environment could be bought off, and that consent was, as one resident remembered, 'up for sale'. It had resonances of an old oil industry joke to the effect that 'the man who can't be bought has "f— all" to sell', according to SIPTU's offshore spokesman Padhraig Campbell. Enterprise Oil had avoided financial inducements when it had first mooted the idea of the project in the locality. At one point, former Fianna Fáil councillor Paraic Cosgrove had suggested that it build a fifty-metre swimming-pool in Belmullet as a commitment to the area, but the company had demurred, believing it would give the wrong signal. 'And I still firmly believe this was a good idea,' Cosgrove said several years later.

Micheál Ó Seighin and company had taken particular issue with the idea of financial compensation in a draft sent to them by Cassells. 'Your list of recommendations proposed are deceptive and causing us difficulty, as they give a wrong impression of our part in mediation. We never made an issue and barely mentioned financial compensation either for landowners or local/regional benefits,' Ó Seighin said, in an email to Cassells, who assured him that the issues raised in the mediation and 'other issues' that arose in consultations 'will be clearly separated' in the report. 'As regards financial compensation, which you never raised, it will not figure significantly in the report if at all,' Cassells wrote, on 14 July 2006. In fact, it was to become one of the dominant features.

Shell said it would 'respond positively and in full' to the Cassells report, while the minister displayed obvious irritation with the Rossport Five. 'I believe they don't fully reflect the overall views of the community generally,' Dempsey said. 'I think that if that [the concerns] had been detected earlier, it would have been helpful. But there are people, and I think it is quite clear from reading the [Cassells] report, reading between the lines and from local knowledge, that it didn't matter what you did . . . they would find an excuse to oppose the project . . . I think that we have done everything we could . . . and it is time to get the project moving.' As for the recommendation on a modified route, this was 'entirely a matter for Shell to decide', the minister said. 'There is no reason why they can't go the current route,' he added.[8]

Leo Corcoran, a Dublin-based consultant to An Taisce, begged to disagree with the minister. He had worked with Bord Gáis from 1976 to 1990, and was senior engineering manager on the Cork–Dublin transmission pipeline. He was a member of the Gas Technical Standards Committee, which was responsible for gas transmission and distribution standards, and also worked in the oil and gas industry before moving into health and safety. He had been asked by An Taisce to examine the design codes for Corrib in 2005. He had criticized the Advantica safety review's limited terms of reference then, and had pointed out that some of the consultants' own findings in their draft review, published in late 2005, demanded that its scope be extended to identify the 'optimum location' for the terminal. In his view, the design of the upstream pipeline was compromised to suit the Ballinaboy location.

Like the Advantica study, the Cassells report addressed the 'effect', rather than the 'cause' of the Corrib gas problem, Corcoran said. This 'cause' was the location of the terminal, which should be by the coast, as was 'normal' for this type of project. Not only was Ballinaboy inland, but it was also within a catchment area for a major water supply at Carrowmore lake. 'The ground conditions are unsuitable,' he said, and the connecting pipeline from sea to Ballinaboy was in breach of safety and environmental requirements in running through protected habitats on the coastline.

Corcoran had also prepared An Taisce's complaint on the issue to the European Commission, which maintained that the ministerial consents for the pipeline were invalid as they did not oblige the developer to comply with a pipeline code of practice and were in breach of EU directives. In Europe, the European Commission's environment directorate indicated that the EU's advocate general was due to issue an opinion within 'months' on whether complaints about Corrib – submitted as far back as 2002 by Monica Muller and by the Irish Whale and Dolphin Group – would form part of a wider infringement case being taken by the European Court of Justice against Ireland over implementation of the Habitats Directive.[9]

A TNS/mrbi national opinion poll for *The Irish Times* three weeks after Advantica in May indicated that 44 per cent of respondents believed the pipeline should be rerouted, as sought by local residents. A total of 20 per cent believed it should go ahead on its current planned route, and 17 per cent felt it should be scrapped altogether, while 19 per cent didn't know.[10]

*

Marathon runner Patricia Murphy had just climbed Mayo's holy mountain of Croagh Patrick in the early hours of 30 July 2006 when she set out on a brisk stroll. It was the second leg of a 300-kilometre Shell to Sea walk from Mayo to Dublin to highlight continued concerns over Corrib, and to mark the centenary year of Michael Davitt's death. Murphy gave her support by participating in the second leg, from Ballinaboy to Bellacorick.

Early in August, as the walkers, including Micheál Ó Seighin's son-in-law, blacksmith John Monaghan, were welcomed to Dublin by Lord Mayor Vincent Jackson and given a friendly Garda escort for the last couple of miles, Shell said it would modify the inshore pipeline route in line with Peter Cassells's recommendation. It also said it intended to resume work at Ballinaboy, which had been suspended since the Rossport Five's imprisonment the previous summer. And it said it had hired local fisherman Pádraig McAvock from Belderrig as the company's fisheries liaison co-ordinator.[11]

Ever since the men had been jailed, informal gatherings had been held, once and twice weekly, in the community hall in Glenamoy village. At the outset, there was no set chair or committee and there was intent to this. In Ó Seighin's view, a recognizable steering group might come under undue pressure. The gatherings provided both a forum and a form of support for the community. Shell's announcement that it intended to resume work at Ballinaboy, and its advertisements of same in the local press, was the subject of heated discussion in Glenamoy during August 2006.

The permanent injunction application was also continuing in the High Court; the company said it could not withdraw unilaterally. Bríd McGarry and Brendan Philbin told one of the Glenamoy meetings that they intended to continue with their counterclaim and focus on this. Bríd had felt for some time that the message emanating from the campaign was too contradictory and too narrow in its remit, and it appeared to be 'wholly focused on the safety distances from the proposed pipeline' rather than on the impact of the gas terminal on the environment.[12] She felt that the inland terminal or refinery and associated upstream pipeline were inter-dependent and 'as a consequence both were totally unacceptable' – this was something she had stressed at the Glenamoy meetings. She felt her concerns were realized later on when Shell proposed an alternative pipeline route that took 'distance factors only' into account.

John McAndrew, a builder living near the Glengad landfall, was exercised over the Cassells report. So exercised that he initiated a new petition, the latest in a series by then collected. 'Mr Cassells said that he had talked to people, but he never talked to us, and I am living within a quarter-mile of the isolation valve for this pipe,' McAndrew said. More than three hundred signatures, which he had gathered by early September in the sparsely populated area, included those of two consenting landowners. Shell said that the petition might be based on 'misinformation', and said that it would be announcing its consultation procedure for modifying the pipeline route shortly.[13]

Terry Nolan, Shell E&P Ireland deputy managing director, gave his first press briefing on this on 21 September 2006 in the Imperial Hotel on the Mall in Castlebar. It was aimed at local press, but both *The Irish Times* and Raidió na Gaeltachta were at the event. Nolan, shadowed by communications advisers John Egan and Christy Loftus, introduced himself as a mechanical engineer from Bagenalstown, County Carlow, who had spent twenty-five years working with Shell, much of that time abroad.

The company would take a 'very different approach' than it had in the past and 'recognized the benefits of dialogue'. As a commit-ment to this, he would be living in the area, Nolan said. He firmly ruled out an offshore processing option. Such an approach would require a structure the size of the 'Empire State Building' in the Atlantic. As an image of something unworkable, it resonated in public relations terms, even though Shell to Sea had been laying emphasis on a shallow-water structure – and not one eighty-three kilometres offshore at the wellhead. In response to reporters' questions, Nolan did acknowledge that the height he was referring to was from the seabed up, but it would still have to stand 100 metres above the sea surface to withstand maximum Atlantic waves. Construction work at Ballinaboy, which had been suspended since the jailing of the Rossport Five, would resume the following week and he was 'very confident' that the onshore pipeline issue would be resolved.

Asked by reporters if he was confident of 'consent', he referred to consents already secured and expressed confidence about further approvals. The reporters' questions related to both community and state consent, but Nolan did not appear to see any difference. He would not be drawn on the company's approach if pickets continued at Ballinaboy. People had a 'right to go to work', he said, and he hoped that the protests would be 'small'.[14]

On 23 September, two days after Nolan's briefing, Shell confirmed that the company had written to Dr Mark Garavan of Shell to Sea, seeking dialogue on a new route for the onshore pipeline. Outgoing managing director Andy Pyle outlined an indicative timetable for the project: preparatory work at Ballinaboy, with a resumption of peat removal in spring 2007; resumption of full construction work in autumn 2007, which would continue for about two years; and gas flowing from the Corrib field by 2009. This would supply 60 per cent of the state's requirement at peak (five years), gradually declining over fifteen years, he said. The project cost was about €900 million, a 'good half' of which had already been spent. Pyle also indicated that he was prepared for protests at Ballinaboy. The company had 'no intention of getting involved in any civil actions' against protesters, and he hoped that protests would be 'peaceful'. Civil actions and associated prosecutions would become the remit of the Garda on behalf of the state.

And the Garda's remit was to ensure there were no arrests, and no 'martyrs', according to Superintendent Joe Gannon, who was assigned to Belmullet in July 2006. In an interview for the *Garda Review* magazine later that autumn, Gannon described a scene at Ballinaboy that resembled a conflict zone. The entrance to the Ballinaboy terminal site had been 'blocked for a year and a half' and 'local people had a veto on who went in and out . . .' At the Shell compound at Rossport, anyone arriving in a jeep – be it someone from the company or 'innocent people going about their business' – would be 'surrounded by these locals and their cars, and questioned for anything up to three-quarters of an hour'.

People would be 'questioned and verbally abused', he said, and gardaí would often have to be called. 'I felt that it was time to take the ground back, as is my responsibility as the district officer,' he told the magazine. 'People have constitutional rights to go about their lawful business unhindered.'

Decades of the rosary in Irish drowned his words as he addressed protesters through his loud-hailer at Ballinaboy early on the morning of 26 September 2006. An estimated hundred people, drawn mainly from the local community but with support from the Rossport Solidarity Camp, were determined to ensure that work did not resume before the pipeline safety issue had been resolved. The superintendent turned the dawn convoys of Shell staff back, to 'cheers' from the group, he said.[15]

'I remember the silence then, and the sound of the rosary,' Mary Corduff said. 'The gardaí were on one side of the road, we were on the other. We felt when Superintendent Gannon turned that he was respecting the fact that we were praying . . . in hindsight, as we got to know him, we realized he could have acted very differently then.'

Jim Mulcair, director of Roadbridge Ltd, was upset. 'Just because it hasn't been done before in Ireland doesn't mean it's not safe,' he said of the pipeline, speaking to *The Irish Times*. 'This is the safest nine kilometres of pipeline that we as a company have ever laid – and our company has laid five hundred kilometres to date. Shell is culpable in that communication breakdown, but society puts its trust in engineers.'[16] Shell's deputy managing director Terry Nolan believed the protesters did not represent the view of the wider community in Erris.

'Clear the street, Joe, clear the street. With truncheons, if need be,' concluded columnist Kevin Myers in the *Irish Independent* three days later. 'It's not often that a Garda officer could have been justified ordering a baton charge against people saying the Rosary, but I think on balance Superintendent Joe Gannon might properly have told his men to draw their truncheons and clear the protesting crowds outside the Shell Corrib building project the other day.'

Myers described the recital of the rosary as 'a gruesome and scandalous sacrilege well deserving of a broken head or two', and said that there was a 'cynical method in this apparent religious madness'. It was a

> reminder of the tribal, sub-intellectual nature of the campaign against the Corrib gas refinery. Shell has been granted planning permission by both Mayo County Council and An Bord Pleanála to build their gas refinery. The nation needs it. So does the local economy. Independent consultants have examined the project and found it safe. But of course, this does not satisfy the handful of *firbolgs* of Rossport . . .

There was a double irony to Myers's references, designed to be insulting: one of five leaders of the *firbolgs* who arrived in Ireland from Greece about 3266 BC had settled in Erris. And, as *Western People* correspondent J. F. Quinn had noted in several of his columns published in the 1950s, Erris had often featured in English literature for all the wrong reasons, as in a set for the 'stage Irishman'.

Myers referred repeatedly in his article to the gas 'refinery' – to the chagrin of Shell's public relations team, which had advised journalists against using this description for its 'terminal'. It 'takes a barking idiot to maintain that building a gas refinery out in the Atlantic Ocean is safer and more ecologically sound than building one on terra firma', he wrote.[17]

For Mary Corduff and her neighbours who had participated in the rosary, the article was both shocking and offensive. 'It suggested that the only way Kevin Myers felt conflict could be dealt with was through violence,' she said. 'I suppose he would say the same for any other community. Reading it, you'd be upset and then people would say, "Well, that's Kevin Myers," so we didn't feel so bad then. Maybe we should have invited him up here, if he would have come . . .'

Nor did Myers's view appear to tally with public opinion. A poll by TNS/mrbi for *Nuacht RTÉ* published on 25 September, several days after Terry Nolan had given his briefing, found that 61 per cent of respondents preferred Shell to build its gas terminal at sea, compared to 23 per cent who preferred the onshore option at Ballinaboy, and 7 per cent who didn't mind either location. Some 4 per cent felt the project should not go ahead anywhere, and 5 per cent had no opinion.

The poll also showed that two-thirds of those surveyed in Mayo, at 66 per cent, supported the stance taken by the Rossport Five in 'their defiance of a court order in relation to the Corrib gas pipeline'. And it asked respondents, did they think that overall Shell had handled the Corrib gas pipeline development well or badly? A total of 78 per cent of respondents believed the company had handled it badly, compared to 9 per cent who believed it had handled it well and 13 per cent with no opinion. NUI Galway lecturer Dr Liam Leonard, who published a book that week also on the environmental movement in Ireland entitled *Green Nation*, forecast that protests would not die down in the absence of proper consultation with the community.[18]

And they didn't, even as it emerged that Shell had offered additional payments of €10,000 annually to landowners on the pipeline route – the Cassells report had referred to the need to address anomalies in compensation – and endeavoured to withdraw its legal action, which had led to the jailing of the Rossport Five.

On 28 September 2006, senior counsel for Shell, Patrick Hanratty, told the High Court that the company had decided to discontinue existing legal proceedings against all defendants and had requested them in turn to discontinue counterclaims that they had lodged. Shell would pay their legal costs if the counterclaims were dropped. Two of the six defendants, Monica Muller and Peter Sweetman, had agreed to drop their counterclaims, but four – Willie Corduff, Philip McGrath, Brendan Philbin and Bríd McGarry – had not indicated whether they were 'agreeable' to this.[19]

On 2 October, 114 gardaí of all ranks were moved on 'temporary transfer' to Belmullet, and by 11 October the special policing operation at Ballinaboy had cost €675,639, excluding salaries.[20] The tension grew, even as Mary Corduff left her farm and Betty Schult left her hostel at Pollathomas before dawn, and Mary Moran Horan put on the tea early each morning in the trailer at Ballinaboy. Up to a hundred people from the community, many of them middle-aged and elderly, joined younger people from the Rossport Solidarity Camp and Shell to Sea protesters from other parts of the country in a daily early-morning walk up to the terminal gates, facing up to two hundred gardaí, with each side bearing video-cameras. Their orders were to ensure safe access to and from the terminal for Shell workers and contractors, who were transported in a bus and jeep convoy from Bangor Erris before 8 a.m. each day.

Aughoose parish priest Father Michael Nallen was often present, as was a visiting priest from England who came from the area, Father Eamon Corduff. Father Sean Noone in Pollathomas chose not to attend the protests, but told *The Irish Times* that the concerns about the project were shared by a 'silent majority'. He wrote a letter to the minister, Noel Dempsey, suggesting alternative locations for the terminal, including Inishkeeragh, a fifteen-acre rock two miles west of the Mullet peninsula, which was on a lateral line with the gas field. There was also extensive Bord na Móna property near Muingmore, Geesala, he said. 'This was a plea,' he said in his letter to the minister.[21]

On 3 October two people were injured at the terminal gates, as up to 170 gardaí supported an operation that involved members of the Garda public order unit. One of the two injured required oxygen from Dr Jerry Cowley, who had to walk more than a quarter of a mile with three medical bags as gardaí would not permit his car through.

Superintendent Gannon's view of events on that day was outlined in his *Garda Review* interview of the following month, where he described how plans had been drawn up in Belmullet Garda Station. A Garda 'protest removal group', which had 'expert training to deal with people who are obstructing the roadway', was called in, along with an additional 150 gardaí from outside the district, he said. Normal strength in the entire area had been twenty-five gardaí at most at that time. 'Equipment' was hired locally, and the Garda press office sent a representative to deal with the media.

The day before, a Garda briefing was held in a community hall in Belmullet, and it was decided, Gannon said, 'that members would go in and take the ground at 3 a.m. on Tuesday, 3 October. We planned to have a traffic cordon around the area; there are three main junctions where we would put up barriers and have static security. At 11 p.m., we had word that they [the protesters] had already congregated down there with their vehicles; that scuppered my initial plans.'

He described how junctions were 'cordoned' at 3 a.m. The protesters were asked to move and refused. 'They had all of their vehicles locked, in gear with the handbrakes on, and the wheels turned – so we couldn't lift them with the additional winching gear. We had a powerful tractor with us, so we drove up to the first vehicle. The owner remonstrated with me, but refused to remove it. I gave the direction to the tractor to back up; he hooked a chain on to it and pulled it down the road with the wheels locked and everything. I proceeded to do the same with about five more; by that stage we had opened up a channel at the gate, so I lined the gardaí up either side and created a pathway in.'

Protesters at the terminal gate were asked to move behind barriers, where they would 'be accommodated in a peaceful and lawful protest. They wouldn't budge, so I gave the order for the gardaí to move them back behind the barriers. They all dropped – on command from one of them – and sat down and all linked hands and arms. At that stage I gave instruction to the protest removal group under the supervision of Sergeant Conor O'Reilly – who was also the tactical adviser.'

Superintendent Gannon described how Sergeant O'Reilly's team – confirmed by the Garda press office as the public order unit – 'unfastened' the protesters, while trained 'lifters' carried them across to the cordoned area behind barriers. It took about thirty minutes to remove eighty people. The next step was to remove about fifty vehicles. The superintendent called out the registration numbers on his loud-hailer to give drivers an opportunity to remove them voluntarily.

The first vehicle owner refused, and the tractor pulled the vehicle down the road. A second driver offered to remove his voluntarily, and 'then a whole flood of them came out, individually, accompanied by a garda'. By 7.15 a.m. the road was clear, and the protesters were 'in the corral flanked by gardaí', he said. The press had by this time arrived, in 'large numbers', and workers were driven in at 7.50 a.m., followed by a convoy of lorries to engage in the peat removal. 'There were no arrests,' Superintendent Gannon said. 'That was part of our strategy; we did not want to facilitate anyone down there with a route to martyrdom. That has been the policy ever since.'

He described how succeeding days brought 'similar scenes', and one arrest. 'It was fairly rough on days,' he acknowledged. 'And as the days passed the tensions and the aggression heightened . . . They picked their place on the road where it is narrowest – at the old Ballinaboy bridge. One of the central figures in the protest – not

Above: Willie Corduff takes in the view over Broadhaven Bay and the Sruwaddacon estuary from Dooncarton Hill.

Above: The Sedco 711 semi-submersible rig at the Corrib natural gas field off the coast of County Mayo.

Left: The Enterprise Energy Ireland CEO Brian Ó Catháin shows the then Taoiseach Bertie Ahern, Mayo TD Beverley Flynn and Fianna Fáil councillor Frank Chambers a model of the proposed terminal at Ballinaboy in 2002.

Left: Erris schoolteacher Maura Harrington raises a victory salute after the initial planning permission for the terminal is turned down in 2003.

Below: Bríd McGarry and Mary Corduff react to the news that Mrs Corduff's husband, along with four other men, have been jailed for contempt of court. The men become known as the Rossport Five.

Below: A message written in turf in Erris.

Left: Green Party TD – and later Minister for Energy – Eamon Ryan shows his solidarity with the Rossport Five outside Leinster House.

Below: Several thousand people attended a rally outside the Dáil or parliament at Leinster House, Kildare Street, Dublin during autumn 2005 after the release of the Rossport Five.

Below: The Rossport Five following their release, *from left to right*: Micheál Ó Seighin, Brendan Philbin, Willie Corduff, Vincent McGrath and Philip McGrath.

Above: The Rossport Five, with part of the illegally welded pipeline in the background, near the terminal at Ballinaboy.

Above: Workers dismantle a three-kilometre section of pipeline after Shell admitted it had been assembled without permission.

Left: Mary Moran Horan, baby Maedhbh Monaghan and her mother Bríd Ní Sheighin in the silver trailer, which was head-quarters for the community protests at Ballinaboy.

Left: Gardaí watch as protestors, chained together with concrete casing on their arms, stage a 'lock on'.

Below: Gardaí scuffle with protestors outside the terminal site in November 2006

Below: Rossport resident Ray Corduff admonishes Shell's Corporate and Social Responsibility (CSR) Committee members on a visit to north Mayo in January 2006. Invitations to meet the group had been sent to 'consenting' landowners only. *From left to right*: Ray Corduff, SEPIL managing director Andy Pyle, communications director John Egan and John Cronin of SEPIL, former Dutch prime minister Wim Kok of Shell's CSR committee and fellow committee member Nina Henderson.

Shell is Committed to your Health, Safety and the Environment.

Above: The garda presence at the Ballinaboy terminal grew over the years, with several hundred deployed at various stages of the project from late 2006.

Left: Residents try to prevent Shell contractors, accompanied by gardaí, from using a private road to transport equipment down to a pier at Pollathomas in June 2007. Twenty people, including two gardaí, were injured and a senior garda was later criticized by the Garda Siochána Ombudsman Commission.

Below: In 2008, the huge pipe-laying vessel *Solitaire* (in the background) was towed into Broadhaven Bay, accompanied by an Irish naval escort in the foreground.

Top: The 'stinger' on the *Solitaire*, used for laying pipes on the ocean floor.

Above: Protestors in wetsuits defy a digger dredging the floor of Broadhaven Bay in 2008.

Left: Pat O'Donnell after a protest at Ballinaboy in 2007.

Above: Jonathan and Pat O'Donnell outside Belmullet Garda station after discussions over the detention of Jonathan's boat by gardaí in June 2009, shortly before the *Solitaire* began laying the Corrib gas offshore pipeline.

Left: Unhappy residents at a meeting of the ministerial forum in Inver in 2009, in the aftermath of the assault on Willie Corduff.

The Corrib gas terminal at Ballinaboy, almost complete in spring 2010.

one of the Rossport Five – bonded scrum-like with his son and his son's friend and propelled the two lads forward into a sergeant who was distracted . . . and he was rammed into a ditch, into a deep drain.

'The sergeant has suffered a broken thumb and some ligament damage; there was a lot of water in there too,' he said. It had been captured on video, and three men were arrested later that day. The video recording was a 'new concept in Irish policing plans', the *Garda Review* noted. Gardaí had been filming three to four hours of video footage a day, and had well over fifty hours by the end of October.

The video footage was 'evidence-gathering', Superintendent Gannon explained. 'The protesters have been recording everything since day one, and then you have the outside influences, the eco-warriors, and it is always part of their operations to have a camera in your face, trying to agitate and get a reaction. They will use it to their advantage; and Indymedia have been down there from day one.'

Colleagues of the superintendent described a 'co-ordinated campaign of intimidation', much of it unreported through 'fear' of reprisal, and the use of the Irish language as a 'tool of the protest'. Gardaí originally based in the area found themselves the focus of personal insults by a minority of protesters. Curiously, the *Garda Review* reported that none of those based at Belmullet Garda Station were allowed to claim a Gaeltacht allowance of 7.5 per cent of pay for using Irish, and the Department of Justice's refusal to permit this was 'beyond belief', several officers told the journal.

Superintendent Gannon expressed confidence that gardaí were succeeding in a 'cat and mouse game' where 'they adopt guerrilla-type tactics and they are watching to see when I will stand down resources'. One of his colleagues, Garda Greg Bourke, said that the protesters were 'totally leaderless' and there is 'no one to reason with. Some are peaceful,' he added, 'and gardaí do interact with the protesters – once the war stops, we can converse with them and things cool down.'[22]

*

At this stage, an unsubstantiated Sunday newspaper report alleged Sinn Féin/IRA links with community objectors to the project. The allegations were almost inevitable, Dr Mark Garavan noted later. The protests at Ballinaboy and televised clashes with gardaí, which continued through late autumn 2006, would be seen as 'feeding into an agenda to portray the campaign as dangerous and radical'. There were many precedents for this, such as during the mid-1980s miners' strike in Britain, when there was a series of press reports, some of which were later retracted, discrediting those involved, such as National Union of Mineworkers' leader Arthur Scargill.[23]

On 12 October, Pat O'Donnell was among several fishermen arrested for public order offences at Ballinaboy and later released – a day after the Erris Inshore Fishermen's Association issued a statement criticizing the government's deployment of gardaí to north Mayo. Two gardaí sustained minor injuries. Shell to Sea campaign spokesman Garavan said that, over the previous fortnight, since Shell contractors had returned to work, the objectors had been subjected to 'a co-ordinated assault designed to delegitimize and criminalize local opposition to Shell's project in County Mayo'.

On 13 October, Maura Harrington's photo made national newspapers, when she was pictured lying on the ground holding a cross bearing the name Ken Saro-Wiwa. She said she had been 'knocked to the ground' by a garda and had sustained neck injuries. She was taken by ambulance to hospital in Castlebar.

The following week, four politicians visited the dawn protest at Ballinaboy at the invitation of Dr Jerry Cowley. Fellow independent TDs Seamus Healy (Tipperary North) and Tony Gregory (Dublin Central), along with Socialist Party TD Joe Higgins (Dublin West) called for work at the terminal to be stopped. It was 'incredible' that a government of Ireland could turn part of Mayo into an 'occupied territory' by sending in a police force, Higgins said, as a garda filmed him.

At the TDs' press briefing outside the terminal gates, Ciarán

Ó Murchú, a former Air Corps pilot and owner of an adventure centre in Elly Bay, west of Belmullet, volunteered to say a few words. He didn't resemble Superintendent Gannon's profile of an 'eco-warrior'; neither did several dozen of the young parents and middle-aged residents present at the protest that day.

Ó Murchú said he had been offered €15,000 the year before by Shell to cover the cost of a climbing wall he was building. 'I was told the company would pay for it and that materials could even be delivered in the middle of the night if it was a problem,' he said. He rejected the offer, but said he believed other local businesses had received similar approaches in a bid to 'divide the community' and undermine objectors. His college, which instructed students in adventure sports through the medium of Irish, employed nine full-time staff at that time and up to thirty seasonal staff during the summer months.

Shell confirmed that meetings had taken place between its community liaison staff and Ó Murchú over his concerns about the terminal outfall pipe in Broadhaven Bay. 'At no point was a payment grant of fifteen thousand euro offered to Mr Ó Murchú for the building of a climbing wall at his adventure centre. The direct or indirect offer of payment as alleged by Mr Ó Murchú is totally at variance with Shell's business principles,' the company said. 'The alleged payment is said to have been offered in November 2005; people will surely question why Mr Ó Murchú is only bringing this allegation into the public domain at this time.'[24]

Ciarán Ó Murchú's revelations were overshadowed by a development back in Dublin that same day. The Department of Communications, Marine and Natural Resources confirmed to political journalists that gardaí were investigating 'death threats' made by phone to the minister that afternoon. The reported calls came after Noel Dempsey had defended the deployment of gardaí in Erris in an interview on RTÉ Radio's *News at One*. Asked by RTÉ interviewer Seán O'Rourke to comment on a call by Mayo

Fine Gael TD Michael Ring for the intervention of Taoiseach Bertie Ahern in a mediation role, Dempsey had wondered what role Ahern could actually play.

'Any man who can get Gerry Adams and Ian Paisley working together might have some prospect of getting Shell and the Rossport community,' O'Rourke had said, referring to the Taoiseach's role in the Northern conflict. 'In comparison to some of the people that I think we are dealing with here, those two are two very reasonable men,' Dempsey replied. There were no prosecutions as a result of the investigation into the reported 'death threats'.[25]

Engineers Ireland, the national body representing the engineering profession, endorsed the Advantica safety review, and called for implementation of one of its main recommendations on establishing an appropriate inspection and monitoring regime for the project. However, on 11 October, Leo Corcoran and fellow engineer Brian Coyle made a joint appeal for a study to determine the 'optimum location' for the Corrib gas terminal site. Ballinaboy was unsuitable due to proximity to Carrowmore lake, they explained. Corcoran contacted Advantica, but Dr Mike Acton, one of the authors of its report on Corrib, said that the company wouldn't be comfortable in accepting a commission from another party on the same project, and wouldn't be best placed to identify alternative locations for the terminal. 'There will be many practical constraints on the possible location of a terminal on the coast, including a limited number of possible approach routes for the offshore pipeline to the shore,' Dr Acton told Corcoran. However, he said his company would be 'well placed' to conduct a safety review of possible alternative sites, once those options had been identified.

University students in the middle of their first term at college were among those who made the long journey by bus to Erris for a protest on Friday, 20 October. It was now three weeks since the dawn pickets had begun, and this was the largest gathering so far. The pattern was proving to be a nightmare for Ballinaboy residents who had opted not to engage in protest action when their own appeal against the gas terminal construction was overruled. When some of the protesters strayed on to the road to delay the convoy of jeeps carrying Shell contract staff, they were moved back by gardaí and several scuffles broke out.

Over the following two to three hours, the protesters tried to delay lorries, insults were hurled, gardaí recorded events on video-camera, and one man was arrested for allegedly causing criminal damage to Garda video equipment. Like a 'pressure cooker' was how Dr Jerry Cowley described the atmosphere, and he called on the government to intervene before someone was 'seriously hurt'.

In an opinion article published in *The Irish Times* the following day, Shell's deputy managing director Terry Nolan appealed for dialogue. Dr Mark Garavan responded. Shell had so far only been willing to engage in 'explanation', rather than 'meaningful discussion', and there was 'no real evidence' that it wanted to change its plans for Corrib.

Garavan referred back to Andy Pyle's olive branch of talks on all options on 4 May, which had then been 'effectively binned' when the company ruled out the offshore option within forty-eight hours.[26] Several days later, the minister contributed to the impasse, stating firmly that there was 'no case' for looking at alternative sites for the Corrib gas terminal.[27]

Fisherman Pat O'Donnell was driving along the 'materials haulage route', upgraded to allow for Shell's transfer of peat from Ballinaboy to Bangor Erris, when he noticed two trucks. O'Donnell had been at the Ballinaboy demonstration earlier that morning. The trucks appeared to be trying to force him off the road. When he pulled up, two drivers tried to haul him out of his vehicle, he said afterwards. A third driver intervened, but O'Donnell was sufficiently shaken to report the incident to Belmullet Garda Station.[28] He was later told that gardaí could not pursue it as it was a civil matter.

Dr Jerry Cowley was concerned about the general security atmosphere. He called on Garda Commissioner Noel Conroy, a Mayo man, to ensure that the force showed 'respect' for the local Erris population engaged in the protests. After all, this corner of Ireland had one of the lowest crime rates in the country. It appeared that a 'minority' of gardaí were showing signs of 'ill discipline' and 'undue aggression' at the protests.[29]

Dr Mark Garavan was also worried. Someone was going to get hurt if there wasn't some movement, and it seemed that dialogue between Shell, the government and the community was the only option.

'We never imagined we would see this in rural Ireland,' Mary Corduff said, of filmed scenes that appeared on the YouTube website. 'Gardaí that we knew on first name terms using batons against local people, tossing them into ditches. We were seriously concerned about people getting hurt.'

Her husband Willie joined Micheál Ó Seighin and Garavan at a Shell to Sea press conference in Castlebar, which was effectively an appeal for help. The group called for an independent and public commission of inquiry to investigate the 'optimum development concept' for the Corrib gas project. The wording had been carefully chosen. This was not an uncompromising demand for an offshore terminal, but a request for a review of all aspects, which would facilitate discussions. The proposed commission would comprise one or more members who were acceptable to all sides in the dispute, and community consent would be the 'critical criterion' employed to determine the most suitable method and location for processing the gas. Other essential criteria would include health and safety, environmental aspects and local and regional benefits.

There had never been a full and independent overview of the project, Garavan explained. There had been a 'truncated consents procedure', and efforts since the release of the Rossport Five to achieve consent were 'entirely flawed'. The Advantica review of the pipeline had not allowed for alternatives in its terms of reference, while the report by government mediator Peter Cassells 'excluded the core problem. Both of these processes failed because they were put in place as part of a policy of persuasion,' he said.

The proposal was endorsed by Labour Party president Michael D. Higgins, his party colleagues Tommy Broughan and Joe Costello, Independent TDs Dr Jerry Cowley and Catherine Murphy, and Sinn Féin Mayo-based councillor Gerry Murray. The three Labour TDs had by then visited the Ballinaboy protests and seen what was going on for themselves. Broughan sought a meeting with Andy Pyle.

The Shell to Sea proposal was dismissed by the minister and by Shell, just hours after it was proposed, while the minister said he 'did not see anything new in the initiative'. Shell said that the project had been through a 'thorough and rigorous consents and planning process', and Dempsey added that it was 'unrealistic' to consider any alternative to Ballinaboy.[30]

Leaflets were prepared, emphasizing a peaceful protest, on 10 November 2006, but there was a sense of foreboding in the wake of the minister's speedy rejection of the Shell to Sea compromise. By the end of that morning, there were bruised arms and legs, torn clothes and uniforms; eight people, including four gardaí, had been injured and two had been arrested – a morning that residents would not forget for a long time. After what Superintendent Gannon described as over two hours of negotiations, and several warnings on his loud-hailer, batons were drawn on protesters at Ballinaboy. He described the use of 'truncheons' to move them from the road as 'slow and methodical'. The altercation was broadcast later on RTÉ and YouTube.

Buses and cars carrying supporters from Dublin, Galway and elsewhere had arrived the night before. The stand-off had begun at around 7 a.m. when roadblocks were erected, preventing protesters from approaching the terminal site. Maura Harrington began driving her minivan towards a line of gardaí and blowing her horn. Her vehicle was then pushed by supporters through two Garda barriers and two gardaí used batons to smash her vehicle's windows and pull her out.

Up the road, American-born resident Ed Collins, a father of three children, was among a group of about forty who made their way to the main road. He said he was pushed into a water-filled ditch and found a *bangharda* on top of him in the mêlée. Collins's account is disputed by the Garda Síochána, but he was taken to hospital with a severe contusion to his lower lumbar spine and soft tissue injuries to his legs, and was to remain on crutches and in chronic pain – pursuing a civil action, when a complaint he filed to the Garda Complaints Board was unsuccessful. The *bangharda* was one of four treated for injuries at the scene.

Several protesters drove to a local quarry and builders' suppliers, both of which were identified as working for the project. Pat O'Donnell was among them. He stressed that the approach was not a picket or protest, but an attempt to persuade the businesses to disengage. The gardaí followed them there and O'Donnell was pulled aside when he attempted to speak to the quarry owner. He sustained three broken ribs, and another man's nose was broken.

The activity was caught on camera by independent film-maker Jim Cahill, who described it as 'very harrowing'. A truck had already left the quarry, and O'Donnell was participating in a 'thin picket line' when gardaí intervened, Cahill said. O'Donnell said he believed he would have been 'left for dead', had it not been for the presence of Cahill and his cameraman and the intervention of another local man.

In a subsequent statement made by Sarah Clancy to the Garda

Complaints Board, she described how she saw O'Donnell being flung to the ground, with 'first two, then one garda kneeling on his back', and 'they pressed his face into the dirt, all the while hitting him with batons . . . There were four gardaí at least involved in this.'

Clancy, a student from Taylor's Hill in Galway who would later become campaigns and regional development officer with Amnesty International, described in her statement to the Complaints Board how she had been pushed and verbally abused by a garda earlier when she was walking along the Bangor–Ballinaboy road. She had seen a woman who was a classmate of hers being punched in the stomach. Her camera phone was kicked away from her hand by another garda when she attempted to take a photo of gardaí wrestling to the ground an activist she also knew from Galway. She was sharing a packet of biscuits with another woman at Barrett's quarry, as part of a group of fifteen protesting peacefully later that morning, when the gardaí arrived, and she was pushed, held to the ground and hit several times. The assaults, 'observed by senior members of the gardaí', had come without any warning or request to move beforehand, she said.

Clancy's complaint was deemed admissible by the Complaints Board, and she was interviewed in January 2007. Several months later, she was told that she might have to testify in court. However, she subsequently received a letter saying that the Complaints Board deemed there had been 'no breach of either discipline or protocol' by the gardaí involved. A complaint lodged to the new Garda Síochána Ombudsman Commission in 2007 by Pat O'Donnell was deemed admissible, but a recommendation sent to the Director of Public Prosecutions was not accepted.

'I believe someone will be killed, given the violence by the state and the low number of trained police,' a distressed Micheál Ó Seighin said afterwards. Glenamoy farmer and Shell to Sea supporter P. J. Moran, brother of Mary Moran Horan, appealed for help. 'If Bertie Ahern had an ounce of cop-on, he'd come down and see for himself what's happening. We're not asking for anything, only for our safety,' he pleaded.[31]

It seemed as if the Taoiseach already had his mind made up. RTÉ north-west correspondent Eileen Magnier caught up with him on a tour of Roscommon and Leitrim the same day. 'We have over the last number of years done an enormous amount of work independent of the investigations of Advantica, Peter Cassells,' Ahern told her. 'Quite frankly, from the government's point of view, that's it. The negotiations are over, the rule of law has to be implemented and the work goes on. And if there are those who try to frustrate that, they're breaking the law and it's a matter for the gardaí to enforce it . . .'

'So you're saying to Shell to get on with it?' Magnier asked him.

'Yes,' the Taoiseach said.

'There's a danger of someone getting seriously hurt,' Magnier continued.

'Well, that's always the danger when you have people breaking the law,' Ahern replied.[32]

That weekend, Fine Gael leader Enda Kenny said he believed 'outside influences' were undermining the legitimate concerns of local campaigners. 'My appeal to law-abiding citizens of my own constituency is to distance themselves from aggressive influences from outside,' he said, at a conference in Dundalk, County Louth.

Labour Party president Michael D. Higgins responded. He criticized the Taoiseach for 'distorting the facts' and said it was 'simply wrong' of Kenny to suggest that the Shell to Sea campaign had been infiltrated.[33] The *Western People* editorial of 14 November recalled the words of Bord Pleanála inspector Kevin Moore in his ruling against the Ballinaboy terminal of 2003, and said that such words had come back to 'haunt the government, Shell and the people of Erris. In truth, the violence that raised its head . . . had been a long time in the making and was as inevitable as it was heartbreaking,' the editorial said. 'The Garda baton charges that occurred on Friday morning [10 November 2006] in Ballinaboy were not the product of Sinn Féin or Provisional IRA machinations; they were the product of abject government incompetence.' Wondering what might have been achieved if the government had listened to An Bord Pleanála in 2003, the leader writer described the controversy as 'the single greatest tragedy that has occurred in this country in modern times', the effects of which would be felt for generations. It was 'not an exaggeration to say that Erris will never be the same again'.

Chief Superintendent Tony McNamara was keen to defend the actions of the gardaí, while emphasizing people's right to protest 'peacefully'. The gardaí had been acting on information that certain republican sympathizers had been invited to the day of action. On 21 November, plain-clothes gardaí and private security staff were on hand in and around the Broadhaven Bay Hotel in Belmullet, when Shell held two open evenings to explain its plans for finding a modified pipeline route. Residents attended, including Maura Harrington, and Chris Tallott, founder of the North West Mayo Action Group. He had condemned the jailing of the Rossport Five the previous year and felt Shell had 'made some stupid mistakes'. He didn't like watching 'what's happening with the gardaí either' but he appealed for dialogue, speaking to *The Irish Times*. After all, fifty permanent jobs in Ballinaboy was the 'equivalent of 500 jobs in Dublin in an area like this', he argued.[34]

RTÉ cameras were also present, for a *Prime Time* programme special timed to coincide with another day of action later that week by Shell to Sea. However, the group had cancelled the protest for 24 November on safety grounds as it said it was 'no longer safe to invite large numbers of people to participate in a peaceful protest'. As Maura Harrington recalled, the decision, which she condemned as a 'failure of courage', was taken after Dr Mark Garavan said he would otherwise stand down as spokesman.

Echoing Garavan's concerns, MidWest Radio journalist Liamy MacNally, who had followed the controversy tirelessly from the outset, had expressed fears in his *Mayo News* column that week that a day of solidarity might turn out to be a day of 'action'. While the Shell to Sea group had 'local people as the backbone', there were 'unfortunately' others 'whose appreciation of local issues and local realities is somewhat different'. These people would 'disappear when the whole Corrib débâcle finally comes to rest', he said, but in the meantime the 'toll on the community mounts higher and higher'. He continued:

> To date, the local community has been rent asunder by claims and counter-claims over the Corrib project. Relationships that have survived generations of emigration, hardship and isolation have fallen victim to a project that has done what neither a colonial government nor a thoughtless national government could ever do. No one bargained for a combination of political incompetence and corporate intransigence. That blend is nothing short of a recipe for disaster. The fruits are currently on show – local person against local person against gardaí.
>
> Some of the gardaí neither want to be on duty in north Mayo, nor wish to be engaged in confrontation with local people. Local people are being vilified by politicians, gardaí and other Corrib commentators.

MacNally took particular issue with recent comments by Minister for Justice Michael McDowell about 'Sinn Féin-led' opposition to the project. This was 'as daft' as it was 'mischievous', MacNally said.

> Some criticism is necessary, especially when local businesses are included in protests [several local shops had been subject to boycotting]. Treating those who engage with Shell, for whatever reason, as a 'sinful blur' will only deepen the hurt that is already evident in an already divided community. The one thing that can be said about the Shell to Sea group is that it is the easiest to

criticize and, for its members, the hardest to control. There is no one voice with one leader. This allows a situation to develop where those with extreme views can do what they wish under the banner of the group in general. The one fact that remains is that the consent of local people is a major issue for this project to proceed.

Shell's 'corporate ethos is well documented in its many places of work, from Nigeria to Russia', MacNally added, and the company 'has enough Mayo people on its payroll for one or more of them to take "a risk for peace" with the Corrib project'.[35]

'Guerilla war threat by Shell protesters', read the headline in the *Irish Mail on Sunday* on 26 November 2006, in a report, labelled 'exclusive', by Warren Swords, who had passed himself off as a supporter of the Rossport Solidarity Camp. He reported that 'Activists from Dublin and Britain plan to target individual companies who do business with Shell in a chilling echo of the tactics adopted by animal rights activists in England, whose campaigns in recent years degenerated into violent assaults and even grave robbing.' Swords also wrote a 'dispatch from inside the protest camp' where he lived on communal meals, which he described as 'mostly rice, pasta, chickpeas and boxty cake'. As one camp leader noted afterwards, the worst part about it was that he had bought Swords a pint.

Bríd McGarry, who had not participated in the Ballinaboy protests, was one of those invited to the RTÉ *Prime Time* programme recorded in Belmullet. RTÉ had recorded footage earlier at Shell's modified pipeline route presentation. She was given a seat at the back, to her disappointment, which gave her no opportunity to participate. This was not the first time this had occurred. She had been invited to *The Late Late Show* with the Rossport Five wives in 2005, and had been conveniently bypassed when it came to her turn to express her opinion on the show.

On the panel there were two men of similar age and similar south Mayo background, but with divergent views: Dr Mark Garavan for Shell to Sea and John Egan for Shell. Shortly before going on air for the pre-recorded broadcast, participants were given the results of a poll jointly commissioned from Red C by RTÉ and the *Irish Independent*.[36]

The results, already set to be published in that newspaper the following morning under the page one headline 'Majority want the Shell Corrib pipeline to go ahead', found that 51 per cent of respondents believed the project should go ahead as planned if it were not an option to change the current proposal, compared to 33 per cent against and 16 per cent in the 'don't know' category. Asked if the current protests should cease or continue, 38 per cent said they should continue but without attempts to impede access to the Ballinaboy site; 32 per cent said they should cease; 14 per cent said they should continue and 16 per cent didn't know.

There were further questions, which indicated a majority agreed with a recent statement by Minister for Justice Michael McDowell that 'a tiny minority of people continue to confront the law' and were 'being supported in this by the Sinn Féin party. Provo tactics won't work.' Some 40 per cent believed the protesters spoke for a minority of the local community, compared to 35 per cent who believed they spoke for a majority, 17 per cent didn't know and 8 per cent felt they spoke for all. Some 53 per cent agreed with a question suggesting that the protesters were an 'intimidating' presence that dissuaded local people from disagreeing with them, but in a separate question, 59 per cent agreed that the protesters were just doing what the respondent would do if the project was on his/her doorstep.

There was another question, which Garavan was ready to respond to when the programme went on air: 'What is your preferred development concept for the Corrib project?' Some 44 per cent of respondents favoured an offshore option, while just 29 per cent favoured the existing proposal. Garavan started off by referring to the finding, but the pre-recording stopped temporarily, and the question was dropped when it resumed.

During the programme, Egan accused Garavan of trying to stop the project altogether: he had found a letter on Mayo County Council's file to this effect, which Garavan had written on behalf of an environmental grouping, Feasta, predating the Rossport Five's jailing and Shell to Sea. However, Garavan, who said he was 'flattered' to find he was the focus of such attention, made it clear that he was speaking for the Shell to Sea campaign and not personally.

Afterwards, Garavan sent a letter of complaint to RTÉ and to the Broadcasting Complaints Commission. RTÉ confirmed that one of the poll questions had had to be 'dropped', but said that this was because 'It became apparent to the programme team during the course of research that the wording of the question contained assumptions which were in fact incorrect.' It said that no one outside the *Prime Time* team had any influence on the decision, and the programme as broadcast was 'accurate and fair'. RTÉ senior press officer Carolyn Fisher also pointed out that the programme had reminded viewers of a poll conducted by *RTÉ Nuacht* in September, which showed that most respondents in Mayo would prefer to see the Corrib gas terminal located at sea.

Garavan was now fully aware that a targeted public relations battle was being waged, one with resources that he and the supporters of the Shell to Sea campaign did not have – and which they could not counteract when there were images of conflict on television news bulletins that undermined, in his view, the campaign's key messages about health and safety. Earlier that month, Shell had confirmed that, in addition to its own formidable public relations team, it used external agencies, including Financial Dynamics in Dublin and Powerscourt in London – the latter providing 'strategic advice'. Formed by Irish business journalist Rory Godson, Powerscourt also listed Independent News and Media among its Irish client list.[37]

*

That last weekend of November 2006, the exhausted Shell to Sea campaigners in Mayo hosted another visitor from abroad – one of a number now expressing concern about the project. Terje Nustad was one of the most influential men in the oil and gas industry, as head of the Norwegian energy workers' union, Safe, representing seven thousand workers. Most of them were employed by Norwegian state company Statoil, but some also by Shell. He told *The Irish Times* that he was 'very disappointed' with Statoil's role as a partner in the project with Shell and Marathon, as he believed the participation was in breach of the company's ethical values.

A conflict like this could not have happened in Norway, because the fundamentals were so sound. Communities up along the coast-line could happily queue for Statoil projects in their area in the knowledge that the company was obliged to pay both local and national tax and contribute to the state pension fund. The terms and conditions developed by a strong Labour government in Norway more than thirty-five years before ensured the 'state take' but also that expertise was built up within government, because Statoil was a state company. At the same time, Nustad, and Helge Ryggvik, a researcher at the University of Oslo who accompanied him on the visit to north Mayo, had been concerned for some time at the partial privatization of Statoil and the company's record abroad.

Nustad, who had scheduled meetings with Irish union leaders, including Jack O'Connor, president of SIPTU, said he would be raising the issue at a political level in Norway. He called on Statoil to stop work on the gas terminal and try to reach agreement with the local community first. 'We support the commission of inquiry into the best development concept which has been proposed earlier this month by Shell to Sea,' Nustad said. The government needed to 'wake up and take action', he added.

Statoil responded to Nustad's call. It said that it was fully committed to the project, to the findings of the Advantica report, and halting work on the terminal was 'not an option'. It said that 'misunderstandings' had been created by the failure of the original developers of Corrib, Enterprise Oil, to communicate with residents. This was a 'big error', according to its spokesman Kai Nielsen, based in London. Yet Statoil had been a partner back then too.[38]

Just before Christmas, Shell announced it was moving from Bangor Erris to a €1-million office development in Belmullet. The complex was in property leased from the state agency for Gaeltacht development, Údarás na Gaeltachta. The gardaí had also been using part of the complex, rent-free, for deployment of extra members of the force to provide security. Given the Gaeltacht agency's role in social and economic development in the isolated area, the business arrangement was perceived by some involved in Shell to Sea as yet another betrayal by the state. The business arrangement was defended by the authority's chief executive, Pádraig Ó hAoláin. It was not clear how many Irish speakers would be working with the company, or whether any such requirements would apply.

The Rossport Five had another reunion – in the Imperial Hotel in Castlebar, for the publication of their book, *Our Story: the Rossport Five*, comprising a series of interviews with Mark Garavan, published by Small World Media. It was a good-humoured gathering on a cold winter's night, but the low-key media attention given to the occasion was in sharp contrast to the adulation showered on the five men and their families in Dublin after their release from jail almost fifteen months before. The following morning, some of the families would be out before dawn at Ballinaboy again. Between seventy and eighty people, mainly from the locality, were maintaining the morning protest. Three days before Christmas, gardaí denied using a truncheon on a protester at Ballinaboy, and denied they were trying to provoke locals with what Mary Corduff described as 'personal remarks'.

It was a 'very sad development coming up to Christmas', she concluded. 'It seems as if the gardaí are getting annoyed with us now, as they haven't succeeded in doing what Shell wants them to do, which is to frighten us away.'[39]

12

THE SPLIT

He wore a green helmet, a luminous yellow 'hi-vis' waistcoat, he had his climbing gear, and he seemed to know what he was about. On 30 January 2007, British police arrested Gareth Hallam, who had ascended to the balcony of the Irish ambassador's office in Grosvenor Place, London. He had unfurled a banner in Irish, before being talked down. He was questioned and released with a caution. One of his associates handing out leaflets in support of the Shell to Sea campaign in Mayo told *The Irish Times* that the action was timed to mark the eve of Royal Dutch Shell's announcements of profits.[1] Just over six days of such profits could build an offshore gas terminal, she said.

Two weeks later, on 16 February, a 'solidarity walk', rather than a 'day of action', was planned for Ballinaboy. Some played samba drums and others beat a similar rhythm with sticks on the security fence: Shell to Sea's Mayo branch wished to emphasize a 'peaceful carnival' atmosphere, and had asked Belmullet gardaí to facilitate this the day beforehand.

And Al Jazeera was in town. A film crew commissioned by the Arabic television network was making a documentary on the dispute – one of many international media outlets that had found Erris on the map since the jailing of the Rossport Five. Shell to Sea had organized buses to Erris from Dublin, Belfast and Galway, and some of the residents had volunteered to steward the event. Chief Superintendent Tony McNamara of the Mayo Garda Division said he was 'delighted to see the campaign interacting with us' in giving information on the nature of the protest. It was estimated that up to 180 gardaí would be present all the same.

Shell staff were escorted to work at the earlier time of 6.30 a.m. The plan was to hold a peaceful walk from Ballinaboy to Glenamoy, where films would be shown. However, a group of about a hundred broke into the terminal site at the junction for Pollathomas, after the main body of the march had already headed for Glenamoy hall. John Monaghan followed the group: he was keen to ensure they came back out. At first, the atmosphere seemed fairly relaxed, but then gardaí began arriving in vans. 'It reminded me of one of those scenes in *Braveheart*,' he said afterwards.

It was just what the television cameras needed. Dublin Shell to Sea supporter Tadhg McGrath's arm was injured, and he described the attitude of some of the site staff as 'hostile'. Shell spokesman John Egan said that the 'trespassers' had vandalized tools and threatened construction staff, who had been subjected to 'extreme verbal abuse and personal intimidation'. As Egan noted to journalists, 'This again shows that the Shell to Sea rhetoric about peaceful protest is simply untrue.'[2]

By now, the cost of policing the Corrib protests was running at €3.14 million, for four months from October to late January, according to Minister for Justice Michael McDowell, and support for the Shell to Sea campaign had cropped up in all sorts of places. Denny Larson of Global Community Monitor, who had visited the area from early 2005, returned with a team to take testimonies from residents who believed their rights had been violated; a report by the group, published later that summer, questioned the training levels of gardaí involved and said there was evidence of 'excessive physical force' – a claim denied by the gardaí.

The 16 February Ballinaboy break-in was condemned also by Mayo TD Dr Jerry Cowley, who said he was 'very disappointed' in the light of efforts to have a peaceful protest. A Shell to Sea spokesman said that the actions of a 'number of individuals' had not been sanctioned by the campaign. Maura Harrington, who was among those involved in the 'breach', said that it had been 'spontaneous' and accused Shell of 'lying about damage to equipment'.[3]

At a subsequent stormy meeting in Glenamoy, the first open discussion about the use of non-violent direct action (NVDA) took place and the approach was endorsed. Dr Mark Garavan withdrew from his role as spokesman for Shell to Sea. Garavan, who had been a Franciscan friar in the mid-1980s and had worked as a social care worker with the Simon Community, was committed to radical non-violence. He believed that NVDA would be counterproductive, as it would undermine the campaign's three key messages about health and safety, the legal process and community consent.

Garavan was mindful of the injuries sustained by Ed Collins, among others, on 10 November, and the risk to locals, Shell staff and gardaí. And 'Because images are always stronger than words, images would result in eroding the campaign's support and justifying criminalization and marginalization,' he said afterwards. No one dared admit it at the time, but it marked the beginning of the end of Shell to Sea as they had known it.

It was a fraught time: the Environmental Protection Agency (EPA) had given a preliminary ruling the previous month in favour of Shell's application to it for an integrated pollution prevention and control (IPPC) licence for the Corrib gas project. Imelda Moran, who had been studying the IPPC application for the campaign since it was lodged by Shell on behalf of the Corrib gas partners in December 2004, was very disillusioned with the interim approval. The EPA had sought additional information in 2006 from Shell on aspects of its environmental impact statement (EIS), which had been found to be 'defective', she said. 'This flawed EIS had already been used by other agencies, including Mayo County Council, to award planning permission. This effectively means that a developer can withhold essential information until it has to provide it,' she said.

Residents were already concerned about aluminium discharges into their public drinking water supply at Carrowmore lake, and about community representation on two local monitoring groups established under the auspices of Mayo County Council and the Department of Communications, Marine and Natural Resources. A meeting hosted by the two state bodies to extend representation on the project monitoring committee and the environmental monitoring group, as advised by Peter Cassells, ended in disarray. Shell to Sea questioned the nomination pro-cedures. 'It's not democratic, but is intended to give that illusion,' Mary Corduff said.

Shell had been set a limit for aluminium discharges from Ballinaboy while removing peat to prepare for the terminal, but on 8 September 2005 the company was forced to take immediate action when Mayo County Council threatened to prosecute over 'exceedances'.

In 2006, reports by the EPA and the North Western Regional Fisheries Board indicated there was no risk, but the problem deteriorated in early 2007, according to data provided by Bord na Móna from 30 January to 8 March for Ballinaboy. Drinking water limits were only exceeded on one date officially, on 23–24 January 2007, and Mayo County Council attributed this to an accident at its own water-treatment plant rather than to Shell.

The developer argued that aluminium could come from many natural sources, as Shell's landholding comprised 'only two per cent of the total catchment area of Carrowmore lake' and the terminal earthworks constituted a 'small part' of the site. The state and developer responses did nothing to assuage the concerns held by

John Monaghan, who had been monitoring the situation for Shell to Sea, and Mary Corduff, who had been ordered by gardaí to 'clear the road' when she tried to report a diesel run-off from a generator in the Ballinaboy site into a local stream one Monday evening.[4] Ironically, several years later Shell staff would acknowledge during a site tour of Ballinaboy that they had had to take action over the high aluminium discharge.

The EPA denied claims by Shell to Sea that it was under political pressure to issue an early ruling on the terminal emissions licence, and said that there could be an oral hearing before a final licence was issued, as there was a twenty-eight-day consultation period.[5] The environmental impact of controversial practices, such as flaring and cold venting, and the risk of polluting Broadhaven Bay and the local water supply at Carrowmore lake, were issues Imelda Moran focused on in her response to the EPA. Shell had by now appointed RPS consultants, one of the state's largest and most successful engineering firms, to come up with a modified route for the onshore pipeline.

The Erris Inshore Fishermen's Association chair, Eddie Diver, also worked on a response to the EPA. His organization, which was still very concerned about emissions into the marine environment, had withdrawn from discussions on this with Shell in January. Diver said that the company had 'never seriously addressed the problem', while Shell said it would be dealt with via the 'relevant legislative process'.

In March, fishermen who were affiliates of the association participated in a training exercise in Belmullet harbour with up to twenty environmental activists who had formerly been members of Greenpeace and were now known as International Sea Solidarity. Orla Campbell, one of its members, said that the group involved 'individuals representing international groups who are determined to ensure the success of the fishermen's campaign to protect their livelihood and the marine eco-system'.[6]

*

The government's deal with oil and gas companies, which Shell to Sea had doggedly tried to highlight, arose in late March at local authority level in Mayo. Two councils – Mayo County Council and Westport Urban District Council – adopted motions urging the government to renegotiate all deals related to the state's natural resources, including the Corrib gas field. Fine Gael councillors supported the proposal at Mayo County Council level, variously describing it as a 'disgrace' and a 'horrible deal'.

The seconder, Labour councillor Johnny Mee, referred to the structure that former Labour minister Justin Keating had put in place.[7] Interestingly, his son, Alan Mee, worked in the oil and gas industry and would be appointed deputy Mayo manager for Shell on the Corrib project the following year.

The motions had not long been in print when Shell and Statoil confirmed that they had begun exploring a further area off the Mayo coast. Independent TD Dr Jerry Cowley expressed disappointment at this, given that a government review of terms given to oil and gas companies was still supposedly in train. He was perplexed by a series of 'unusual coincidences'. He and six other Shell to Sea supporters, including Garavan, believed their telephones might have been tapped. A Dáil reply by Michael McDowell on 28 March 2007 said it would be 'contrary to public interest' to disclose if authorization for surveillance of telephones had been granted.

The justice minister referred Cowley to section 9 of the Interception of Postal Packets and Telecommunications Messages (Regulation) Act 1993, which provided for a complaints referee to examine complaints of contravention of the legislation. Judge Carroll Moran of the Circuit Court was the current referee. Cowley said he intended to ask him to investigate.[8]

Meanwhile, Garavan had decided to run for one of the six university seats in the Seanad. Although he was not elected, the campaign was a platform for highlighting his concerns about the Corrib project. A general election was also held that year, and his move was part of a strategy to broaden political support for Shell to Sea, already expressed by Labour, the Green Party, the Socialist Party, Sinn Féin and independents led by Cowley.

In late March, Garavan also submitted a petition to the European Parliament's petitions committee, with the support of Kathy Sinnott, Member of the European Parliament for Munster. He contended that EU directives on environmental impact assessments, Seveso II on transfer of hazardous substances, water quality and public consultation had all been breached in relation to the Corrib project, and he urged the committee to initiate a 'comprehensive inquiry'. Shell responded that it was confident that all proposed operations were in line with all relevant Irish legislation, which in turn conformed with EU directives. Pipelines were explicitly excluded from the Seveso II provisions, it noted, and the IPPC licence, which it was seeking from the EPA, would ensure that it complied with 'all relevant water quality legislation'. Garavan's complaint was deemed admissible by the EU's environment directorate in July of that year.[9]

The EPA's oral hearing into Shell's application for the IPPC or emissions licence, relating to discharges into the atmosphere and marine environments, got off to an unhappy start although, interestingly, it referred to Ballinaboy as a 'refinery' in its documentation, and not a 'terminal', as the developers preferred. There was the issue of an interim approval, which had upset residents. There was the fact that EPA director Dr Mary Kelly had appeared on a promotional video made by former Corrib gas developers Enterprise Oil some time before she joined the EPA – the agency made it clear she would not be involved in any decision regarding the licence application.

There was a significant Garda presence at the Broadhaven Bay Hotel in Belmullet, which was offputting. Key documentation had not been supplied in Irish, and this was a Gaeltacht area. Imelda Moran, one of thirteen appellants including Shell itself, also asked

EPA inspector Frank Clinton why other agencies, such as Mayo County Council, the Health and Safety Authority and An Bord Pleanála, were not represented. After all, she pointed out, this was an 'integrated' pollution prevention licence.

Most of the EPA hearing's opening session was dominated by the carefully delivered testimony of Agnes McLaverty, Corrib gas environmental adviser, who was one of eight witnesses on a Corrib gas team. The Norwegian chemical engineer, who had previously worked for Enterprise Oil and for Marathon, said that the overall objective was to ensure that discharges would cause 'zero' harm to the environment.

The issue of a Garda presence was raised by Shell to Sea's John Monaghan, also an appellant, and the inspector agreed to examine the issue. Bríd McGarry had prepared submissions on behalf of herself and her mother, Teresa, living in Gortacragher. Their home, McGarry explained, was two kilometres downwind from the refinery, yet Bangor had been named as the nearest village in an incomplete environmental impact statement, she said.

She told the inspector that the first ever environmental impact statement prepared for the developers, then led by Enterprise, back in 2000 had contained 'revealing data' that was omitted later. A second such statement had reduced environmental emissions, she said, and this showed that the applicant had 'no credibility' and it reflected a 'disregard' for local residents. She was concerned about a 'close political corporate climate', which existed and which, she said, meant that the EPA was under intense pressure, like An Bord Pleanála, to approve the project. It involved use of 'primitive technology', the applicant had not carried out feasibility studies on processing gas offshore and it did not have community consent, she said.

'Pristine waters' would become a 'contaminated sump', and there would be a noise factor, McGarry said. She stressed that she and her mother were not against processing of natural gas in theory, but 'best standards' must apply, and the EPA must take other development concepts on board. As part of the submission, the hearing also watched a DVD she had submitted, entitled *Beauty and the Beast*, about the impact on residents in South Africa of a refinery run by Shell and BP.

Consultant engineer Leo Corcoran represented An Taisce. In his submission, he quoted the text of an email from the Scottish Environmental Protection Agency to the effect that it would 'normally recommend against' placing 'major industrial facilities' at 'locations which could affect public drinking water sources'. The Ballinaboy refinery was too close to the Carrowmore drinking water catchment, he argued. Curiously, the inspector did not include Corcoran's email from the Scottish agency in his subsequent report, and thus his board would make its decision without having this significant piece of information highlighted.

Just several days into the EPA hearing, there was a development in the High Court. On 18 April, it agreed to allow Shell to end its long-running legal action against opponents of the pipeline, including McGarry, Brendan Philbin, Willie Corduff and Philip McGrath. Most significantly, the company also said it would not be relying on compulsory acquisition orders made over defendants' lands, and it intended to seek new orders relating to a modified pipeline route. McGarry sought to have this taken into account by the EPA. *The Irish Times* of 19 April said that the company faced a legal bill of more than €1 million in legal costs. McGarry, Philbin, Corduff and McGrath had decided to continue with their counterclaim.

A delegation of nine politicians led by Dr Cowley had set up a meeting in Dublin with Statoil; the Garda Complaints Board had confirmed that three senior officers were investigating a series of complaints about Garda behaviour during Corrib gas projects; the total cost of policing has risen to €5.4 million; and the EPA hearing was reaching its conclusion in Belmullet when Willie and Mary Corduff arrived in convoy from Knock airport. The couple had flown in from San Francisco and were bearing a trophy and a cheque.

Willie had become Ireland's first winner of the Goldman Environmental Prize. Known as the Nobel of the environmental movement, the award of a trophy and US$125,000 (€92,000) to six international nominees had been endorsed by more than a hundred heads of state. The late Nigerian writer Ken Saro-Wiwa had received the award in 1995, while the 1991 winner for Africa, Wangari Maathai, had been awarded the Nobel Peace Prize in 2004. Sharing the platform with Willie that year were five others, including Icelandic businessman Orri Vigfusson who had campaigned for a ban on driftnetting for wild salmon.

Mary Corduff was somewhat uncomfortable with the award for all sorts of reasons. Under the rules, it had to be given to an individual – yet Willie was still known as one of the Rossport Five, and felt he was a 'back page, rather than a front page, man and we are very much part of a community'.[10] The organizers had sworn them to secrecy beforehand. 'I couldn't tell my kids, I couldn't even tell my own mother,' Mary recounted afterwards. 'I remember we were going together to Castlebar to look for clothes for a wedding, and I was trying to keep an eye out for something for the trip to San Francisco without letting on. It was awful.'

The couple had only flown once before, when they had visited their son in England two years previously. 'I'd rather be out at home on the bog and with the horses and cattle,' Willie said. 'As for the money, I've always said – and I said it to three groups of lawyers helping us to fight our case against Shell, some of whom tried to persuade us to settle for compensation – that it is not about that. Money has ruined people, and you're never going to be left over the ground when you die. Even if you're penniless, someone will bury you. So this will be going to someone or some group who will help us with our case.'[11]

On their return the Corduffs were elated but exhausted after a hectic schedule that had included international press interviews and meetings with a number of US Senate and Congress members, including Nancy Pelosi, Speaker of the US House of Representatives, over on the east coast in Washington DC. Pelosi described the 'conviction and courage' of all six Goldman award winners as an 'inspiration'. Willie was moved by the stories of some of his fellow recipients. 'It would seem as if our situation is only going to get much worse,' he observed, on arriving home.

Goldman funds were assigned for campaigning work, such as a later trip to Brussels and to Norway two years later. Yet gardaí would taunt Willie Corduff regularly afterwards, and there was some enmity within the community over the money, Mary acknowledged afterwards. Philip McGrath felt that Willie should have divided out the cheque among the five men. 'If it happened again, I'd take the trophy and paperwork, but not the cheque,' Mary Corduff said.

The following month Willie was in a plane again, when he and four supporters of the Shell to Sea campaign travelled to the Royal Dutch Shell headquarters in The Hague, Holland, for a meeting of its shareholders. The Shell to Sea campaign had been donated fifty shares anonymously more than a year before. The campaign had been working with a Shell shareholding group, the Ecumenical Council for Corporate Responsibility (ECCR) in Britain, with which Sister Majella McCarron and Maura Harrington had made contact. The Oxford-based organization also held Shell shares on behalf of its members, comprising representatives of mainstream Christian denominations, church bodies and individuals, as a means of exerting some influence on corporate policy.

The ECCR was compiling a report on Corrib. It had already analysed the records in global corporate responsibility of Shell and other multinationals, such as Rio Tinto mining, British Petroleum, GlaxoSmithKline and Marks & Spencer. Harrington had already been to meetings at Shell's headquarters in London, and knew the form. However, an ever-optimistic Willie Corduff was very disillusioned by the trip. He felt that Shell shareholders, many of them middle-aged intelligent people, just didn't want to know about life as it was unfolding in north Mayo.

General election fever was in the air, and the Irish Offshore Operators' Association had written to politicians to counter 'negative views regarding the terms under which oil or gas can be produced offshore Ireland'; Micheál Ó Seighin wrote his own letter to deputies. Shell unveiled the first in a series of financial initiatives to win local support for the project, as recommended in the report by Peter Cassells. A programme of ten third-level scholarships worth €4,000 a year was offered to students from four secondary schools in the Erris area. One of the schools was Coláiste Chomáin in Rossport, where Ó Seighin had formerly taught and where principal Micheál Ó hÉalaithe was an ardent supporter of the gas project.

However, the company's good news was short-lived: it had to issue an apology to a resident of Pollathomas, across the Sruwaddacon estuary from Rossport, early the following month. It wanted to carry out a marine survey of the estuary as part of its investigations into modifying the pipeline route. On 8 June, it wrote to Mayo County Council notifying it of its plans to use Pollathomas pier for crews involved in the survey, and it proposed to erect a temporary 'office', including a small generator and a portable toilet, in an eight-by-four-foot container at the base of the pier.

On the same day the local authority replied, stating that it gave permission for same for a 'period not exceeding two months'. It emerged that the local authority or the company or both hadn't recalled that the public right of way to the pier had been washed away in the Pollathomas landslide of 2003. A Shell contractor in a HIMAC digger, along with a jeep towing a trailer, which in turn carried a Portakabin, arrived under Garda escort on 11 June and sought to gain access from the main road through a farm gate down a lane to a pier on Sruwaddacon estuary. Landowner Paddy McGrath said he had not given permission. Ironically, it was in his pub that the first Enterprise Oil meeting about the project had been held when an early pipeline route via Pollathomas was being considered.

A Garda broke the lock on his gate to allow the contractor to reach the pier with the container. In some distress McGrath telephoned around, and neighbours and Rossport Solidarity Camp members rallied. He also contacted his solicitor, Peter Flynn of Clarke and Flynn solicitors in Ballina, who tried on several occasions to reach Superintendent Gannon by phone. Afterwards, Gannon told *The Irish Times* he was not aware that the solicitor was phoning him, although video footage taken by one of the estimated forty people who arrived in support suggests otherwise. The main confrontation occurred at the gate as the HIMAC advanced; there was a steep slope on the northern side of the lane, which posed a serious risk.[12]

'Our only role was [in protecting] the right of access, and people sought to confront us,' Superintendent Gannon said afterwards. 'We pushed them out of the way.' He told the newspaper that one of his sergeants required eight stitches and a second sustained a knock on his head. The confrontation resulted in twenty injuries, including those of the two gardaí, and John Monaghan was arrested. Gannon said that his own shirt was ripped, along with one of his epaulettes, and he had had 'muck fired' at him. He said he had tried to speak to local residents in a community policing capacity the previous September, and confirmed to the newspaper that he now made a video of everyone he spoke to in the area, with their consent, for his own 'protection'.[13]

Footage from the scene was shown on television the next day, along with reports on an extensive blaze that fire units from Belmullet, Ballina and Crossmolina had fought throughout the night of 10 June and well into the following day on Coillte-owned forestry at Bunahowna, up to within four hundred yards of the Shell terminal site at Ballinaboy. Glenamoy farmer and Shell to Sea supporter P. J. Moran said that the blaze was close at one point to the original Corrib onshore pipeline route and 'highlighted once again the serious health and safety issues surrounding the project'. The risk of bogland fires during dry weather had been raised in inspector Kevin Moore's April 2003 report on Ballinaboy.

An *Irish Times* editorial on the Pollathomas confrontation, written before all the facts had become public, described the situation as like a scene from the 1949 film *Whisky Galore*.[14] John Monaghan said that the events had marked 'the lowest point yet in the authorities' handling of the Corrib gas project' and people had been left 'both physically injured and severely traumatized'. Paddy McGrath was hospitalized afterwards and his family said he never quite recovered from the shock.

Shell's public image was once again at a low point after the Pollathomas débâcle. However, there was a brief interlude when press reports in July 2007 referred to a fish managing to do what the Rossport Five had done – halt work on the project. A ling nick-named 'Larry' had blocked robot-guided drills out at the rig, preventing the suction piling from working properly – at a cost of 250,000 euro.

In September 2007, even as RPS consultants were still continuing their work on modified pipeline routing, residents of the 'landslide area' of Pollathomas were worried about a possible return of pipeline routing to their side of the estuary. They signed a petition for Shell, stating, 'You have a duty of care to notify us in relation to possible further landslides consequent to you or your agents carrying out work in Sruwaddacon Bay. We the residents do not consent to anything that we have not been consulted with.'

Shell suspended plans for its marine survey, removed the container 'in the interests of harmony' and told *The Irish Times* of 14 June that it apologized for the distress caused to Mr McGrath. Company representatives had visited him, it said. However, it told the newspaper it was 'not apologizing for its handling of the preparations' on 11 June. Subsequently, the Department of the Marine said that a Shell vessel could not undertake survey work until it was passed by its marine survey office.[15]

The events of June 2007 at Pollathomas would be investigated by the Garda Síochána Ombudsman Commission, and two years later it would recommend disciplinary action against a senior garda.[16] The case against John Monaghan, who had been charged on public order offences, was dropped on the advice of the Director of Public Prosecutions with no explanation issued.

After the May general election of 2007, it seemed that the Green Party would be kingmakers in a coalition government with Fianna Fáil. Its leader Trevor Sargent had long been a supporter of the Rossport Five, had participated in Shell to Sea protests, and had formally congratulated Willie Corduff on his Goldman award win at a photocall in Galway's Eyre Square before the general election. His party colleague Eamon Ryan, who had also visited the five in jail, was a founding member of the Campaign for the Protection of Resources, intended to lobby for a review of the oil and gas tax and licensing conditions.

In November 2006, Sargent had said he had 'serious questions' about the manner in which the government had granted permission for the location of the Ballinaboy terminal and said it had a 'responsibility to facilitate, without preconditions, a transparent inquiry'. In February 2007, the Green Party had endorsed the Cowley/Garavan proposal for establishment of an independent commission to examine the best development concept for Corrib if the party was elected to government. The resolution was passed unanimously at a Green Party conference.

Sargent had made it clear that he would resign as leader, as he had earlier promised, if his party entered a coalition with Fianna Fáil. However, his party colleagues clearly felt otherwise, so he stepped down and was replaced subsequently by John Gormley. In early June, as talks on forming a government were in progress, Sargent phoned P. J. Moran, with whom he had stayed when visiting Erris.

He pledged that the Green Party's opposition to government handling of the Corrib gas project would be discussed with Fianna Fáil. 'He assured us that he had not forgotten about us and that the issue was very much live,' Moran said afterwards.[17]

'An ideal opportunity for a fresh and positive outlook to be taken to the Corrib gas conflict' was how Dr Mark Garavan welcomed the appointment of Eamon Ryan as the new coalition government's communications, energy and natural resources minister. Green Party colleague John Gormley had taken the environment portfolio. Padhraig Campbell of SIPTU's offshore oil and gas committee was delighted, as Ryan was well familiar with the Corrib conflict.

However, there was also some upset in the Shell to Sea camp, as Dr Jerry Cowley had lost his seat in Mayo; rivals put it down to his close association with Shell to Sea, which, they said, was proving too divisive in the county. The Corrib gas developers perceived it as proof that the project had widespread support. The day after the election result, Cowley was back in his GP clinic.

Eamon Ryan's first move was tough, if not quite as tough as Shell to Sea campaigners had hoped for. He announced that the government would impose an additional resource tax of up to 15 per cent on the oil and gas sector for all new licences issued from 1 January 2008. The tax rate would depend on profit ratios. His decision had been informed by a report from Indecon consultants, commissioned by his predecessor, Noel Dempsey, as a result of a private members' motion moved by Cowley.

The department analysis challenged the arguments put forward by the Irish Offshore Operators' Association, which warned that offshore Ireland remained a 'high risk, high cost activity'. The analysis suggested that the potential of the Atlantic margin had been played down, and that there could be at least 10 billion barrels of 'oil equivalent' resources – crude oil or gas – off the west coast, which would be worth US$600 billion (€455 billion) at a rate of US$60 a barrel.

Ryan didn't rule out a windfall tax on profits for those fields, which would not be affected by his new terms – including Corrib. At the time, some thirty companies were working on various aspects of mineral exploration and production. 'Ireland's oil and gas is a resource of the people,' he said.

Cowley gave the changes a qualified welcome. The Labour Party's energy spokesman Tommy Broughan said it was a step in the right direction, but regretted that Ryan and his consultants had rejected suggestions for a petroleum-sharing contract regime, similar to that run by states like Denmark. Up in Mayo, Micheál Ó Seighin was not impressed. The gains to the taxpayer were 'decades away', given that the changes were not retrospective, he said. A 'two-tier' situation could 'be easily manipulated by the industry' to avoid paying new taxes.[18]

However, Ryan's next move was far less welcome: he confirmed that the government would not be reviewing the entire gas project. Reminded of his party's unanimous resolution on Corrib the previous April, Ryan told *The Irish Times* that, even though this had been passed, he was now a minister and had to adopt a 'neutral position'. In any event, his legal advice was that Bord Pleanála approval of the Ballinaboy refinery 'could not be reversed'.

Ryan reminded the newspaper of his record on Corrib. He was critical of the manner in which the onshore pipeline had been given consents, and was also concerned about bog movements. As his party's spokesman, he had also suggested other locations for the terminal in the past, such as the old Asahi factory near Killala and at Belmullet (where he had then been lobbied by local interests who did not want it in their area).

However, he believed that Noel Dempsey's Advantica safety review of the pipeline had recognized and made allowances for many of the community concerns. It had 'validated' those concerns and had recommended 'significant' changes, such as recommending reduced pressure.[19] Up in north Mayo, the perception was that a decent man was now in a position of power and had changed his mind.

During the month of August 2007, when he was on holiday, the new minister called on residents in Erris, including members of the Rossport Five, along with Monica Muller and Bríd McGarry. McGarry called over to Brendan Philbin's house to meet him. 'I remember asking him before he left if he would at least see to the slick in the Glenamoy river opposite our house,' McGarry recalled, two and a half years later. 'The slick had been there since 2001, when Enterprise was testing in Broadhaven Bay, and an amount of oil was discharged and was washed up into the Glenamoy river.' The Department of the Marine had initially blamed it on spent mushroom compost from a mushroom factory behind Belmullet, she recalled, but this had to be withdrawn when the mushroom factory objected to the 'misrepresentation'. Even a boom at high tide would help to collect it, Bríd suggested. The minister nodded on leaving, but the request was not carried through. 'The slick is still present in the area.'

Shell to Sea welcomed Ryan's attempts at dialogue, given that Dempsey had never visited the area, but it was cautious and had requested a formal meeting with the minister.

Summers were always eventful for those involved in the dispute, and this one was no less so than usual. Pat O'Donnell, his son Jonathan and Enda Carey were convicted at Swinford district court in mid-July of assaulting a Garda sergeant the year before and were released on bail when they lodged an appeal.

A protest in support of the three fishermen took the form of a 'lock-on', when two local men chained themselves to each other and to a van, parked sideways on the road at Ballinaboy. It took five hours for gardaí, fire brigade and ambulance staff to cut them out of the chains, and the pair were then arrested and taken to Belmullet Garda Station.

Mayo County Council initiated legal action to remove the Rossport Solidarity Camp from its most recent base at Glengad, following a complaint lodged by Monica Muller that it was a special area of conservation. And Shell to Sea called on Shell to disclose details of all its sponsorships of local organizations, following a row over its funding of the annual Féile Iorrais festival. Poets Rita Ann Higgins, Cathal Ó Searcaigh and musician Andy Irvine, formerly of the group Planxty, had been invited to participate and expressed dismay when they found out that Shell had given €2,000 towards the event.

Higgins considered withdrawing, but was advised by Vincent McGrath that this was not necessary. The festival was an important cultural event, and a withdrawal might cause further divisions, he said, even though the organizing committee should have made the sponsorship clear. The Shell logo had appeared on the festival website, but not on its programme. Festival committee member Seán Ó Coisdealbha said that there was no 'deliberate attempt' to conceal Shell's involvement.

Higgins opted to travel, but wrote to the festival committee to say she wouldn't take a fee. On taking to the stage in Belmullet, she was heckled from the audience by several people who were obviously critical of her public stance. She said afterwards that she was shaken by the experience, but had received a number of kind comments from others who, she believed, were 'intimidated by a vociferous minority clearly in favour of the Corrib gas project'.

Shell's spokeswoman Susan Shannon explained that the company didn't consider it appropriate to disclose details of community funding, amounting to grants of between €2,000 and €10,000, without the permission of recipients. Such recipients included the Humbert Summer School in Ballina, which Shell had supported, and which had been happy to make this public. Among at least forty applications to the company for funding, a third had come from local sports groups, a third from voluntary organizations and a third from organizers of local events. A number of applications for the third-level scholarships were being assessed.[20]

A long-term local investment fund – which, the company acknowledged, was tax deductible along with all its financial supports – was formally initiated by Shell the following month. It had been recommended by Peter Cassells, the company said. As a gesture of good faith, it was giving a preliminary grant of €150,000 to Belmullet GAA club. The local lifeboat committee was also being sounded out, as it was seeking support to raise money for an inshore vessel.

However, Erris Inshore Fishermen's Association, whose members were volunteers for the Royal National Lifeboat Institution, said it would not be party to any development grant until the pipeline discharge issue was resolved. The company should be addressing real issues, rather than trying to 'buy support', its chair, Eddie Diver, said.[21]

Separately, RPS said its work on a new onshore pipeline route had been delayed due to 'landowner opposition'. Several land-owners had returned letters received from RPS; Shell claimed that they had been 'intimidated' into doing so.

Protests were still continuing at Ballinaboy, and on 13 September four English supporters of the Shell to Sea campaign, Kate Kirkpatrick, Anna Rudd, Emma Jackson and Julie Ryder, along with Eoin Ó Leidhin from Ballinskelligs in Kerry and the Rossport Solidarity Camp, were sentenced to a hundred hours of community service for blockading the entrance to the Bord na Móna peat depository site near Bangor Erris on 5 June 2007. The five had locked themselves together with climbing gear, and put their hands into pipes laced with concrete at the entrance to the site, holding up some two hundred trucks transporting the peat, Judge Mary Devins heard at Ballycroy district court.

Judge Devins had recently revealed that she had received 'hate mail' from members of Shell to Sea; the solicitor defending the five, Alan Gannon, asked if she should not therefore disqualify herself from the judgment. Judge Devins replied that she was not a coward, and controversy had never influenced her thinking in her last nine years in the district court.[22]

*

Minister for Energy Eamon Ryan reiterated that he wanted Irish waters to be 'fully explored', but with a 'proper return to the state' if oil and gas were discovered, when he announced terms for a new exploration licensing round in the Porcupine Basin in October 2007. The terms for the 63,500 square kilometre area would be subject to the new tax regime as announced by him in August, he said; the next exploration round would take place in the Rockall Basin in 2009. Dependence on imported oil and gas had grown to more than 85 per cent in the past decade, but Ireland had become 'much more attractive to oil and gas companies' as 'we approach a peak in oil finds', Ryan added.

Fergus Cahill of the Irish Offshore Operators' Association agreed that it was 'in everyone's interest to have this area explored, and the high price of oil, coupled with technological development, will help'. However, there had been a lot of effort for very little yield in Irish waters so far, he said, with only two of nine exploration wells drilled over the past decade proving substantial: Shell's Corrib, discovered by Enterprise in 1996, and Dooish in 2002.[23] Later that month, Ryan's cabinet colleague, Minister for Justice Brian Lenihan, confirmed that the approximate cost of Corrib policing was now €8.1 million for the past year.[24]

Meanwhile, Colin Joyce, a Galway graduate who had worked in marketing and public relations, most recently with Fianna Fáil East Galway TD Joe Callanan, had joined the public relations team at Shell. He had spent five years on the Fianna Fáil national executive. Joyce was enthusiastic and keen to improve communication. He believed that there was 'broad support among the communities' for the project, as the work continued at Ballinaboy, and on new pipeline route surveying; some 75 per cent of staff were from the region, the company had pointed out.[25]

The assertions by the gardaí that opponents were mainly from outside the area appeared to have some substance when three men arrested on 9 November 2007 at Ballinaboy and charged with obstruction of a peace officer gave their addresses: Dominic McGlinchey, Dunmore Road, Tuam, County Galway; Cathal Larkin, Turner's Cross, Cork; and Robert Jackson, Lenadoon Avenue, Belfast. McGlinchey was the son of the late paramilitary of the same name with the Irish National Liberation Army, while Jackson, formerly of the IRA, had served twenty years for explosives offences and was released under the Good Friday Agreement.[26] Each was subsequently given the Probation Act. Brendan McKenna, a veteran of Catholic protests against Orange Order marches on the Garvaghy Road and member of Éirigí, was also spotted by gardaí who said he was advising protesters.

At one point, a prominent Derry republican, Gary Donnelly, who was involved with the 32-County Sovereignty Movement, political wing of the Real IRA, was also reported as having attended a Corrib protest.[27]

On the evening of 9 November, two quarries supplying to Shell reported damage to lorries, to a jeep and to a computer. Shell to Sea condemned the action, accepting that it did not have control over who turned up at protests. Many locals knew this was becoming a problem, but also an inevitable consequence of their having been abandoned by officialdom and those in power.

That November, Shell was given permission by An Bord Pleanála to retain a temporary road at Glengad, which it had built initially without authorization at the pipeline landfall. It also secured its emissions licence from the EPA. The agency said that the licence was subject to ninety conditions. The board noted that cold venting of gas was of 'major concern' to residents, but it had accepted the recommendation by the oral hearing chairman that this was the 'best environmental option in the circumstances, due to the small volume of gas released'. It also accepted that controls needed to be strengthened to ensure protection of drinking water supplies at Carrowmore lake.

Shell described it as a 'significant milestone', while Mary Corduff and John Monaghan described it as a 'very sad day for Erris'. They were not confident of the strength of any controls by state authorities, although the actions of Minister for the Environment John Gormley had been decisive the previous month when Shell consultants RPS drilled two boreholes between fifteen and twenty-five metres deep on the Glenamoy bog complex special area of conservation. The consultants had secured landowner permission, but had not consulted with the National Parks and Wildlife Service as they were statutorily obliged to do.

Only six weeks before, the Rossport Solidarity Camp had been evicted finally from the same area for a similar reason – special habitat status. A National Parks and Wildlife Service officer was despatched to the area after the borehole drilling, and confirmed that the work was unauthorized. Gormley ordered that the SAC should be restored, although the actual direction for this work was not issued until December. It appeared that no significant damage had taken place, but the minister said he was still 'very disappointed'.

On 14 November 2007, at least twenty commonage holders at Rossport and adjoining townland Muingnabo, including Monica Muller, objected to RPS consultants entering their lands to carry out tests for the new pipeline route. Judge Mary Devins ruled at Belmullet district court that they had not been properly notified, neither was their consent sought under the Gas Act, and she issued an order prohibiting entry to the commonage by Shell or its agents for site investigations or otherwise.

It seemed that there was still a firm resistance on the ground. 'The opponents, portrayed as baddies by Shell, are the prophets when it comes to our environment,' Father Sean Noone of Pollathomas observed. 'Some people have been attracted by money, and the divisions this has caused will take more than a generation to heal.'[28] On 16 October, he and two parish colleagues, Father Michael Nallen and Father Michael Gilroy, wrote to Minister for

Energy Eamon Ryan, questioning how 'promotional material' distributed by the Corrib gas developers – such as a newsletter prepared by the Shell press team – could 'validly claim community consent'.

The three priests said that they wished to 'reiterate that we believe most people are not opposed to the gas coming ashore' as 'benefits for community and country are something that most people welcome'. However, 'it should be possible to achieve this goal in an environmentally and community friendly way without the flaws which are linked to the Ballinaboy site'. The *Irish Independent* of 25 October reported on 'local anger at priests' attack on gas project'.

The response from the minister referred to the Advantica safety review of the pipeline, and his technical advisory group, and the fact that consultation was taking place on a new pipeline route. He said he was committed to doing 'everything possible within my regulatory role' to ensure that there was 'appropriate consultation'. The priests wrote back on 14 November, with a more specific proposal: relocation of the refinery to Glinsk, on the coastline near Belderrig. The location above a cliff-face had no housing within several miles of an exposed area of bog, and had been identified by Shell consultants RPS the previous summer as a potential landfall during initial work on the pipeline route modification. The consultants had since ruled it out of a shortlist of three possible routes. Shell would later explain that Glinsk had only been included by RPS at Father Nallen's suggestion to make sure the route selection process 'had been given a fair hearing'.

Locating the refinery at Glinsk would avoid a high-pressure onshore pipeline carrying unrefined gas – there would still be a link to the Bord Gáis grid. It would also obviate the risk to Carrowmore lake, they said. And the priests believed that running a high-pressure pipe up a cliff-face more than fifty metres high had been done before, was 'technically and economically feasible' and would 'comply with the codes of practice and EU directives'.[29]

*

In November 2007 Gary White Deer, a representative of the Choctaw Native American community, visited north Mayo to speak at a 'hedge school' run by the justice and peace organization Afri and to give US$8,000 to the Shell to Sea campaign. The Choctaws had held Ireland close to their hearts since the Great Famine, and had sent US$710 for famine relief in 1847. This contribution had 'no strings attached' and represented the unspent balance of some US$40,000 given in Ireland and the USA to Choctaw victims of Hurricane Katrina in 2005. White Deer appealed for dialogue in the dispute and said he believed there was a 'major opportunity' for talks at that time.

13

'FIVE "GREEN" MEN . . .'

Poets Seamus Heaney, Paul Durcan, Nuala Ní Dhómhnaill and musicians Donal Lunny and Andy Irvine were on Vincent McGrath's mailing list in autumn 2007. A celebratory function for the Goldman environmental award had been organized for Belmullet, even though there were already rumblings in the community about the size of the Goldman cheque. McGrath felt it was a good time to round up some messages of support.

'From my heart I send you my warmest congratulations on this night of celebration and laughter and defiance in the face of injustice,' Durcan, himself a Mayoman, replied. Mayo actor Seamus Moran, best known for his role in the RTÉ television soap *Fair City*, was master of ceremonies for the night, with participants including Galway poet Rita Ann Higgins and Lelia Doolan. Messages of support were read out from Nicaraguan poet Claribel Alegría, Irish writers Colm Toibín, Éilis Ní Dhuibhne, Nell McCafferty, Alan Titley, Gabriel Rosenstock, artist Dorothy Cross, film-maker Bob Quinn, musicians Lunny and Irvine.

And there was a 'congratulations' from Seamus Heaney: 'It's not the most distressful country after all when five "green" men stand their own ground,' the Nobel laureate wrote in his response to McGrath.

Such messages were valued within a community worn out by the dawn picketing at Ballinaboy. In Dublin, Shell to Sea supporters managed to climb on to the roof of an office block on Adelaide Road. It housed Eamon Ryan's department, and the three were wearing masks depicting Ryan, along with Minister for the Environment John Gormley and Taoiseach Bertie Ahern. The energy minister had become a 'spokesman for Shell on the Corrib gas issue' and a 'spokesman for the industry on natural resource licensing agreements', Tadhg McGrath of Dublin Shell to Sea said. The trio, who had been supported by fifteen others in their ascent, left eventually at the request of the gardaí.

Back up in Erris, the North Western Regional Fisheries Board had taken issue with some of Shell's publicity. The Corrib gas partners had announced a contribution of €130,000 towards upgrading the salmonid fishery on the Glenamoy river. 'We are proud to be associated with the Glenamoy Community Angling Association and the North Western Regional Fisheries Board, as well as the many other sports clubs and voluntary organizations that we support,' deputy managing director Terry Nolan was quoted as saying.

Vincent Roche, fisheries board chief executive, said that at no time had the board 'sought nor received any funds from Shell in respect of the Glenamoy river, or any other project', though it had provided some technical advice and direction under its statutory remit.[1] The fisheries board had had its issues with Shell in the past. It had issued a warning of legal action over discharges of silt by the company into the Ballinaboy river in June 2005.

There was another glitch to the local funding in 2008 when the Corrib gas partners confirmed they were 'reviewing' their grant of €10,000 to the Dooncarton landslide committee in Pollathomas. Gerard McDonnell, also a community representative on the Corrib environmental monitoring group, had been chair of the committee

– formed after the devastating mudslides of September 2003. Committee member Gerry Sheeran, whose family had been affected at the time, told *The Irish Times* that his group was unaware that a grant for safety barriers on the mountain had been sought from Shell.

The application had been made to the Corrib gas partners by one committee member, without authorization from the rest of the group, Sheeran said. Any financial aid for safety barriers on the hill should be a matter for Mayo County Council, he pointed out.

It wasn't the sort of weather for another landslide, or a power cut. On the May bank holiday weekend of 2008, Betty and Fritz Schult's Kilcommon Lodge Hostel in Pollathomas had been booked out by divers who regularly visited the area. On Saturday morning, the electricity went off in the village, and remained so for much of that day. Betty Schult was driving along the road between the Ballinaboy terminal and Glengad when she realized the reason for the cut. Five electricity poles had been swept off the road, there was a large crater of about three metres deep, and the movement of peat extended for 200 metres. Local authority workers on overtime to upgrade the L1202 Ballingelly to Pollathomas route were reported to have had a narrow escape, as they were with ten lorries and a digger at the time.

The 'slide' had occurred just eight kilometres from Dooncarton mountain, location of the catastrophic slide of 2003. The Electricity Supply Board confirmed that the event had occurred near Aughoose, and said it was working to restore supplies to an estimated 1,600 customers affected by the power cut. However, Mayo County Council denied that it was a bog slide. It argued that only one pole was dislodged in a 'movement'. It said it had closed the road for several hours to facilitate what it called a 'diversion' of one power line.

Over previous months, Schult had engaged in much correspondence on behalf of a group of twenty-eight concerned

households with Mayo County Council about these road works on
the L1202. The route widening had been planned to facilitate the
Corrib gas project as part of a €3 million upgrade of the road
network paid for by Shell. There had been no environmental impact
statement for the work, in spite of the fact that the route followed
the southern shore of Sruwaddacon Bay's special area of
conservation.

She was never quite sure if her long-held concerns about the
environmental impact of the Corrib gas project meant that Mayo
County Council didn't take her seriously. Two petitions with 125
signatures were collected in the village, and submitted to the local
authority. However, its response to the twenty-eight households in
the residents group was that the roadworks were required for
'health and safety reasons'.[2]

Schult, a trained nurse, and her husband Fritz had been steadfast
in their concern about Corrib. Schult had worked in Norway at the
time the first oil was found in the North Sea, and had witnessed
how the Norwegian authorities had secured the best arrangement
for their people. Travellers from all over Europe were attracted to
the family-run hostel overlooking Sruwaddacon, which Fritz
Schult had opened with his former partner when he first moved
from Hamburg. Sruwaddacon offered a tranquil and over-
whelmingly tempting contrast to city life in Germany. 'We never
had to go far for our own holidays, with Benwee Head and the
Atlantic landscape here on our doorstep,' Betty Schult explained.

When the pickets started at Ballinaboy, she was down most
mornings at 7 a.m. and sometimes earlier, in spite of having
children to prepare for school and a business to run. Guests, some
of whom were sympathetic to the campaign, learned to wait for
breakfast; if there were arrests, they knew it would be late. Schult,
who was fastidious about remaining within the law, sometimes came
home shocked and upset after being 'pulled and punched, and
screamed at by gardaí'. She witnessed neighbours being punched
and bruised. Countless times she and others were told, 'Go home,

look after your families and do a day's work,' as if they weren't doing that already.

The previous year, she had had one particularly unpleasant experience, when she caught sight of the flash of headlights outside the hostel early one morning. Once again, the hostel was booked out by divers, and she was setting tables for breakfast when she saw the car. It was a Garda patrol vehicle, and it was up at the guest car park, just above the hostel and close to the road. Gardaí appeared to be filming several of the guest vehicles. The patrol car had moved out on to the road by the time she reached the top of the avenue, and it was continuing to film from there. Schult put two and two together. Visitors' cars had been loaded with diving equipment, and some of it, such as oxygen cylinders, was clearly visible. A call had gone out the previous week from Shell to Sea for another day of action at Ballinaboy. 'Perhaps the gardaí thought the diving gear was something else!' she concluded afterwards.[3] She made a complaint to the Garda Síochána Ombudsman Commission, but was told her complaint was not admissible and the local Garda station was informed that she had contacted the Garda Ombudsman.

Several months before the 'slide', she wrote to Superintendent John Gilligan at Belmullet Garda Station and subsequently met him. A group of North American students staying at the hostel with their academic director in February 2008 had been filmed by the gardaí without reason, questioned on the length of their stay, and asked if they were drinking. It was not the first time that visitors had been questioned, she told Gilligan, and it was 'counterproductive to our immense efforts to retain and improve the developments in our vicinity'. She told Gilligan he was free at any time to inspect her visitors' book. He was cordial and assured her that the hostel was not under surveillance. The situation with visitors eased, but she was never sure if and when the surveillance stopped.

The Aughoose bog slide was not the only movement in Erris that spring. In late February, a week after Shell consultants RPS began restoring the Glenamoy bog complex where it had drilled without full authorization, Shell reported that several 'thousands of euro' worth of materials had been lost in a fire that broke out in a store compound about 1.5 kilometres from the Ballinaboy site. About a hundred timber mats used by contractor Roadbridge for temporary roadways were reportedly destroyed, and gardaí said they were investigating.

Shell to Sea in Mayo issued a statement in which it said it condemned 'without reservation the apparent criminal damage to materials'. It also said it condemned 'ongoing breaches' of ministerial consents at Glengad – it had written to the Department of the Environment about this, questioning why there was no departmental supervision of works there.

The Pro-Gas Mayo Group, through its spokesman, retired garda Brendan Cafferty of Ballina, condemned the 'wanton destruction' caused by the fire. The group claimed that forest fires had 'started last year not far from the perimeter fence of the Ballinaboy compound'. There had also been 'wanton destruction' of property belonging to quarry owners working for Shell some months ago, it said. This was 'damaging to the reputation of Mayo and Erris' and 'damaging to future job prospects in the region at a time of difficult economic circumstances'.[4]

On 28 February, Shell had to deal with another fire – but this time it was Shell UK, and the location was Bacton in Norfolk, location of a facility that handled a third of Britain's gas supply and involving a number of companies. The blast was said to have occurred in a waste-water system, sending an enormous column of smoke into the air. Shell said that the emergency plan worked well: staff were evacuated and there was no threat to the local community or to gas supplies. North Mayo freelance television cameraman Richie O'Donnell phoned Willie and Mary Corduff. Would they travel over with him for a documentary he was making on Corrib?

For the couple, it came as something of a culture shock. 'We met communities living around and outside the refinery, pleading with the companies for money for facilities, and not being given a cent,' Mary recalled. 'We went to a parish council meeting and they told us about safety issues, and safety plans that were designed primarily for staff rather than residents, and police monitoring for the oil companies. The place was like a fortress, and the community very deprived. The local garage and the shop had closed, and a lot of places were for sale, as refinery staff were spending no money in the area.'

At one point, O'Donnell decided to film a clip of the couple with the refinery in the background. 'Next thing, we had the police questioning us. It was a freezing cold day, I remember, and they kept us for forty-five minutes. We had a sense that this was what was ahead of us back in Erris.'

Back in Erris, Englishman Paddy Briggs had paid his first visit to the area. Briggs had retired from Royal Dutch Shell in 2002 after a thirty-seven-year career in marketing and communications with the company in Holland, Scotland, Hong Kong, Dubai and at corporate headquarters in London. He had heard Willie Corduff speak at the Shell AGM in The Hague in May 2007, and was struck by his sincerity. Briggs had little knowledge of Corrib at the time, but believed it had echoes of the Brent Spar debate of the mid-1990s where Shell 'seems to have the technical high ground but has lost the hearts and minds of the people'. Shell must ensure Corrib is a 'model project', in every respect, without shortcuts, Briggs wrote in a subsequent twenty-page report in March 2008, in which he endorsed the views expressed to him by Micheál Ó Seighin. 'If Shell does not believe that it can honourably defend all aspects of the health, safety and environmental integrity of the project', the company should 'pack its bags and leave', Briggs concluded. In any case, consent was the key, as without consent the project was 'doomed'.[5]

In April 2008, Caitlín Uí Sheighin was on her second plane to

Norway in less than three years. She had first travelled there with Dr Jerry Cowley in 2005 when her husband Micheál and four neighbours were in Cloverhill prison. It was widely believed then that the Norway factor – as in intervention by Statoil as a Corrib gas partner – had led to the men's release.

This time, fellow Rossport Five members Willie Corduff and Vincent McGrath were among a twelve-strong delegation to Oslo, along with three politicians – Labour Party president Michael D. Higgins, who had already met Statoil representatives in Dublin, Galway Green Party councillor Niall Ó Brolcháin and Sinn Féin Mayo councillor Noel Campbell. The three-day trip had the support of the three priests from Kilcommon parish, who had written twice to Minister for Energy Eamon Ryan to propose Glinsk as an alternative location for the refinery.

The plan, as drawn up with Terje Nustad of the SAFE oil and gas workers' trade union, was to meet senior executives of Statoil, as it was now known. Nustad had also scheduled talks with the Norwegian Centre and Socialist parties and non-governmental organizations representing civic society. The Centre Party was part of the coalition government: it held the key oil and energy portfolios in cabinet and it was due to send a fact-finding delegation to Ireland that same month to inform its members on a series of issues, ranging from Corrib gas to the EU Lisbon Treaty.

Initially, Statoil management refused to meet the group. Michael D. Higgins wasn't going home without a discussion, however, and neither was Terje Nustad going to let him. When the talks did take place, they had a 'different atmosphere' from discussions held with the company in Dublin before, Higgins said afterwards. He was convinced that the Oslo headquarters had not been fully informed on certain aspects of the controversy in which it was involved, as a Corrib gas partner.

'Close to zero' were the words of Helge Hatlestad, Statoil's vice-president of exploration and production in Western Europe, to *The Irish Times*, however. There was just about no chance that the Corrib gas partners would move the refinery from Ballinaboy, but it was 'very unfortunate' that the concerns of the north Mayo community had not been listened to during the planning stages in 2000 and 2001, he said.

'We've learned in Norway that there is a need for these sorts of discussions, for consultation and communication, before a project is sanctioned,' Hatlestad said. 'It becomes commercially unviable to do something different once a project has started.' It was also 'very unfortunate' that the government had opted to deploy gardaí in north Mayo, he said. 'Nobody feels happy with the involvement of, or necessity of, using police, but it is up to the relevant authorities to make this decision.'

The Corrib gas project had come to a complete halt for an eighteen-month period after the Rossport Five were jailed, Hatlestad reminded the newspaper, and a proposed modified onshore pipeline route was close to identification. He understood that a majority of people in Ireland accepted the project and welcomed it, but a 'not insignificant minority' opposed it – a description the Corrib gas lead developer, Shell, might not have agreed with. The priests' proposal for Glinsk was not viable, in his view. There were 'no immediate plans' to expand or extend the refinery, he added, as no additional finds had been made off the west coast. However, Statoil was due to drill shortly on the Cashel field, seventy kilometres north of Corrib.

Terje Nustad was insistent. Statoil was 'breaching its own ethical guidelines in north Mayo'.[6] It wasn't good enough to blame Enterprise Oil for making mistakes early on. It might have been lead developer of the project then, but Statoil had also been a partner with Marathon from the outset – and it had been there before Shell.

The group was barely back in Ireland when the final modified onshore pipeline route was published by RPS, on behalf of Shell and its partners. There was no doubt that the original route was still the best one, RPS acknowledged, but the amended route ran for 40 per cent of the original track. The landfall was the same, at Glengad, and it crossed Sruwaddacon estuary into Rossport. However, it then veered north-east to avoid the shoreline, through commonage shared by more than sixty landowners, including Brendan Philbin, Monica Muller and Willie Corduff.

The route cut into the Glenamoy bog complex – the special area of conservation – before crossing the Sruwaddacon estuary for a second time and travelling south to the refinery at Ballinaboy. The new route was 'twice as far from occupied housing' as the original, RPS said, pointing out that this fulfilled the recommendation to avoid housing made by government mediator Peter Cassells almost two years before. It would be submitted for planning approval – in contrast to the original route, which had been exempted from planning.

The planning permission would be sought directly from An Bord Pleanála under the new Strategic Infrastructure Act. Consents would also be sought from Minister for Energy Eamon Ryan under the Gas Act, from Minister for Agriculture and Fisheries Mary Coughlan in relation to foreshore licences, and, if An Bord Pleanála directed so, from Minister for the Environment John Gormley, because parts of the route would be through special areas of conservation.

Compulsory acquisition orders would still have to be issued for access to land, but Shell and RPS were confident that they would have the support of most landowners. Shell managing director Andy Pyle said that the announcement showed 'we have made every reasonable effort to address the concerns expressed by local people', and the Ballinaboy gas terminal was now 30 per cent complete.

Shell to Sea in Mayo was not happy. Its spokesman, John Monaghan, said the amended route 'exposes not just the people of Rossport, but the people of the entire parish of Kilcommon, to unprecedented and unacceptable risk. We do not give our consent to this and will resist it through every legal, political and campaigning means open to us, even though this could lead to more years of unnecessary conflict. This conflict can be resolved if there is a genuine willingness on the part of Shell and Statoil to reach agreement and secure consent. The tragedy is that there has always been a better way. What we need is resolution and agreement, not the forced imposition on an unwilling community of an unwanted and unsafe project.'

Pro-Gas Mayo said it welcomed the route, which should 'allay fears which householders in the Rossport area had'.[7] Labour's Michael D. Higgins had another viewpoint: he had received legal opinion that suggested there were problems with the statutory instrument used by former marine minister Frank Fahey for the consents on the original pipeline route. This would suggest that the validity of the original route was in question and that the company could have been aware of this when agreeing to a modification.[8]

Nine days after the new pipeline route had been published, seven north Mayo residents who had been part of the Norwegian initiative decided to drop their demand for a gas refinery to be built offshore. The seven, who described themselves for press purposes as community leaders in the Kilcommon parish, were key members of the Mayo Shell to Sea campaign – Caitlín Uí Sheighin, Mary and Willie Corduff, Philip and Vincent McGrath, Glenamoy farmer P. J. Moran and Porturlin fisherman Pat O'Donnell. Norwegian contacts had explained during their trip north that offshore refineries did pose serious safety issues, and suggested that they needed to reassess their stance on processing gas offshore.

The group, which included several landowners who had been on the original pipeline route, but were not on the modified one, called for serious consideration of a compromise proposal – construction of the refinery or terminal at Glinsk. The coastline location near Belderrig had been proposed by the three priests and had been identified as a possible landfall in the initial RPS consultancy engineering work for Shell.

The compromise move was welcomed by Michael D. Higgins, Fine Gael Mayo TD Michael Ring and the Bishop of Killala, Dr John Fleming, in the bishop's first public statement on the issue. 'This deserves an immediate and generous response,' Higgins said. The group, which called itself Pobal Chill Chomáin after the Kilcommon parish, was making a 'very genuine effort' to 'restore peace in the area and get on with their lives'. He believed Minister for Energy Eamon Ryan and his officials should 'not let this opportunity be lost'. Michael Ring said it was a 'move in the right direction' and he contended that alternatives should be looked at, and he appealed to Shell to 'sit down and talk – but only if is prepared for compromise', while the minister should also recognize the opportunity, he said.[9]

Maura Harrington was floored. At several meetings in Glenamoy to back the compromise, she verbally abused neighbours, prompting a walkout led by Willie Corduff and P. J. Moran. She readily admitted later that it took her a long time to recover from the blow. There were other local tensions over the group's claim to represent the community. The McGarrys owned land near Glinsk. Had they been consulted? However, there was also a keen awareness that any public suggestion of a 'split' could be highly damaging to the overall campaign.

Later that month, as Terry Nolan took over the managing directorship of Shell E&P Ireland from Andy Pyle, the company ruled out Glinsk. Nolan wrote an opinion piece on the issue in *The Irish Times*, while a *Mayo News* leader and an article by Kevin Myers in the *Irish Independent* sang from the same hymn sheet. Leo Corcoran, consultant to An Taisce, supported the Pobal Chill Chomáin call: Glinsk was technically feasible from an engineering perspective, he insisted.

The Mayo branch of the Green Party weighed in behind the compromise proposal to move the refinery, without specifying

Glinsk, and suggested that this was the only way to resolve the Corrib gas issue. In June, a motion proposed by the Mayo Greens at the Green Party national council was passed by a two-thirds majority, but the members were persuaded to postpone issuing a press statement on this until discussions had been held with energy minister Eamon Ryan. A press release was eventually issued on 6 July.

Pobal Chill Chomáin, steered by Vincent McGrath, began working on various legal and other moves designed to distance the group from situations that might lead to clashes with the gardaí. It asked Leo Corcoran to apply his engineering expertise to a formal complaint lodged with the European Commission, focusing on the consents and licences granted for Corrib by the minister, by the Environmental Protection Agency for the design, construction and operation of pipelines and the operation of a gas refinery at Ballinaboy. It called on the government to implement the Aarhus Convention on public access to environmental information – Ireland was one of the few countries in Europe that had not transposed it into national law. And McGrath and colleagues prepared a submission, which was despatched later in the summer of 2008 to the Organization for Economic Co-operation and Development (OECD), contending that the project violated OECD guidelines for multinational companies. The submission was facilitated by Afri, the peace and justice group, and Sherpa, a French non-governmental organization.[10]

The complaint to the European Commission sought interim measures to stop the work, and maintained it was in contravention of European law. The interim measures, equivalent to an injunction, had been issued by the Commission's environment directorate within the previous two years in relation to spring hunting in Malta and Italy, and the threat posed to a peatland habitat in the Rospuda valley in Poland by a road-building project.[11]

It represented the latest in a series of appeals to Europe from or on behalf of the community. The previous year, the EU's petitions committee, which already had Dr Mark Garavan's complaints about Corrib, had called on the government to halt work on part of the M3 motorway through the Hill of Tara landscape. A two-thousand-year-old archaeological site had been found in April 2007 at Lismullin that was not identified in the original environmental impact statement for the project.

If Europe and the Green Party couldn't save the Tara landscape – and it seemed they couldn't – there was little hope for those concerned about the Erris environment, unless health and safety aspects could be upheld. A Europe heavily dependent on the provision of energy by the world's largest gas producer, Russia's Gazprom, might otherwise not be so sympathetic to the concerns of a tiny community living in a westerly corner of Europe's most westerly island.

Developers had begun preparing in earnest for laying the offshore pipeline when the cost of Corrib gas policing was confirmed at €8.9 million to the end of April 2008. It emerged that the Garda Síochána Ombudsman Commission had sought permission unsuccessfully from Minister for Justice Brian Lenihan the previous year to examine the management of 'crowd protests and civil disobedience' in north Mayo. The submission for the modified onshore section had been submitted to An Bord Pleanála in late April, and no decision was expected before the end of the summer. It was also expected that there would be an oral hearing.

Out at the Glengad landfall, nets had been placed by Shell contractors on the cliff-face to stop sand martins nesting, and to facilitate construction work. Willie Corduff, a lover of all wildlife, had been so upset that he and others had gone down to try and remove the netting when it was first put in place. However, the gardaí, who were obliged to investigate complaints of breaches of the Wildlife Act, were now describing this passion to protect the coastal environment as 'criminal damage', which they were investigating – and the press was briefed informally of same.

Michael D. Higgins, who knew a thing or two about EU directives from his time as minister for arts, culture and the Gaeltacht in the Fine Gael-Labour coalition government of 1993 to 1997, questioned the legality of placing netting in a sensitive habitat. He said it would appear that Shell contractors were 'playing fast and loose with the EU Birds Directive'. There was no public response from Minister for the Environment John Gormley, in spite of his concern the year before about unauthorized work by a Shell contractor on a sensitive habitat at Glenamoy bog. And Shell had no comment to make.[12]

In late May, gas had come at last to Mayo. It was 'switched on' in Mayo towns as part of the Bord Gáis 'GasWest' project, which supplied the fuel to eleven towns in Mayo and Galway along the Mayo–Galway transmission pipeline. The demands of north Mayo residents had ensured that this took place. However, as the *Western People* noted, the fuel was not from the Corrib field but from the national supply extended west.

In early June, a group with which Sister Majella and Maura Harrington had maintained links, the British church-based Ecumenical Council for Corporate Responsibility (ECCR), called on Royal Dutch Shell to revisit the compromise proposal aimed at resolving the Corrib gas dispute. It said that Shell stood to gain 'long-term reputational benefits' from reaching a settlement with 'the vast majority of the local Rossport community', and that it was notifying British and Irish church investors, trade unions and other pension funds, and the wider responsible investor community, and asking them to join in calls on Shell to accept relocation of the refinery for 'the greater good'.[13]

In Norway a little later that summer, trade union leader Terje Nustad called for some form of 'humility, dialogue and ethical behaviour' from Shell and Statoil. Nustad wrote an article on his union's website in which he asked the Norwegian parliament's energy and environment committee to undertake a fact-finding visit to this 'vulnerable and beautiful region of Ireland'.[14]

In late June 2008, Minister for Agriculture and Fisheries Brendan Smith issued approval for offshore pipelaying and associated fore-shore work at Glengad. The consent for offshore pipelaying had already been granted by former energy minister Noel Dempsey. As reports appeared in the press in early July that the world's largest pipelaying vessel, *Solitaire*, was returning to Broadhaven Bay, Vincent McGrath and fellow members of Pobal Chill Chomáin grew increasingly concerned about protests and safety.

McGrath contacted Superintendent John Gilligan in Belmullet. He proposed that they meet to discuss community policing, and Gilligan suggested that McGrath send him a list of nominees. However, the superintendent wasn't happy with the list as submitted, as he said that some of the names were of people 'known to the gardaí'. At this stage, parts of the entire Kilcommon parish could argue that they fell into this category due to their presence at protests, driven by their continued health and safety concerns.

Tensions rose again as summer approached. Some residents in Pollathomas, close to Glengad, refused to give Mayo County Council consent for access to their property for continued road-widening work. Shell began putting pressure on the Erris Inshore Fishermen's Association to come to an agreement on compensation in exchange for suspension of fishing activity to facilitate the off-shore work. However, a fishermen's flotilla protest, involving inshore boats at Ballyglass, called on the government to protect the constitutional rights to fish, and Eddie Diver voiced the frustration felt by his members in a stirring address at the pier.

The dialogue with the fishermen would run into early August, when Shell issued letters to them, promising to use an alternative method of discharge for treated produced water from Ballinaboy, subject to statutory approval. It described it as a 'significant good-will gesture' and offered compensation, but with a deadline of acceptance forty-eight hours after receipt of the letters. Liamy MacNally reported in the *Mayo News* that the Corrib 'democratic deficit' continued, when he described how Shell had also purchased a share in the Rossport commonage, over which it intended to run its amended onshore pipeline.[15]

14

OPERATION GLENGAD

Glengad resident and musician Colm Henry was not having a very good start to his summer. Not only was there much activity at Glengad beach, but a new private security firm had been employed by Shell. Henry believed that staff hired by Integrated Risk Management Services (IRMS) were filming him and his grandchildren as they strolled down a field to Glengad beach. It had been a 'paradise', he told *The Irish Times*, when he and his wife Gabrielle had built the house on some of her family's land thirty years before. Their site overlooked Sruwaddacon and the wide sweep of Broadhaven Bay embracing diverse marine and shoreline flora and fauna.

The Henrys had been troubled by the security since April 2008; unlike the gardaí, the young men in 'high-vis' jackets, helmets and sunglasses wore no identification numbers and at that point were not legally obliged to. They took out binoculars when Gabrielle went to the washing line. Her husband was sure that they were being photographed and filmed every time they came to and left their home. Calm and clear about a situation which was costing him sleep, Henry explained that he supported the gas project, if conducted safely; he had been sympathetic to the plight of neighbours on the northern shore of the estuary but had not been an active supporter of Shell to Sea.

However, he was so concerned about video footage of his grandchildren that he had submitted written complaints to the Garda, the Department of Justice and the Department of Health and Children. The gardaí had said the issue was a 'civil matter'. Significantly, an environmental impact statement of April 2008 prepared by the developers for a modified pipeline route had omitted the Henry house from accompanying maps.

The golden beach below Henry's home now resembled an extended security zone, with fencing attached to 1,620-kilogram blocks erected down the public beach for about half a mile. He and his family had watched a pod of dolphins herded away by a speedboat hired for the Corrib gas work offshore. Father Michael Nallen, down the road in Aughoose, also believed he was under surveillance. He said his car registration details were taken every time he passed the Shell compound at Glengad – he felt it was a form of intimidation and harassment.[1]

On 22 July, a group of up to twenty locals, including Colm Henry and his neighbour John Monaghan, went down to the beach to question the authorization for the work. The previous day, the residents had spoken to the National Parks and Wildlife Service about the issue, given the impact of the work on sand martin nesting sites. The Wildlife Service indicated that its role was advisory; residents said afterwards that there was no mention of authorization. Some thirteen people were arrested and later released without charge.[2] Several days later, Minister for Energy Eamon Ryan had to issue an apology when it transpired that details of the work authorization had not been published on his department's website. Had those details been available, thirteen people might not have been detained. Ryan promised that all information would appear on the website in future; it was particularly embarrassing for him as minister for communications. Dr Mark Garavan had already called that summer on Ryan to 'start acting like a minister on the Corrib issue'.

Ryan's colleague John Gormley, who had acted in relation to the unapproved borehole work on a special area of conservation at Glengad the year before, was also now embroiled. It emerged that his department had exempted a section of the pipeline onshore at Glengad, above the high-water mark, from Bord Pleanála approval. Fault would be found with this exemption later on by An Bord Pleanála.[3]

Arthur Boland, a letter writer to *The Irish Times*, found the picture 'very depressing'. He described as 'startling' the newspaper's description of a pod of dolphins being herded out into the bay while sand martins were prevented by netting from nesting. 'May I ask exactly what terrorist black-list dolphins and sand martins are on?' he wrote. 'What threat are they to Shell, whose website proclaims that it operates in "environmentally and socially responsible ways"?'[4]

In early August, An Bord Pleanála wrote to the Corrib gas developers for extensive additional information on the modified pipeline application. It had a series of questions for RPS consultants about the pipeline's impact on ground stability, on habitats, and the 'cumulative impact . . . having regard to the other elements of the Corrib gas field development'. Significantly, it also asked for details on the impact of any extension of the life of the wellfields.[5]

Later that month, there were press reports that the Department of the Environment had criticized the 'deficient' and 'inadequate' nature of the environmental impact statement submitted to An Bord Pleanála. The *Mail on Sunday* reported that Julie Fossitt, a National Parks and Wildlife Service senior ecologist, had criticized the Corrib gas developers for keeping crucial facts on the pipeline from the experts assessing the company's application under the Strategic Infrastructure Act, and that protected habitats likely to be affected were also excluded from maps provided by the company. Her critique was part of a submission made by the Department of the Environment to An Bord Pleanála.[6]

Shell and the Erris Inshore Fishermen's Association were now talking again in the lead-up to the crucial offshore pipelaying plan for summer 2008, but Shell said the discussions were 'confidential'. The association, which didn't take that view, proposed that the discharge pipe from the refinery be moved to a location twenty kilometres west of Eagle Island. It insisted that the company's planned location twelve kilometres out posed too much of a pollution risk, given the tidal movements in Broadhaven Bay.

'This is the best deal we could get after seven years. We were between a rock and a hard place,' Eddie Diver said, after his members had voted for compensation from Shell several days later without securing their full demands. The fishermen had been given forty-eight hours by the developer to respond by noon on 8 August 2008. 'We were never against the gas project, but against the methods being used, which would harm the marine environment,' Diver told *The Irish Times*.

Shell had undertaken not to discharge 'treated produced water' from unrefined gas through the outfall pipe. A legal agreement committing the developer to this would be drawn up. Compensation offered by Shell would amount to €15,000 for registered and licensed vessel owners who 'do not normally fish in the prescribed pipeline corridor' and up to €30,000 for owners who could demonstrate that they fished regularly on the offshore pipeline route from Broadhaven Bay out.

Shell pointed out in its letter to the Environmental Protection Agency (EPA) that, although 'international experts' had 'confirmed that the treated produced water will cause no harm to the marine environment', it had made this concession as a 'goodwill gesture'. Imelda Moran, who had opposed the EPA's issuing of an emissions licence for Ballinaboy, studied the deal. It would require a change to the existing emissions licence, she said – and she was right. It would be a good eighteen months before the company would submit the application to do so, and only after several deadlines given to the fishermen had passed.

Pat O'Donnell was resolute. He and his son Jonathan were not going to be party to any deal between Shell and the Erris Inshore Fishermen's Association: he did not trust the company to amend its plans for its discharge pipe into Broadhaven Bay. In 2008 the skipper and managing director of his own shellfish company was now in regular contact with his solicitors. There had been his three experiences in 2006 – his arrest at a Ballinaboy protest on 12 October; the alleged assault by two truck drivers on the 'haulage' route to Bangor Erris; and the injuries he had sustained at Barrett's quarry in November, when he was also hospitalized. A subsequent investigation by the Garda Síochána Ombudsman Commission on Garda actions at the quarry was forwarded to the Director of Public Prosecutions, but that office said there was no case.

O'Donnell was back in hospital again with bruising and broken teeth after an altercation with gardaí at Ballinaboy on 19 January 2007, but he, his brother Martin and two other men who had been prosecuted were subsequently acquitted. During the protracted trial, it emerged that seven minutes of Garda videotape acquired by the defence for the trial was missing. O'Donnell's solicitor initiated a civil action due to the extent of injuries sustained.

O'Donnell questioned the authority of a private company to put pressure on him to leave publicly owned fishing grounds that he was licensed to work on. 'I believe the health of the marine environment for future generations is more important than short-term compensation,' he said.

However, *Solitaire* was now in Donegal and waiting for orders to sail south to Broadhaven Bay. In Norway, Terje Nustad described the ship's hiring for the offshore work, while so many uncertainties remained on land, as a 'provocation'.[7] There was, after all, still no state approval for the onshore pipeline link.

Focus for protests about the project would now move from Ballinaboy to Glengad, just across the estuary from Rossport. After the failure to engage in talks with the gardaí, Vincent McGrath of Pobal Chill Chomáin wrote to three government departments expressing concern about the Corrib work at Glengad, as it was clearly beyond the high-water mark and therefore beyond the scope of the foreshore licence issued by the agriculture and fisheries minister, Brendan Smith, earlier in the summer. It was also 'contrary to the environmental management plan'.

The fisheries department said its engineering division was 'monitoring the works' and had been 'requested to provide a report on the storage of the material'. The energy department intimated to the press that the offshore pipelaying in advance of any onshore approval was at the developer's own risk.

'Risk' was one word to describe the scenes captured by John Monaghan of Pobal Chill Chomáin and photographer Peter Wilcock on 21 August, when Shell to Sea supporters in wetsuits paddled out to sea at Glengad. Three of its kayakers had already delivered a letter to the captain of *Solitaire* in Donegal Bay, asking him not to steam into Broadhaven. A large digger was photographed scooping up sand from the seabed as the wetsuited figures approached – two were within feet of its bucket, one with his hand in the air in an apparent gesture of alarm. Four gardaí in a rigid inflatable were photographed standing by.

For the media, it was a 'David and Goliath' image; for Superintendent John Gilligan, it represented 'the most dangerous' action yet, and he appeal to demonstrators to 'take a step back from the water's edge'. He had 'serious concerns about the safety of protesters, workers and members of the Garda water unit' during the demonstration. Shell said it recognized people's right to protest, but it urged 'everyone to take sensible health and safety precautions in the vicinity of vessels and the site at Glengad'. Eight people were arrested and later released.

On the same day, Pat O'Donnell and those fishermen who did not support the association's compensation deal with Shell received an ultimatum to take their fishing gear out of the water by noon.[8] O'Donnell ignored the letter, and he and his son Jonathan took to sea in two of their boats, the ten-metre *John Michelle* and *James Collins*, to watch over up to eight hundred crab pots laid in Broadhaven Bay. O'Donnell appealed for state protection to ensure his gear was not removed by Shell contractors. McGarr Solicitors, representing the O'Donnells, said its clients had been put under pressure at sea by Shell contractors, and also pointed out that the Garda water unit had failed to come to Jonathan O'Donnell's assistance when requested the week before; Jonathan was at sea when he said he was approached in an intimidating fashion by a Shell contractor vessel. The gardaí confirmed they had received complaints from both parties.

McGarr Solicitors said that the Corrib gas developers had no legal authority to remove fishing gear to facilitate laying of the off-shore pipeline. The Department of Agriculture and Fisheries confirmed the constitutional dilemma: it said that the 'exercise and enforcement of rights and entitlements lawfully arising is a matter for the respective licence holder' and 'any dispute that cannot be resolved between parties acting reasonably may be determined in court'.[9] In other words, both had the law on their side.

By now, the Garda bill for policing Corrib was running at €11 million, and Minister for Justice Dermot Ahern was consider-ing asking Shell to contribute, according to the *Irish Independent*.[10] *Irish Times* security correspondent Conor Lally reported several weeks later that gardaí had sought the assistance of Interpol to identify what was described as an 'influx' of non-Irish activists to the Corrib protest.[11]

On 26 August, O'Donnell received a registered letter at his cliff-top home in Porturlin. The state's deputy chief surveyor was warning him under the Maritime Safety Act 2005 to keep a safe distance from the Shell contractor vessel, *Highland Navigator*, and any other vessels displaying signals indicating that they had restricted ability to manoeuvre under international maritime collision prevention regulations.

'Shell calls in the Navy' ran the large headline on page one of the *Irish Daily Mail* on 30 August 2008. 'This is the first time we have been used against civilian protesters', ran the strapline underneath. A 'gunboat' had been sent to Broadhaven Bay to protect laying of the offshore pipeline, the newspaper said. There was much press focus on Erris again after a period of disinterest. The international oil and gas industry journal *Upstream* carried a two-page special report on the controversy in late August, which highlighted some of the residents' issues. Corrib needs 'real and decisive leadership' urged an editorial in the *Mayo News*. Shell's west Dooish exploration, which might lead to an expansion of the Corrib refinery, was focused on in the business pages of the *Irish Independent*.[12]

The irony of another event was not lost on Erris residents: Minister for Energy Eamon Ryan had refused a prospecting licence for gold mining; there had been a long and successful campaign some years before to halt such work on the holy mountain of Croagh Patrick.[13] Meanwhile, the EPA confirmed Imelda Moran's claim that the Corrib gas partners would have to reapply to change their emissions licence for the refinery, following the agreement brokered with the Erris fishermen on the discharge pipe.[14]

The pressure was increasing on Pat O'Donnell and his son as they spent days and nights out in their small boats, monitoring gear, some of it laid along the offshore pipeline route, which they refused to move. Two Navy ships loomed behind, with a third further out beyond the bay, and Garda and Naval Service RIBs approached with officers, some of them wearing balaclavas. O'Donnell, chainsmoking, was under enormous stress; it reminded Mary

Corduff of the pressure she and her family had felt in the lead-up to and after the imprisonment of the Rossport Five in 2005. Onshore, there was a large Garda presence, backed up by Naval Service personnel who said their attendance had been requested as an aid to the civil power.

David Kirwan, an east-coast fisherman who had formed a group to campaign over difficulties within the industry due to rising fuel prices and falling stocks, sent O'Donnell a message of support. 'Fishermen trying to make a living are being hounded out of it in north Mayo,' he said.[15]

Pat and Jonathan O'Donnell were arrested at sea under the Public Order Act on 10 September 2008. However, McGarr Solicitors challenged the order successfully, and the two were released. On the same day, a 'technical problem' arose with the stinger on *Solitaire*, according to Shell, which was anxious to ensure television news cameras filmed the problem at sea. The ship would have to return to Killybegs in County Donegal for repairs. Coincidentally or otherwise, the fishermen's legal case was such that it appeared the Shell contracting firm, Allseas, might be worried about liability. It declined to comment.

'You don't build trust with gunboat diplomacy,' wrote Fintan O'Toole, assistant editor, in *The Irish Times* on 16 September, while Labour Party president Michael D. Higgins called on Minister for Energy Eamon Ryan to provide 'leadership, in relation to resolving the controversy, and asked the minister to explain why the Naval Service was providing protection for several vessels contracted to Shell, when these vessels were flying flags of convenience in breach of the United Nations Law of the Sea Convention'. This had been confirmed by the Irish branch of the International Transport Federation, Higgins said.[16]

Mayo Fine Gael TD Michael Ring learned in a Dáil reply that the Corrib gas field was now valued at €9.5 billion – almost five times its monetary value when it had been declared commercial in 2001. Therefore, if delays were costing the developers money, they were still set to be generously compensated in the increasing value of the resource.

Father Kevin Hegarty focused on Corrib in his weekly column for the *Mayo News*. He dismissed the proposal to relocate the Corrib gas refinery to Glinsk as 'unrealistic, even outlandish' and 'as likely to happen as Cristiano Ronaldo playing for Erris United in next season's Mayo Super League'. Referring to the three priests who had proposed Glinsk initially, he said that he was aware that they were 'motivated by their responsibilities for the pastoral care of the Catholic community of the parish, which, of course, includes those who support and oppose the gas project'. However, 'as priests, their academic competency lies in theology, canon law and liturgy. Being able to wade studiously through the intricacies of the *Summa Theologica* of St Thomas Aquinas does not qualify one as an expert on the location of gas terminals.'

He urged Pobal Chill Chomáin, which had 'freed itself from the increasingly empty mantra of Shell to Sea', to accept that Shell had got 'legal consent for the project so far and enter negotiations with the company and the relevant government agencies'. The protesters could already take credit that the project was safer now than as originally envisaged by Enterprise. 'Surely it is possible to agree on a group of independent experts who would monitor the safety of the project in its construction and operation,' he wrote. The 'search for honourable compromise must always be part of the democratic process'.[17]

In both the *Mayo News* and a reprinted version in *The Irish Times* the previous November, Father Hegarty had been more forthright, while also stating that he had already been placed in the 'ranks of the unclean' by some Shell to Sea supporters, for accepting a nomination to the board of Shell's third-level scholarship scheme for Erris students, chaired by former *Mayo News* editor Sean Staunton. He referred then to Shell to Sea having lost support, with Dr Jerry Cowley's 2007 election defeat being a reflection of this, with the number of daily Ballinaboy protesters, which had been swollen by 'serial environment protesters from far places', now dwindling.

He referred to there being 'no community consent' for the 'verbal abuse and intimidation – sometimes subtle, sometimes overt – of those who work on the project or do not support the cause'. And there was no community support for the 'vandalizing of the property of local businesses providing services to the project'.[18]

His articles reflected a marked change in the position of a very influential voice in the area, who had called, back in the April–May 2001 edition of *Céide* magazine, for an independent environmental, health and social impact study of the entire Corrib gas project. He was also one of ten Erris priests who issued a statement in the summer of 2005 that called for the 'restoration' of the Rossport Five to their families, and expressed hope that all genuine worries about health, safety and the environment could be addressed.

Galway film-maker and environmentalist Lelia Doolan took issue with him in the letters page of *The Irish Times*, noting that a recent poll in the newspaper had shown an 84 per cent vote in support of Shell to Sea's arguments.

She said there had 'never been a blanket rejection of the use of the gas for the betterment of life', but there 'were and are other routes, other options. No wonder Shell is trying to mend its hand with various inducements and sponsorships and scholarships. By the time this affair is settled, it may regret the financial losses suffered by its refusal to undertake the cleaning of its raw gas at sea.'

And Father Hegarty was 'naïve on the topic of peaceful resistance', she said, when 'the power of a state to frustrate and abuse its citizens' legitimate entitlements through full-time, permanent, pensionable, official and bureaucratic means is daily practised and well known in every aspect of Irish life.

'Considering the provocation, most would accept that the Shell to Sea campaign has shown remarkable forbearance,' she continued, and quoted a line from Sean O'Casey's character Fluther Good: '"A few hundhred scrawls o' chaps with a couple o' guns an' Rosary beads, again' a hundhred thousand thrained men with horse, fut' an' artillery . . . an' he wants us to fight fair!"'[19]

To Maura Harrington, it didn't matter whether the press coverage was good or bad; she was just happy that the media were taking an interest. In fact, she was making her own stand in support of Pat O'Donnell. She had asked her Shell to Sea supporters to deliver a letter to the captain of *Solitaire*, informing him of her decision to go on hunger strike on 9 September. She pledged that she would not come off the strike until his ship left Irish territorial waters and could prove it in writing. This was to be confirmed by fax, she said.[20] She spent her fifty-fifth birthday parked in her car down at Glengad. Her birthday also marked her official early retirement from her post as principal of Inver national school.

However, also on 16 September, an event occurred in Dublin that was to distract attention from activity on the ground in Mayo. An 'explosive device' was reported on the steps of Shell E&P Ireland headquarters at Corrib House in Dublin's Leeson Street, and the army bomb disposal squad was called in to defuse it. A senior Garda source told *The Irish Times* that the device comprised a drinks bottle filled with petrol, which was attached to a battery and a clock. It was also attached to a can of paint, which would have sprayed out if the device had exploded. The sources said that while all the components for a viable device were present, the ensemble was not wired properly and therefore could not have exploded. Gardaí said that they believed the incident was linked to the Corrib gas protests.

Shell described it as a 'sinister development', and stated that the work 'currently being undertaken' on Corrib had 'all the necessary consents and permissions required by the various statutory bodies which oversee the project'. It believed that this was 'a time for calm assessment', and said, 'We remain open and willing to talk to any individuals or groups who continue to have concerns about our project.'

Vincent McGrath and John Monaghan, of Pobal Chill Chomáin, and Ciarán Ó Murchú, the former Air Corps pilot turned adventure-centre managing director who now chaired a new Erris business group, Pobal Le Chéile, used similar language in a joint statement: 'It is time for everyone to take a step back and for Minister for Energy Eamon Ryan to live up to his responsibilities.' They condemned 'unreservedly' and 'totally' what they described as an 'appalling action', and called for an 'immediate suspension of activities on every side to allow for a peaceful and diplomatic solution to Corrib that doesn't put lives recklessly in danger'.

Shell to Sea and the Rossport Solidarity Camp, which had relocated to a house in Pollathomas since its eviction from Glengad the previous year, said that the device was 'nothing to do' with them, and Niall Harnett, spokesman for both groups, said they were 'not into the politics of condemnation'. Separately, Dublin Shell to Sea said it rejected an 'unfounded insinuation' by Shell that the device was 'made and placed by Shell to Sea supporters'. The Pro-Gas Mayo group said that the device was a 'further sinister attempt at intimidation and proof, if proof were needed, that this campaign has a large element of subversive activity attached to it, which some sections of the media do not address'.[21]

In mid-November, there were press reports that four people, including the son of a republican, had been held over the device found at Shell offices in September, and later released . . .

On 17 September 2008, *The Irish Times* reported that Shell had abandoned plans to lay the offshore pipeline that year, and the following day Shell said in a statement that *Solitaire* was to leave Killybegs for Britain. The ship was to be assessed in deep-water berthing facilities in Glasgow, it said, and an eighty-tonne section of her pipelaying equipment, which had reportedly become detached the previous week, would be retrieved from the seabed in

Broadhaven Bay. On 18 September, the two community groups, now known as the 'Pobals', said in a joint statement that the announcement was 'inevitable after it became clear that Shell had no legal entitlement to remove fishing gear belonging to Erris fishermen in Broadhaven Bay'. There were also 'serious legal doubts' over the arrests on two occasions of Pat O'Donnell and other fishermen, the groups said, and the suspension of activity provided a 'critical window' for a resolution of the dispute. 'Our community has had enough of conflict and trauma.'

Maura Harrington was making no comment, but she had come under 'serious pressure', as she described it, to drop her hunger strike, which was now in its second week. She came off her fast on 19 September. She had been six stone nine pounds when she started, and she said she was down to six stone. Pointedly, she also rejected the Glinsk option for the Ballinaboy refinery as backed by the two community groups. 'Any alternative location for the Corrib gas infrastructure will not build new schools, new hospitals or contribute to the national pension fund,' Harrington said. 'Until we the people benefit from what is rightfully ours, any attempt to extricate Shell and the government from the mess that is Corrib remains doomed to failure.'

Shell had nothing to say about her action, or her statement, but it was well aware that there would be a perception that the community opposition was now officially split. In fact, as always, there had never been any leadership, any real structure or any real appetite in a rural community for public rifts.

A week later, Liamy MacNally reported in the *Mayo News* that Shell had requested that Bord Pleanála inspector Kevin Moore should not handle an appeal to amended planning permission it had been granted for Ballinaboy. The 'minor modifications', as described by the company, had been approved by Mayo County Council, and then appealed. The company had initially sought the changes under the fast-tracking Strategic Infrastructure Act directly from An Bord Pleanála the year before, and Moore had been involved in the pre-consultation discussions. However, the appeals board advised that the submission did not meet the strategic infrastructure criteria.

Moore had chaired the oral hearing on Ballinaboy, and had issued the devastating report advising against permission in 2003. An Bord Pleanála confirmed that it had received the request in writing from Shell's agents. It said it hadn't assigned Moore to the appeal, but this decision was not influenced by the Shell request. It said it had full confidence in Moore and its planning staff 'to deal in an impartial and professional manner with any case assigned to them'.[22]

Richard Kuprewicz of Accufacts Inc., the US pipeline expert who had analysed the Corrib pipeline plans for the Centre for Public Inquiry in 2006, was back in Ireland. He had been invited to attend a hedge school weekend in Erris hosted by Afri, and had undertaken a report on the Glinsk option, which he described as 'vastly superior' on 'health, safety and environmental grounds' to Ballinaboy. He said that 'serious questions should be raised as to why this site was not evaluated when identifying site alternatives for possible consideration from the Corrib gas field'.

Kuprewicz, who confirmed that a meeting with Minister for Energy Eamon Ryan was on his schedule, said he was 'saddened' to see there had been no resolution; the only obstacle was 'trust' and Shell's willingness to spend money on the safe design of existing plans, or relocation of the gas refinery to a coastal site. 'It is not hard to move a gas plant – this is not an oil refinery,' he said. 'The only cost is money, which Shell is not short of.'

Asked to respond, Shell said that this showed a 'lack of understanding of Irish planning processes and the statutory scrutiny which underpins every application for major infrastructure', and moving from Ballinaboy was 'not an option'.[23]

At the same Afri weekend, former UN assistant secretary general Denis Halliday, who resigned his post as humanitarian co-ordinator in Iraq in mid-1998 over the impact of UN-imposed sanctions on the Iraqi people, criticized the 'ugly methods' used by gardaí in north Mayo to control Corrib gas protesters.

On 29 September, the government made its first direct intervention in the conflict since Noel Dempsey's term in office. Minister for Energy Eamon Ryan and Minister for Community, Rural and Gaeltacht Affairs Éamon Ó Cuív travelled to Ballina to meet various delegations over the issue, along with local politicians. The community Pobals welcomed the initiative, and said they would work with 'all parties', including Shell, Statoil and Marathon, to help deliver a safe solution for the Erris people. Council for the West chairman Sean Hannick, a Killala businessman, accompanied representatives from the Pro-Gas Mayo and Pro-Erris Gas Groups. Maura Harrington, who was photographed in the following day's newspapers engaging with Mr Ó Cuív during a press briefing, described it as an 'exercise in futility', while Shell welcomed the move.

Early the following month, the two ministers announced that a consultative forum would be established. Speaking at a press conference in government buildings, the 'two Eamons', as the pair became known in Erris, said that the forum would be chaired by former Department of Justice secretary general Joe Brosnan, a Kerryman who was a member of a number of influential bodies, including the Independent Monitoring Commission to oversee decommissioning of weapons in Northern Ireland.

'Undoubted mistakes' had been made with the Corrib gas project, including a failure to consult adequately with the local community at the outset, Eamon Ryan said. A 'window of opportunity' had allowed the government to step in, and this was 'not an initiative for its own sake', he told *The Irish Times*, in an interview in which he defended his decision as minister to support the Corrib gas project, having been an opponent when in opposition. 'I have been intimately aware of and involved in the development of this project for many years,' Ryan said. However, he said that his safety concerns had been allayed by the Advantica safety review.

'Now my position and that of the government is that this project is good for Ireland,' the Green Party minister said, but he added that north Mayo had significant energy potential beyond it and, presumably, fossil fuels. He referred to a study recently completed on building a grid energy connection to exploit wave energy west of Belmullet, and the enormous wind energy potential in the area. 'In my deliberations on this, it is clear that all sides never sat down together, the government needed to intervene to provide leadership and facilitate healing, and the benefits to the local community needed to be considered.'

Ó Cuív also admitted to 'mistakes' and said that the new forum aimed to 'obviate the need for deployment of gardaí in north Mayo'. The bill for Garda overtime and expenses in policing Corrib had now reached €11 million, according to official figures. The forum initiative was 'worth a chance', Ó Cuív said. 'From a community point of view, it's an effort to bring the government closer to the people in this situation. Eamon Ryan brings a huge contribution to this – everyone would trust his bona fides and his credibility,' he added. The forum would 'not be semi-detached' and would report regularly to the two ministers.[24]

The forum was going to be very limited, however, as the two ministers made clear that it would not revisit any consents or planning permissions already granted. And it stopped well short of an independent review of the optimum development concept as originally proposed by Dr Jerry Cowley, former TD, and Dr Mark Garavan, former Shell to Sea spokesman, back in 2006.

Vincent McGrath and Ciarán Ó Murchú were disappointed: they believed that the two Pobals would not be able to participate under its current terms of reference. Maura Harrington, who initially said that Shell to Sea would consult 'widely', also found difficulties with its remit.

*

Shell also hosted a trip to a number of gas sites in Holland. Selected press were invited, along with Pro-Gas Mayo's Paraic Cosgrove and a consenting Rossport landowner Padraig McGrath. A video of the two men's tour was subsequently put on the Corrib gas website. On 18 November 2008, both Shell and the state asked the High Court to halt the action by four local residents – Bríd McGarry, Brendan Philbin, Willie Corduff and Philip McGrath – alleging that ministerial consents granted six years before for the Corrib gas pipeline breached European law and the principles of natural and constitutional justice. The residents were still seeking a High Court declaration that Shell had no interest in lands acquired under compulsory acquisition orders issued in 2002 by Frank Fahey, and declarations that the then minister had acted in breach of the principles of natural and constitutional justice and in breach of European law by giving consent for those orders.

The four residents also claimed that both Shell and the state were attempting to prevent issues being determined by the High Court in an effort to 'protect illegal acts' from judicial scrutiny. They claimed that Shell failed to comply with a condition of ministerial approval that a full assessment of the effect of leaks or a rupture should first be carried out, and an environmental management plan should also have been drawn up and approved by the minister before, rather than after, consent was given. The state denied the claims. Ms Justice Laffoy said she had to consider the applications; in March 2010, she said that two of the four residents who were still pursuing the legal challenge were entitled to have the validity of their claim determined by the courts, and Shell promptly sought to challenge this.[25]

In early December 2008, Chief Superintendent Tony McNamara and Shell managing director Terry Nolan were among those asked to a pre-Christmas dinner hosted by Mayo county manager Des Mahon in the Mulranny Park Hotel. Significantly, on 12 December, *The Irish Times* reported that Shell had withdrawn its modified onshore pipeline route application from An Bord Pleanála. Sources close to the company told the newspaper that a revised application would be submitted in the new year and there would be 'no overall delay' to the project. It seemed as if the application as submitted would have fallen short of planning requirements.

As the year drew to a close, forum chairman Joe Brosnan said that 'flexibility' in the terms of reference for the ministerial forum should allow community groups to participate, and he was prepared to talk informally to anyone outside the forum confines without prejudice. 'There was a concern among some people that it would be a talking shop,' he told *The Irish Times*. 'But my view is that in talking about it, the positions of the various parties become clarified – such as relating to what kind of regulation, monitoring and adherence to national and international standards will be in place. There cannot be any prescribed outcome of this in advance. It is a question of getting involved in discussion, and to have open-ended dialogue.'

The two Pobals said they were 'willing to help the forum, the government and the developers to bring this sorry saga to a successful conclusion', and referred to the Glinsk compromise option. 'Our groups will have great difficulty in partaking in a process that effectively has no power to change anything,' they said. They quoted one of their supporters as saying that 'We do not need more money and we do not need group therapy.'[26]

Shell confirmed that there was 'no pause button' on Corrib gas.

15

NO QUARTER

When Susan Shannon was first assigned by her employer Royal Dutch Shell to the Corrib public relations team, the word was that a red-haired beauty had been hired to slay the Rossport Five ... or perhaps their ginger-haired spokesman, Dr Mark Garavan. Shannon's petite demeanour belied serious experience. As a fluent Russian speaker, she had worked on Shell's controversial Sakhalin project. Yet, as one seasoned Corrib observer noted, she seemed the type that one might find on the set of BBC Television's *Dr Who*, or perhaps between the pages of *Harry Potter* ...

During her time on Corrib, Shannon maintained a professional respect for journalists, and a detachment that contrasted with the messianic approach of some of her superiors. However, by early 2009, she was on the move to a new post in The Hague – having spent her last year setting up a €5-million investment fund in Erris as part of the company's commitment to the Cassells report. The first named beneficiaries would be Belmullet GAA and the local branch of the Royal National Lifeboat Institution (RNLI).

This was something of a coup for Shell. Enterprise's effort to woo the Church and the educators had backfired. The GAA and the lifeboats still represented two highly influential institutions in a coastal community. Still, the €200,000 contribution to the RNLI Ballyglass station came as something of a surprise. The RNLI had been an apolitical organization with a distinguished 185-year history in sea search and rescue, and with a reputation for community building.

The Shell money would meet half the costs of supporting a new inshore lifeboat, purchased and launched the year before with fundraising through the *Western People*. When they had first got wind of it, Erris fishermen had opposed Shell support for the appeal. Questions were now being asked as to whether the RNLI's local branch had formally applied for the money in spite of the potentially divisive consequences.

RNLI area fundraising manager Tony Hiney told *The Irish Times* that Shell had 'responded to a general appeal sent out to all businesses and individuals in the region', and the Ballyglass lifeboat committee was 'delighted' with the money, which, it said, would help to pay for Land Rover costs and a dedicated station in Belmullet town.[1]

The RNLI denied that several volunteers had tendered resignations when the contribution became public. However, Pat O'Donnell said that fishermen were 'very upset', given the ongoing issues with Corrib and its impact on the marine environment. 'If we are involved in an incident at sea, and need the lifeboat, will the response time be the same if it is known that we have spoken out over Corrib?' O'Donnell said. The RNLI countered that it would 'never be compromised' by any donation or fundraising appeal, stating that it had 'no legal or ethical reason' to refuse the donation.[2]

In the preceding weeks, there had been a flurry of business press reports about the economic importance of Corrib gas. Chambers Ireland issued a statement in early January 2009 on behalf of business groups across the state, and Minister for Foreign Affairs Micheál Martin told an Oireachtas European Affairs committee that the project delay was 'extraordinary'. An *Irish Times* editorial of 7 February 2009 did point out that when the gas was finally brought ashore, it would supply 60 per cent of energy needs for only five years at peak supply, reducing gradually during its projected fifteen- to twenty-year life. Shell to Sea pointed out that the gas would be sold at full market price, with no benefit to the state for many years under tax write-offs.

In early spring, the Organization for Economic Co-operation and Development (OECD) said that it could play a role in mediation between Shell and Pobal Chill Chomáin, having admitted a complaint lodged by the community group. However, there was a separate potential breakthrough when the ministerial forum's chairman, Joe Brosnan, managed to persuade Pobal Chill Chomáin and Pobal Le Chéile to engage in discussions with the two ministers and a Shell team, which included Shell E&P Ireland managing director Terry Nolan, and John Gallagher, vice-president (technical) of Shell E&P Europe. These would be the first direct talks between the community and the developer, in spite of Peter Cassells's efforts in the year after the jailing of the Rossport Five.

Not everyone was happy, though. Maura Harrington's husband, Naoise Ó Mongáin, staged a picket outside the discussions at the first meeting in 'Craggy Island' – the nickname for Ó Cuív's community, rural and Gaeltacht affairs department on Dublin's Mespil Road – on 20 March 2009. 'Even if we were invited, we would not talk to Shell while Maura is in Mountjoy jail,' he told *The Irish Times*.

On 11 March, she had been sentenced to twenty-eight days for assaulting a garda at Pollathomas on 11 June 2007 – during the fracas over Shell's bid to place a Portakabin on the pier. Judge Mary Devins also directed that Harrington should undergo psychiatric assessment – a move criticized by Senator David Norris. In sentencing, Judge Devins told Harrington that she was less inclined to believe in her passion for her cause, having 'witnessed the enjoyment she seems to get in being in the public limelight'.

The day after Harrington's transfer to Mountjoy, Pat and Jonathan O'Donnell received letters from Superintendent Michael Larkin in Belmullet. The letters stated the Director of Public Prosecutions had recommended no prosecutions following their arrests the previous September. It was a very significant decision, according to the fishermen's legal team, which had challenged the legality of their arrest at a time when *Solitaire*, the pipelaying ship, was in the bay.

The Brosnan-chaired direct talks were described as 'full and frank'. Along with the six Pobal Chill Chomáin representatives were Ciarán Ó Murchú and fisherman Anthony Irwin, for Pobal Le Chéile. Joe Brosnan wasn't present for the second session on 6 April, however. The talks broke down after more than five hours, due to a 'fundamental disagreement' between the two sides. Shell's delegation at this second session had included Mike Wilkinson, Royal Dutch Shell vice-president for sustainable development.

Vincent McGrath was glad that Wilkinson was present, along with John Gallagher and Terry Nolan. 'We were keen to ensure that Royal Dutch Shell heard our concerns directly, and we believe that Shell headquarters is fully aware now of our position,' McGrath said. 'We pointed out the benefit to Shell in public relations terms of coming to an agreed solution, given that the legal case against Royal Dutch Shell for human rights abuses against the Ogoni people in Nigeria is due to be held next month,' he added.[3]

Shell, which had long denied it had played any role in the execution of writer Ken Saro-Wiwa and eight Ogoni colleagues in 1995, was still haunted by those events. It was now facing a suit taken by the Center for Constitutional Rights in New York on behalf of the Ogoni litigants, based on an eighteenth-century law that allowed foreigners to pursue corporations in US courts.[4]

'Serendipity' was one commentator's reaction to the timing of Maura Harrington's release from Mountjoy, the day after the direct talks collapsed. Beaming, she was photographed with supporters, reporters and a bouquet of red roses outside Dáil Éireann, where

Minister for Finance Brian Lenihan was preparing a Budget speech – the economy was now officially in trouble. She said she had put on the half-stone she had lost during her hunger strike of the previous year, and she was as resolute as ever. She had believed the mediated discussions would go nowhere, but there was always 'a danger that someone within the community might approve a deal'.

The Irish Times opinion page editor Peter Murtagh wrote of the 'half truths behind the Corrib gas controversy' on 16 March 2009, in which he referred to the support Maura Harrington of Shell to Sea was prepared to accept from 'several extreme organizations and individuals whose beliefs and track records would make them un-acceptable to most right-thinking people'. Incorporating a personal criticism of Harrington, he referred to the 'sustained abuse and intimidation of anyone associated with the Corrib project'.

'Of course, Maura Harrington is a troublesome individual,' Lelia Doolan responded, in a letter commenting on Murtagh's piece published in *The Irish Times* on 18 March 2009.

> Of course, her civil disobedience is awkward and inconvenient, and of course there are other methods of opposing bad policy and illegal infringements . . . But when a landscape and a community are being torn asunder and the human rights of its inhabitants set aside over many years in pursuit of questionable material gain, heroic self-control may not be the chosen path for all.
>
> The whole matter of human rights, natural resources and the environment in an adolescent democracy such as ours deserves proper public discussion. The people of Rossport deserve a visit, and support, from anyone interested in opposing the desecration of a beautiful area of our country. Is it too much to hope that current fiscal embarrassments might put a stop to the gallop of Garda overtime, and their own brand of muscular arm-wrestling, in Erris? Peter Murtagh's bad-tempered opinion

piece of 16 March offers the old Irish attitude of attacking by
association, while ignoring the core issue and the fact that
sensible alternatives have been advanced by the campaign over
many years . . . I know Maura Harrington and I greatly value,
and sometimes shake my head at, her rebel spirit. By the way,
she weighs about ninety pounds and towers to a height of
almost five feet.

Shell lodged a further application to amend planning permission
for the terminal in spring 2009. Down the coast, An Bord Pleanála
approved permission for a liquefied natural gas (LNG) pipeline to
serve a terminal planned by an Irish subsidiary of the global Hess
energy corporation on part of a state-owned land bank on the
Shannon estuary. Irish US-based actor Pierce Brosnan and his wife
Keely had given their support to a Safety Before LNG campaign,
which had opposed the development. However, Shannon LNG, the
developing company, had learned from mistakes made with Corrib
and engaged in extensive community consultation.

Several days after the forum talks collapsed, Minister for Energy
Eamon Ryan approved an environmental management plan for the
Corrib offshore pipeline. Mary and Willie Corduff and other
members of the two Pobals were dismayed at the speed of his
decision. It would appear to have put paid to any further attempts
at a mediated resolution. The minister said he had no choice but to
issue the approval at that time.

With this approval, the Corrib gas developers moved back in
force to the pipeline landfall at Glengad, supported by the I-RMS
security firm. On Good Friday, the community in Rossport held its
annual peace walk. Afterwards, at Ballinaboy, P. J. Moran hitched
up the silver trailer, which had served as campaign canteen during
the 2006/7 pickets, and towed it home. There was nothing particularly symbolic about this, as his sister Mary recalled: 'He needed it
back! You know, PJ's kids grew up thinking that I lived in that
trailer – they even knew which key unlocked it on my keyring . . .'

*

Shell communications advisers couldn't quite believe it when a report from Bolivia flashed on to newsroom computer screens in Dublin. On 16 April, a young Irishman from County Tipperary had been shot dead in a hotel in Santa Cruz, along with two colleagues, Árpád Magyarosi, thirty-nine, a Romanian of Hungarian descent, and Eduardo Rózsa Flores, forty-nine, a Bolivian with Hungarian and Croatian passports, who had at one point led a group of foreign fighters in support of Croatian freedom during the Balkan wars.

The Irishman was Michael Dwyer. He was twenty-four and had spent the previous summer working on Shell security in north Mayo. Described as a 'very pleasant' student by Galway-Mayo Institute of Technology's building and civil engineering department, he had graduated the previous year in construction management. It emerged that his parents, Martin and Caroline Dwyer, in Ballinderry, County Tipperary, only learned of their eldest child's violent death in phone calls from journalists.

Initial reports suggested that their son had been involved in a plot by 'mercenary terrorists' to kill Bolivian president Evo Morales. There had been some unrest in the main wealthy Santa Cruz region over the Morales-led government support for land reform, nationalization of gas pipelines and redistribution of gas revenues. Bolivian police said that they been greeted with gunfire when they raided the Las Americas Hotel in Santa Cruz, where Dwyer was staying. Yet there was no convincing evidence of weapons in the rooms. The men appeared to have been sleeping, and the hotel owner denied that there was any shoot-out.

Dwyer, it was reported, was 'recruited' to travel to Bolivia by colleagues he had worked with in Erris during a six-month contract in 2008 with I-RMS on the Shell project. When his contract had ended in October, employment prospects for a construction management graduate were uncertain. He told his parents he was going to Bolivia in November for two months to complete a bodyguard course. The course never materialized, but in emails home he said he had been employed as a personal bodyguard for a guy who 'really just has me for show, mostly'. He had a girlfriend, Brazilian medical student Rafaela Moreira, and was planning to return to Ireland in late April.

Minister for Foreign Affairs Micheál Martin called for an international panel of inquiry, with Irish involvement, to investigate Dwyer's death. As the Department of Foreign Affairs liaised with the Bolivian authorities to ensure Dwyer's body was released and returned home, his family had to cope with press and Internet reports describing their son as a member of a racist right-wing separatist group. There were photographs of him in fatigues holding what resembled a gun. Images were reproduced of badges of duty bearing such slogans as 'Operation Glengad Beach' and 'Operation Solitaire Shield', which one of Dwyer's travelling companions had advertised on the Internet. A clip appeared on YouTube of Dwyer trying to remove Rossport Solidarity Camp supporter Eoin Ó Leidhin from fencing at Glengad in 2008.

'"Mercenary" was Corrib gas guard,' read the front-page headline on the *Western People* on 21 April 2009. Hungarian Tibor Revesz, who had worked for I-RMS as confirmed by the security company, was reported to have been the link between Dwyer and Flores. Both Flores and Revesz were members of the Szekler Legion, a Hungarian group seeking independence in Szeklerland, home to ethnic Hungarians in Romania. Many press reports agreed that it was unclear how much Dwyer knew of what he was involved in, given his language restrictions.

The press spotlight also turned on I-RMS, owned by former army rangers Terry Downes and Jim Farrell, and on the company's vetting of employees. The I-RMS website came down shortly after Dwyer's death. Cached material recorded that the company offered 'international armed and unarmed security', and there was reference to providing security in 'hostile environments'.[5] Dwyer and company were reported by film-maker Richie O'Donnell to have adopted the name 'Foireann Cahil/Charlie team', as per an

I-RMS security grouping, for their trip to Bolivia. The company was already the subject of complaints over alleged surveillance of the Henry family and Father Nallen, and this latest information caused serious anxiety among some of the residents living along the Sruwaddacon estuary. On 2 June 2006, *The Irish Times* reported that one of two Hungarians and a Slovak who had travelled with Dwyer were still working with I-RMS.

In early May, Judge Mary Devins said in Ballina district court that she could not ignore what she had read in the newspapers about the firm. Before her, once again, was Maura Harrington – this time on charges of assaulting a security guard at the Shell Glengad compound, and entering the building with intent to commit an offence. At previous court sittings, three Polish I-RMS staff had given evidence that Harrington had entered the compound and 'kicked one guard in the groin'. Judge Devins said she wanted to seek the view of the Director of Public Prosecutions as to the credibility of the company's witnesses.[6]

At the subsequent inquest in the Dublin coroner's court, I-RMS managing director Terry Downes said he had known nothing about the bodyguard course that his 'former employees' were attending until after news of Dwyer's death had broken. Bolivian ambassador to Britain and Ireland Beatrice Suveron claimed that the group leader, Flores, who had died in the police raid, was being paid by Santa Cruz businessmen who supported independence from Bolivia for Santa Cruz. She said that they had already caused an explosion at the home of Santa Cruz's Roman Catholic cardinal.

The Bolivian authorities' post-mortem conclusion that Dwyer had been shot six times was disputed by state pathologist Dr Marie Cassidy. She pointed to shortcomings in tests that appeared to show gunshot residue on Dwyer's hands, and she said that he had been shot once through the heart with a bullet designed to cause massive internal injuries. He was 'most likely' in bed at the time, she said.[7]

For the Dwyer family – his parents and three younger siblings – the inquest was closure of sorts, according to Carl O'Brien, who interviewed them afterwards for *The Irish Times*. They felt the full truth would never be established unless there was an international investigation. No one in the family had been able to sleep properly, and their struggle to grieve and to come to terms with their loss had been made all the more difficult by the fact that it was 'so public'.[8]

Continued media coverage, and a particular focus by Shell to Sea on the issue, would spur *Sunday Tribune* journalist Michael Clifford to write an impassioned piece three months later in which he criticized both Indymedia and the Shell to Sea campaign for their approach. He questioned how a 'citizen journalist' website, which 'does much that should be commended' and regularly 'decried' the 'right-wing corporate media', could cite two such mainstream publications as source material for a 'long article' that was 'littered with suggestive links between Dwyer and fascist organizations. For the sake of cheap expediency, elements within the protest and its wider support base are using a dead citizen to further their own agenda, eschewing any standards of decency towards the bereaved family,' he wrote. 'That lonely outcrop, the high moral ground, is in danger of being totally abandoned in the dispute over the Corrib gas field . . .' Clifford suggested that more 'humane elements' within the protest who would 'distance themselves from such stuff' should 'forcibly' make their views known.[9]

Mary Corduff would have been among the 'humane elements', but she was with Pobal Chill Chomáin, which had already tried to distance itself from Shell to Sea – and she had much on her mind in the weeks following Dwyer's death. Her own husband Willie had been hospitalized after the alleged assault at Glengad on the night of 22/23 April, the night of his protest under the chassis of a Shell-contracted delivery truck. The events of that night had occurred just a week after the shootings six thousand miles away.

It had begun with a row over authorization for Shell work at Glengad. Corduff and neighbour Gerry Bourke weren't happy with what they had been told, and chose to protest under the parked vehicle. A factsheet given to Mary Corduff at Belmullet Garda Station said that landfall fencing works at Glengad were required temporarily for construction activities on a site 'pursuant to a development which is exempted development'. However, it said that 'no works will take place on the onshore pipeline until the process is complete, notwithstanding this [*sic*] the original onshore pipeline consent is still valid'. Monica Muller, who, like Bríd McGarry, had become an expert on the legal technicalities of the Corrib project, didn't agree.

Nor did An Taisce in Dublin. As the organization pointed out, Bord Gáis had applied for planning permission for ancillary works when constructing the Galway–Mayo gas pipeline. An Taisce's reading of the situation was that ancillary works could only be undertaken 'pursuant' to planning permission, which Shell was applying for under the Strategic Infrastructure Act. However, the developer appeared to be relying on the fact that the original onshore pipeline had been exempted development.

'They must have come up from the beach, and they were wearing dark clothing and night-vision gear, because I neither saw nor heard them,' Willie Corduff said, several days later. 'First thing they did was to hit me with a baton, which dazed me. Four of them pressed me to the ground so I couldn't breathe, and one guy had his knee in my ear and had my hands up behind my back. They'd release my hands every so often to see if I was getting weaker. They were highly trained, knowing how to cause maximum pain with minimum marks to the body.

'The bruises only came up hours later. And it was only when I stopped struggling and lay still that they stopped. I heard one fella say to the other, "Leave him now, he's nearly gone." They had a paramedic with them who put his face to me and repeatedly asked me if I was refusing his offer of help. I was afraid of him – he smelled of drink. It was then that I knew that the gardaí were there – perhaps someone had called them. I heard one of the gardaí saying, "Willie, you're at your old tricks again." My son was outside and roaring to be let in to help me. They wouldn't let him near, until the ambulance came.'

The 'breaking news' on RTÉ Radio's *Morning Ireland* just before 9 a.m. on 23 April, and on *The Irish Times* website, was not focused on Corduff. It focused on a Garda report that up to fifteen people, wearing balaclavas and carrying tools, bars and chains, had vandalized the compound at Glengad, starting up a mechanical digger and damaging a fence and gate, and injuring a security guard's wrist.

Chief Superintendent Tony McNamara of the Mayo Division was quoted as stating that the incident appeared to be 'well planned' and was 'an almost military-style operation'. Later radio reports would upgrade this to 'paramilitary precision'. The incident was said to have started at about 11.30 p.m. when there was a 'small number of security staff on duty' and they 'withdrew because they could not deal with the numbers in the masked group'.

The group were said to have been carrying bolt cutters, iron bars and chains. 'This is a serious escalation. We've never had people wearing balaclavas and so on breaking into the site in a group with implements,' one Garda source told *The Irish Times*. Superintendent Michael Larkin of Belmullet said that arrests could be 'expected'.

The superintendent, who said that he was appealing to local people to 'take a step back' and use the ministerial forum for any complaints, spoke to Teresa O'Malley of MidWest Radio on 23 April. He could not confirm if gardaí at Glengad were present during what he described as 'thuggery' in damaging equipment, but said that no arrests were made at the time as 'circumstances didn't permit'. Asked by O'Malley to comment on injuries inflicted on Willie Corduff, Larkin said he could not comment on specifics. However, he said that a 'well-known protester' had been escorted from the Glengad site early in the morning and was transferred to hospital as he was complaining of feeling 'unwell'. The Garda press office statement of 23 April said that 'A protester who was present at the site yesterday [Wednesday, 22 April] was this morning [Thursday, 23 April] removed from the compound by security staff' and 'was taken to hospital by ambulance as a precaution'.[10]

Pobal Chill Chomáin spokesman John Monaghan, living near Glengad, disputed Garda assertions that there had been an incursion by 'armed' people. He said that it was his understanding that a 'handful' of local people were gathered outside the Glengad compound gate on the night that Corduff was inside and under the truck. His brother-in-law Peter Lavelle was also inside the compound to provide moral support. The group outside had 'unravelled' some fencing, Monaghan said.

The damage to fencing, clearly visible on television cameras, might have been linked to the noise Corduff said he had heard some time during his long night vigil. In a subsequent television documentary, Niall Harnett of Shell to Sea acknowledged there had been some action. The various statements begged many questions about how long the gardaí were present, where the security staff were, and how they knew about, but failed to prevent, a reported break-in conducted with 'military' or 'paramilitary' thoroughness.

Solitaire had been booked again to make another attempt at laying the offshore pipeline, and clearly the developer didn't intend to wait for a Bord Pleanála hearing outcome to conduct the preparatory works on and offshore. If legally questionable, it was also morally dubious. A company insisting on its willingness to talk to and work with the community appeared to be far less interested in doing so on the ground. Notice of the return to work at Glengad had only been given to the Department of Energy on Tuesday, 21 April . . .

As the fall-out from Michael Dwyer's death continued, the Corduff family was also seeking answers. Willie Corduff was reluctant to make a formal complaint to the gardaí about his assault, as he no longer trusted their independence. Chief Superintendent Tony McNamara said that they would be happy to investigate, if he would make a complaint. On the evening of 23 April, former UN assistant secretary general Denis Halliday sent a message on his BlackBerry phone from Jakarta in Indonesia to Joe Murray of Afri in Dublin:

I applaud the courage and commitment to non-violent resistance shown by Mayo farmer Willie Corduff in respect of corporate giant Shell stomping on the human rights of his community concerned with environmental damage and the welfare of their families threatened by natural gas-linked danger. As in Nigeria, Shell appears to show no respect for humanity when greed-driven profits demand feeding.

Sadly, the continuing situation in Mayo raised doubts about the ability and commitment of the government to find a solution that best serves the interests of all Irish people. I do not believe those hoping to benefit from north-west natural gas would wish to do so, were they informed of the burden Shell seems prepared to place on this Mayo community.

As statements go, it was too strong for most newsrooms. Afri incorporated some of his sentiments in its own comment, reminding the public that Corduff had been given the Goldman environmental award two years before. The alleged break-in, if proven to have occurred, it said, was 'totally unacceptable and damages what is a credible, popular and peaceful campaign. However, there is an important difference between the actions of protesters and the duties of security forces charged with ensuring public safety, even in extreme circumstances. One does not cancel out the other.' It called on the government to initiate a 'thorough investigation', and to ensure in the meantime that security guards who might have been involved were removed from their post in Glengad, pending the investigation's outcome.

Nobel Peace Prize laureate Archbishop Desmond Tutu, a patron of Afri, expressed his support for a 'national and international investigation', and for steps to be taken to ensure that 'all parties, particularly those representing the state, demonstrate a commitment to non-violence and the protection of human rights'. His statement would be followed several days later by an appeal to President Mary McAleese, Taoiseach Brian Cowen and Norwegian Prime Minister Jens Stoltenberg to intervene in the Corrib gas conflict, as articulated in an open letter sent to all three by twenty-six recipients of the Goldman international environmental award. Signatories included Sven 'Bobby' Peek, who won the accolade in 1999 for environmental activism in Durban, South Africa, and 2001 winner Jane Akre, a North American journalist who had lost her job with Fox News following her reports on genetically modified growth hormones injected into cattle.

After a local meeting on the night of 24 April, four groups that had differed over the past twelve months in their approach to Corrib – the two Pobals, the Rossport Solidarity Camp and Shell to Sea (Mayo) – issued a joint statement. They said that it was 'with great regret' that they had witnessed a 'disproportionate response' from An Garda Síochána 'in relation to the savage and unprovoked attacks on several local people, including local farmer and Rossport Five member Willie Corduff' at Glengad. 'Rather than gardaí taking immediate action on an incident that could have led to Willie's death, the public were treated to a circus of contradictory and inconsistent statements . . . none of which were willing to take seriously the sinister assault on Mr Corduff. Our community has now lost total faith in the ability of An Garda Síochána to discharge their duties in a fair and unbiased manner when it comes to the Corrib gas project,' the four groups said, and they called on Minister for Justice Dermot Ahern to 'immediately facilitate an internationally supervised investigation into this incident, so that peace and order may be restored to County Mayo and to Kilcommon parish'.

In his Galway constituency, Minister for Community, Rural and Gaeltacht Affairs Éamon Ó Cuív said he was keeping an 'open mind' on the circumstances surrounding the events of 23 April. He intimated that he knew enough of the Corrib situation to be very careful about any information given to him, official or otherwise. Both he and cabinet colleague Eamon Ryan agreed to face the community at what turned out to be a highly charged meeting in Inver on the night of 30 April.

Several nights before, satirist and musician Paddy Cullivan, better known as Clint Velour of the resident Camembert Quartet on RTÉ Television's *The Tubridy Show*, warmed up an audience for a Leviathan political debate at Galway's annual Cúirt international festival of literature. Cullivan presented his view of a future Ireland from space. Galway would be one big motor roundabout, Killarney would be a hotel, he said, to much laughter. And Mayo would be known as the 'Shell exclusion zone . . .'

Several nights later, *Mayo News* reporter Áine Ryan wrote of 'palpable' community anger at a meeting with the two ministers in Inver community hall:

> One after the other – members of the 300 and upwards throng – categorically told two senior government ministers what they thought about them, the Corrib project, Shell, the gardaí, and I-RMS. They also warned Ministers Éamon Ó Cuív and Eamon Ryan that if they didn't do something to stop the controversial development, they would. This was despite repeated attempts by the two ministers to state that, as government representatives, they must always act within the law.
>
> Feelings of isolation and betrayal by the powers of the state were emotionally expressed.[11]

People who had never been used to raising their voice did so now,

such was the level of concern. Colm Henry of Glengad described
the official response to his reports of Shell security surveillance of
his family and home. Speaking directly to Ó Cuív, Henry said, 'My
grandfather was in jail with your grandfather to protect the very
thing that you are giving away . . .'

Willie Corduff dropped his trousers to show the two ministers
his injuries. There was a misunderstanding between Ó Cuív and
Naoise Ó Mongáin, who thought the minister was making some
sort of sexual allusion about his wife. As a Green Party minister,
Eamon Ryan was singled out for particular criticism, and he was
asked why the Aarhus Convention, affording the public access to
environmental information, had still not been ratified.

'The issue of us stopping the project is a legal impossibility,'
Ó Cuív said.

Local eco-tourism operator Anthony Irwin of Pobal Le Chéile
summed up the damage that governments over the past nine years
had wrought through their handling of Corrib. 'For decades, two or
three gardaí policed this community without any trouble. And now
three hundred gardaí can't control it.'[12]

The two ministers were in Belmullet the following morning for
another meeting of the forum – yet again closed to the press. Shell
and the various organs of state, the Garda, several county
councillors, the Pro-Mayo Gas group and others were present, but
the two Pobals, Shell to Sea and the Rossport Solidarity Camp continued their boycott, believing it to be a 'talking shop' with no
statutory function and no remit specifically linked to the project's
safety. Chairman Joe Brosnan made it clear that the door was always
open.

Some fifty locals were down at Glengad beach, however, for a
spontaneous protest. They were outnumbered by gardaí and Shell
security. Speaking to *The Irish Times* outside the forum, Ó Cuív
compared the situation to Northern Ireland during the 'troubles'. It
was 'as difficult a problem as I've ever seen' and with 'as deep a
division as I've ever seen'.[13] Ironically, far south several days later,

Shell's head of its Africa division said that the company was 'ready to work with' the community in Ogoniland.

The following weekend, seven people were arrested during a three-hour protest by up to a hundred people at Glengad. Once again, gardaí and security outnumbered protesters, who included James Monaghan, one of three Irishmen convicted in Colombia of training Farc rebels. Monaghan, walking on his own, ignored the photographers. He appeared to have aged since the many images published of him during that time.

Inside the Glengad compound, it was like a military base. A Garda command and control unit was backed up by twelve Garda security vans, two four-wheel drives and a transporter. Uniformed gardaí had also set up checkpoints on access routes to Pollathomas and Glengad, taking names and preventing some drivers, including local residents, from using the road for the duration of the protest. Eoin Ó Leidhin of the Rossport Solidarity Camp said that he was punched in the stomach by a garda, and another man suffered minor injuries, when a handful of protesters, including two young women, tried to secure a rope to the compound fencing. The protest petered out in the evening sunshine.

Early the following morning, Sunday, 10 May, an event took place that stopped everyone with any view on the conflict in their tracks. Garda Terence Dever and two of his colleagues were travelling from Achill Island in two private cars for a 6 a.m. shift at Belmullet Garda Station when they were involved in a head-on collision with a vehicle travelling in the opposite direction.

Garda Dever, a father of three who was very involved in community, sport and music in Saulia, Achill, died, and one of his colleagues, Garda Eneas McNulty, also from Achill, was injured. The driver of the oncoming car also died. Stephen Conway, aged nineteen, the youngest son of Val and Kathleen Conway of Inver, had recently finished school and was working in a restaurant in Belmullet. He was a nephew of Terence Conway, the popular 'wood butcher' or carpenter, who had returned to Ireland from the US in 2000 and had become a stalwart of Shell to Sea when he returned to Mayo in 2004.

Initial radio reports suggested some link with the previous evening's protest. However, Garda Dever had not been policing the protest at Glengad, and Stephen Conway had not been protesting as he might normally have done. His uncle recalled that he had said he didn't want to risk getting arrested at Glengad, as he might then miss a good Saturday night out with friends.

Garda Dever and Stephen Conway were buried in the same county on the same day. Stephen Conway's brief but fruitful life was recalled at his funeral mass in Inver by parish priest Father Edward Rogan, while Garda Dever was described as an 'extraordinarily gifted community builder' by Dr Michael Neary, Archbishop of Tuam, at his mass on Achill Sound. He had been a member of the 'A Team' at Belmullet Garda Station, the large attendance was told. His colleague, Garda Eneas McNulty, attended in a wheelchair with medical support.

A preliminary session of An Bord Pleanála's oral hearing into the Corrib gas onshore pipeline was deferred as a mark of respect. When it opened on 19 May 2009, in the Broadhaven Bay Hotel in Belmullet, there was no sign of the Garda presence that had characterized the Environmental Protection Agency's sessions two years before. However, there had been protest the night before at Ballyglass pier. Plans by Shell E&P Ireland to resume attempts to lay the offshore pipeline with *Solitaire* in advance of onshore approval had hit another obstacle, with failure to secure a new agreement with Erris fishermen.

Erris Inshore Fishermen's Association chairman Eddie Diver was not at all happy. He said the company had not so far lived up to its promise in relation to discharges from an outfall pipe, as no application had been lodged to revise its emissions licence. 'This is not about compensation or money – there is a principle involved here,' Diver told *The Irish Times*. Fishermen would remain at all times within the law, but would continue to fish in Broadhaven Bay, as was their legal right, he said.[14]

In the lead-up to the key hearing, Shell had published an engineering review, which it said it had commissioned the previous year from consultants Arup. The review rejected the Glinsk option for a coastal refinery as proposed by the three priests and the two Pobals.

Once again, there was a large team on behalf of the Corrib gas partners, including senior counsel Esmonde Keane, at the opening session of the hearing, chaired by inspector Martin Nolan. Also present for some of it was Denise Horan, former editor of the *Mayo News*, who had taken up a post as Shell senior communications adviser in place of Susan Shannon. Horan, a distinguished sportswoman, had represented Mayo on four All-Ireland football teams. She was the latest of a number of journalists recruited by the company, but the appointment still came as a surprise to many of her colleagues.

In one corner of the ring sat the developers and a plethora of consultants and advisers. An Bord Pleanála had received forty-five submissions, and seven comments on the onshore pipeline application. Pro-project submissions ranged from Goodbody Economic Consultants to Chambers Ireland to Engineers Ireland. Local supporters included Pro-Gas Mayo, the Council for the West, Belmullet GAA club, Erris Chamber of Commerce and consenting Rossport landowner Padraig McGrath, who had written a joint letter with former Fianna Fáil councillor Paraic Cosgrove endorsing the project.

'We believe that Shell are doing their utmost to guarantee absolute safety,' McGrath and Cosgrove wrote. 'We have visited and studied other similar pipelines in the Netherlands and are satisfied that the safety distance of the pipeline from housing is more than adequate.' They did not state that Shell had hosted this visit.

In the other corner, a number of groups and residents included Bríd McGarry and Brendan Philbin, Monica Muller and Peter Sweetman, the Henrys, Niall King, Imelda and Ed Moran, Pobal Chill Chomáin appellants Vincent McGrath, John Monaghan and Micheál Ó Seighin, their consultant Leo Corcoran, Pobal Le Chéile's Anthony Irwin and Ciarán Ó Murchú, and members of the Rossport Solidarity Camp, including St John Ó Donnabháin and Eoin Ó Leidhin. An Taisce had also lodged an appeal, as had Maura Harrington, who was in jail for the opening forays.

The inspector introduced himself, and his team, which included senior inspector Stephen O'Sullivan, British pipeline expert Nigel Wright and geotechnical engineering expert Conor O'Donnell. The Health and Safety Authority had submitted a letter saying it had no remit in relation to the pipeline's safety. The agency informed the appeals board that off-site gas pipelines were not controlled by the Control of Major Accident Hazard Regulations 2006. The residents were upset, but not one bit surprised at this. Once again, it would appear that no state department had overall responsibility for residents' safety. If they were all employees working on the project, it would be a different matter.

Several of the appellants sought unsuccessfully on that first morning to clarify the extent of An Bord Pleanála's jurisdiction. An onshore section at the Glengad landfall had been deemed exempt from planning by the Department of the Environment in 2008. Yet the developers had said that all of the onshore section would be submitted for planning approval – the first time that the high-pressure pipeline would be subject to such scrutiny. Philbin and McGarry in particular wished to know if the entire onshore pipeline was then under An Bord Pleanála's remit for the planning application.

Micheál Ó Seighin stressed the importance of clarification. It would decide whether or not people lost their land to the pipeline in a few months' time. Philbin and McGarry weren't happy with the inspector's response, and withdrew. The strain of the past nine years had begun to take its toll: Bríd McGarry, who had spent many midnight hours on her preparatory work for the hearing while trying to run a farm, left in tears. Yet the subsequent Bord Pleanála ruling would vindicate her.[15]

RPS technical director Ciarán Butler told the hearing's opening session that the route met all relevant codes and standards, and selection was open and transparent. It ran a minimum 140 metres from occupied housing, twice the minimum separation distance than the original route, he said. Describing the selection procedure, Butler said that it included a more complex land valve installation at Glengad.

A 'fail safe' isolation valve had been designed to comply with all relevant codes and standards as part of a commitment after the Advantica report publication in 2006 to limit pressure in the pipeline to 144 bar. Overall the proposed new pipeline route achieved the 'optimum balance of community, environmental and technical criteria', Butler said, and this was backed up by his RPS colleagues.

Glengad residents Colm and Gabrielle Henry weren't so impressed with Butler's 'optimum balance'. Henry expressed serious concerns about infringement of basic rights and privacy, along with health and safety impacts of the land valve installation, which was due to be built less than a quarter of a mile from the couple's home. Their house had only been included in a second environmental impact statement submitted to An Bord Pleanála.

The area had already been subject to severe landslides from Dooncarton mountain, just above Glengad, in 2003, Henry reminded the hearing. Planned tunnelling by Shell under an estuary crossing below the mountain could 'further destabilize' it, he said. And the land valve installation would include 'flood lighting, security fencing, CCTV equipment and around-the-clock armed security presence' just down from his house.

It was on the sixth day that Pobal Chill Chomáin produced a witness who would question the RPS safety assertions. A retired army bomb disposal officer, Commandant Patrick Boyle, had served in Ireland and with the United Nations in Lebanon. He warned of 'horrific' consequences for civilians in and around Rossport if an onshore gas pipeline serving the Corrib gas project should rupture and explode. A hospital with a specialized burns unit and a fire station would have to be located in the area, due to the potential for fatalities and serious injury in a very isolated area, he said.

Boyle said that many recent pipeline accidents had occurred with pressure loads of 70 bar – half that proposed for the Corrib gas onshore pipeline. He cited as an example the July 2004 explosion in Ghislenghien, Belgium, in which twenty-four people had died and more than 120 were injured. Most of those killed were police and fire-fighters responding to reports of a gas leak in a pipe operated by Fluxys, a pipeline operator owned by Royal Dutch Shell.

Explosions caused by released fuel mixed with air had a multiple factor, Boyle said. The Ghislenghien explosion was equivalent to forty-one tonnes of TNT, and similar to the impact of smaller tactical nuclear weapons. He said that a separation distance of at least five hundred metres from dwellings would be more appropriate than the currently proposed 140 metres.

Former Bord Gáis engineering manager Leo Corcoran, also representing Pobal Chill Chomáin, said that the project was a case study in 'escalation of commitment' by developers who continue to build 'beyond the point of failure'. Corcoran, who had been a member of the Gas Technical Standards Committee, identified a number of difficulties with the current application, and reiterated, as he had so many times before, that no code of practice for a high-pressure gas pipeline was mandated in the ministerial consents for the onshore and offshore pipelines issued in 2002. The site selected for the gas refinery in Ballinaboy was also 'in breach' of the code, and located within a drinking water catchment area of ten thousand people, Corcoran said. The site for the land valve installation at the Glengad landfall, linking the offshore and onshore pipelines, was an 'insecure location' and 'socially unsustainable', as it was close to housing. Without the support of the local community, it would be 'difficult to enforce the invasive measures required to secure this facility', he pointed out.

At this stage, Micheál Ó Seighin knew the project inside out, and had earned the grudging respect of his opponents. Once again, he was able to engage humour with devastating effect. When Peter Cassells's report was cited, Ó Seighin remarked that the prospect of Cassells holding technical/engineering expertise in high-pressure gas pipelines was as likely as Shell consultant/RPS chief executive P. J. Rudden holding the professorship of medieval music at the University of Heidelberg. Rudden found it hilarious; he told Ó Seighin afterwards that his family would find it funny too, because he hadn't a musical note in his head.

All pipelines were designed and built to be as safe as possible, Ó Seighin told the hearing. 'But pipelines still fail – *Lloyd's List* is just one record of such, with 150 failures from July 2002 to July 2008 – for reasons known only after the event,' he said. There to back him up was ninety-year-old former diver Des Branigan of Marine Research Teo in Dublin, with a decade of accidents recorded by Lloyd's Intelligence Unit.

The *Lloyd's* statistics from 1999 to 2001 made depressing reading. Two ten-year-old boys and a teenager had died from injuries sustained in Washington, USA, in June 1999, when a pipe rupture fuelled a fireball in a park. An eighteen-year-old fisherman died after fumes overcame him in the same incident. There was the damage caused by a 1.29-million-tonne oil leak from a pipe run by Petrobras into Guanabara Bay, Rio de Janeiro, on 10 February 2000, which wiped out the fishing operations of five local villages. There was the gas pipeline fire at Lancaster, Pennsylvania, in July 2001, which fuelled flames of up to a hundred feet high. There was the evacuation of twenty thousand people in southern China in September 2001 when a fire ignited by a welding accident at an oil facility injured six workers, and witnesses described 150-foot-high flames. There was the gunman who shot a hole in the trans-Alaska oil pipeline in October 2001, causing a major leak and halting the supply of up to one-fifth of North American domestic production. In total, the Lloyd's Intelligence Unit data submitted by Branigan recorded 1,200 deaths in the past decade to 2008, as a result of pipeline fractures in fifty-eight countries.

The hearing had convened again after the June bank holiday weekend when Shell and its team of consultants took the 'stand' again, but this time to answer questions from one of the board's two experts, Nigel Wright. The consultant and former British Gas engineer was a big man, who appeared to tower over the men in suits from Shell, RPS, J. P. Kenny and Det Norske Veritas. There were few enough visitors to the hearing at this stage, apart from the appellant, and you could have heard a pin drop as Wright began his analysis.

Wright had a long list of technical queries. The relevant expert would reply. Wright would rephrase the question. The consultant would reply. For every second response, it seemed that Wright wasn't too happy. Eventually he would rephrase the response, mindful of a non-technical audience. The picture created in the discourse was a little unsettling. The consultants conceded that safe shelter in the event of a rupture and explosion had not been identified for residents living close to the proposed Corrib gas onshore pipeline. What was more, houses within 230 metres of the pipeline could 'burn spontaneously' from heat radiation if gas in the pipe was at full pressure when a rupture occurred. Residents would have just thirty seconds to escape from thermal radiation, the team acknowledged.

Houses within 171 metres would be at risk if the gas pressure was at 144 bar – the level agreed by the developers after the Advantica review. Both 171 metres and 230 metres were greater than the minimum separation distance of 140 metres from houses allowed for in the pipeline route, Wright observed. The Corrib project team told him the 'assumption is that there will be shelter'.

The consultants conceded that this particular pipeline was 'unique'. And why had the onshore pipeline beyond the landfall been designed for 345 bar when it had been agreed to limit pressure to 144 bar onshore? This was 'appropriate to prudent design', they said. The integrity management system also allowed for 'small leaks; past valves in the network'. Wright outlined a leakage scenario. If pressure crept up in the system, would flaring to release it be used? It would, the team acknowledged.

Wright referred back to the Advantica safety report. Why was a recommendation in that study on the use of an isolation joint at the land valve installation at Glengad not being applied? A consultant said that such joints created 'inherent weaknesses'. Wright wanted to know more about the security of the beach valve, and why potential damage in a terrorist attack had been excluded from the quantified risk assessment (QRA) submitted. This information was 'not part of a normal QRA', he was told.

Wright referred to evidence given by the former army bomb disposal expert Commandant Patrick Boyle on the impact of the July 2004 gas pipeline explosion in Ghislenghien, Belgium. Classification society Det Norkse Veritas, for Shell, confirmed that modelling for such accidents was drawn from British data. Actual testing had not been carried out for cost reasons. Questioned on state responsibility for the safety of staff at the land valve installation, the team said that a document would be sent to the Health and Safety Authority (HSA). However, the HSA had already informed An Bord Pleanála that it had no remit in relation to the pipeline's overall safety.

The body language said everything. The consultants were not comfortable. It was to prove to be a significant day. 'One of the big challenges in discussing the Corrib gas issue is separating fact from fiction,' Dr Mark Garavan wrote in a letter to *The Irish Times*.

The Corrib project in its present configuration should be opposed because it constitutes an unacceptable risk to the health, safety and environment of a small community in County Mayo. How can we prove that it does constitute such a risk? The answer lies with Shell's own evidence to the Bord Pleanála oral hearing presently under way in Belmullet.

There are houses within this zone with families and children. This is dramatic evidence finally confirming facts previously denied by Shell and their apologists. I recall then Minister Dempsey stating definitely in 2006 that people were safe beyond three metres in the event of the pipe rupturing.[16]

There would be more long days, and then one of the longest of all. On 11 June, the two community groups, Pobal Chill Chomáin and Pobal Le Chéile, informed the inspector they would have to withdraw temporarily from the hearing in protest. Fisherman Pat O'Donnell and a crewman had been rescued early that morning. Their boat had been boarded, they said, and sunk while they had been tending gear in Broadhaven Bay.

16

'A VERY DANGEROUS GAME'

'I was there watching my gear since about eleven or twelve o'clock. We'd gone out a few nights before because I heard there was some boat going to come in, and damage my gear. It was only a rumour, but you see they arrested me last summer but they couldn't prosecute me.[1] We, myself and Martin McDonnell, were there for about two hours and next thing I heard a noise behind me. There were four of them, masked and in wetsuits. Two of them had guns and they told us to be quiet.'

It was a couple of days after Pat O'Donnell and his crewman, Martin McDonnell, had taken to a life raft in the early hours of 11 June 2009, and O'Donnell was speaking to Áine Ryan of the *Mayo News.*

'Two of them went down below – what they were doing I don't know . . . looking for sea cocks or busting out a plank. Of course I was frightened. I was worried for my life and for Martin's. I was thinking of my family, but I don't want to talk about that. When they came back up they stayed with us for a while. If they wanted to drown us, why didn't they take the life raft, or why didn't they

313

throw us overboard? I could say they wanted us dead, but I don't think so. I think they just wanted to give me one hell of a fright.

'I noticed the boat's roll slowing and getting heavy. When the bandits left our boat, I looked below and there was water covering the engine. I told Martin to get his lifejacket on. I had to go to the top of the wheelhouse and cut off the dinghy. I had a problem launching it and I also had to cut the paddles out of their plastic cover. I had to grab a handheld VHF and lifejacket for myself. I knew we'd better get away from the boat and paddle like hell because I was afraid we'd get sucked under by the boat sinking.

'After I made the mayday call and gave an estimate of my co-ordinates, I rang Belmullet Garda Station and told a female garda what had happened and that they had escaped north. Next thing, the gardaí had arrived and [skipper] John Healy had arrived in *Rachel Mary* [an O'Donnell-owned boat]. John Healy had nothing to do with the protest. In fact, we never talk about it. *Rachel Mary* had been steaming from Ballyglass harbour to her fishing grounds further west of Erris, as she does every morning during the summer, weather permitting. I could be fishing over there at Glengad because I have a licence to fish there. But I won't make a fool of myself. I haven't been to sea since. I'm shook up and I'm frightened. I need another few days to pull myself together.'[2]

Minister for Community, Rural and Gaeltacht Affairs Éamon Ó Cuív supported a call, initiated by Green Party justice and marine spokesman Ciarán Cuffe, for a full investigation into the sinking of O'Donnell's boat, *Iona Isle*, off Erris Head.

Coincidentally, there had been some damage reported at a quarry supplying the Corrib project the day before the sinking. Tyres had been slashed and hydraulic pipes cut on vehicles, it was reported, with damage estimated at around €9,000, on the night and following morning of 9 and 10 June. There was no evidence of any protester involvement in this, but by now some of the community suspected there might be an *agent provocateur* in their midst.

On the morning of *Iona Isle*'s sinking, gardaí had arrested six of a group of fourteen protesters in kayaks, mainly from the Rossport Solidarity Camp, who had taken to the water from Glengad at about 4.30 a.m. to try to board dredgers working in Broadhaven Bay. The camp had only planned this the night before, and Eoin Ó Leidhin recalled afterwards that they had no idea of the activity further out to sea, which had led to the sinking of *Iona Isle*.

O'Donnell and McDonnell were taken to Mayo General Hospital after they were brought ashore, latterly by the Ballyglass lifeboat. They had been given dry clothing by the lifeboat crew. In response to press queries about a Garda investigation, Chief Superintendent Tony McNamara said that the skipper was 'not cooperating'. O'Donnell countered that he would be cooperating through his legal representatives, but pointed out that he and his crewman were still in a state of shock when they were met at the hospital in Castlebar by four detectives and a uniformed Garda who had asked them to hand over their clothes.

Shell E&P Ireland issued a statement on 11 June in which it said it 'emphatically' rejected 'the allegation that people employed on the Corrib gas project were involved in any way in the incident, which led to the sinking of the *Iona Isle* fishing vessel in Broadhaven Bay early this morning'.

> The location where the fishing boat sank is approximately 10 miles away from our worksite at Glengad. All boats and personnel – both operational and security – working in the bay on SEPIL's [Shell E&P Ireland's] behalf in the early hours of this morning are accounted for and none was in the vicinity of the *Iona Isle* when it sank. In fact, three vessels working for the project responded to the distress signal.
>
> A number of malicious allegations have been made against SEPIL and its security contractors in recent weeks. These claims

are designed to cast doubt on the integrity of the project and the personnel working on it, but have no basis in fact. To our knowledge no complaints have been made to the gardaí.

In addition the Corrib project and some local suppliers have in the past two months been subjected to acts of criminal damage and lawlessness by protesters, most of whom are from outside the Erris area. The increasingly reckless attempts to frustrate the works are putting staff, contractors, the gardaí and the pro-testers in potential danger.

The Corrib gas project, which is currently employing more than 1,500 people in Mayo, Donegal and Dublin is of strategic importance to Ireland. Despite the repeated attempts to disrupt, SEPIL is committed to completing work on the Corrib gas project, for which all the necessary permits and consents have been granted, in a safe and environmentally sustainable manner.

Eddie Diver of the Erris Inshore Fishermen's Association and Ciarán Ó Murchú of Pobal Le Chéile weren't happy with the response by both Shell and the state. Diver described the sinking as 'sinister'. In a statement, he said it was 'not for us to apportion blame, point the finger in any one direction or pass judgement on who is or was responsible for this drastic act. Suffice it to say that it is not acceptable to us that a vessel has been sunk; that its owner, a member of our organization, is at a considerable loss and that two lives have been recklessly endangered.'

Ciarán Ó Murchú spoke to the newspapers and on Raidió na Gaeltachta. Shell was 'trying to sow the seeds of doubt' in relation to responsibility for the sinking, and similar 'seeds of doubt' had been sown in relation to the April assault on Willie Corduff, Ó Murchú emphasized. 'If you had to choose between believing a company which has been found guilty of misleading its shareholders, and members of my community who never had criminal records before this project, I know who I would believe,' Ó Murchú said.[3]

Seeds were being sown – Martin McDonnell wasn't a regular crewman and had panicked during the sinking, it was said. It was rumoured that the *Iona Isle* had been 'stripped down' before it went to sea – in fact, two of the boat's regular crew were on holidays and O'Donnell confirmed that a pot hauler had been taken off. He was aware of talk of insurance scams in an industry struggling to survive. 'If I had said it was an accident, I would have an insurance claim,' he said later. The twelve-metre boat had been insured for €60,000, but there was no cover for loss defined as an 'act of terrorism'. As far as residents like Bríd McGarry were concerned, a fisherman with Pat O'Donnell's reputation would never sink his own boat, and it would be considered very bad luck to do so.[4]

Pat O'Donnell travelled to Dublin to make his statement to gardaí, accompanied by his solicitor. Labour Party president Michael D. Higgins expressed his concern about the 'deteriorating situation' in Erris, with a fisherman's boat sunk, and with the developer bringing in a large offshore pipelaying vessel while an oral hearing on the onshore section was still deliberating. Higgins pointed out that O'Donnell had expressed fears about his own safety more than a month before, when he had refused to sign up to an agreement eventually brokered between Shell and the Erris Inshore Fishermen's Association.

'Accurate information' was required for 'establishment of fact', Higgins said, and he asked Taoiseach Brian Cowen for information on 'the status of the existing legislation on the practice of private security companies and what, specifically, is their relationship with the Garda commissioner'.

Minister Éamon Ó Cuív, who had met members of Galway Shell to Sea earlier in the month on their concerns about the private security issue, confirmed that the chief executive of the Private Security Authority Association of Ireland, Geraldine Larkin, would attend a session of the government's north-west Mayo forum in Belmullet. This was not what Galway Shell to Sea had looked for in the discussions with Ó Cuív, as they had difficulty with the forum's terms of reference. However, it seemed that this was the best they were going to get.

The forum convened on 22 June in the Broadhaven Bay Hotel, and for the first time it was open to the press. Journalists were placed on a platform overlooking the cast seated at the long rectangular table – Shell's Terry Nolan, Chief Superintendent Tony McNamara of the Mayo Garda Division and Belmullet superintendent Michael Larkin, Mayo county manager Des Mahon and several councillors. A slightly hesitant security authority head said that the company providing the security for Corrib, which she did not identify, was the 'most monitored' in the state by her organization. It had about 156 staff based currently at Corrib. The authority was mandating all security staff in the state to display badges from September on, but I-RMS had agreed to use the badges 'before this'.

None of those present raised the death of young Michael Dwyer, and the controversial links of some of his former I-RMS colleagues. However, as if in anticipation of a question, Larkin said that staff who had been living more than six months outside Ireland were required to provide a criminal record certificate for every jurisdiction they had been in. An individual could work for six weeks while awaiting vetting approval, she added.[5]

Before the end of the month there was another significant development, when US oil company Marathon sold its share in Corrib to Canadian rival Vermilion in a deal potentially worth US$400 million. As Barry O'Halloran reported in *The Irish Times* on 26 June 2009, the deal valued the Corrib gas field at that point, even before it had produced gas, at just over US$540 million. Even as all this activity was taking place, the *Western People* noted that a Pollathomas village-enhancement scheme had been disbanded because of Shell's involvement; and that oil had been reported off the west coast by a British company, Serica Energy.[6]

*

If *Iona Isle*'s sinking had been designed to scare him, Pat O'Donnell wasn't buying it. Shell wrote on 24 June to four fishermen – Pat, his brothers Martin and Tony and son Jonathan – asking them to remove their shellfish gear temporarily to a safe location. Otherwise, it might be obliged to do so, as *Solitaire* was en route to lay the offshore pipe. The O'Donnells had set up to three thousand crab pots in Broadhaven Bay. The letters from Shell offered 'fair and equitable' compensation for disruption to fishing activities. O'Donnell's solicitor was clear: while Shell held a foreshore licence for its work, the O'Donnells held fishing licences and the company had no legal authority to remove gear. Pat O'Donnell contacted the gardaí and asked them for protection – a request confirmed by Chief Superintendent McNamara.

The Irish Times reported a 'major security operation' was due to be initiated on 25 June for the pipelaying work with 'a fleet of Garda boats and vessels staffed by Shell's security contractor I-RMS'. This was the first confirmation that I-RMS had staff on the water, working alongside the gardaí. Once again, the Naval Service had been asked to assist.[7] That same day, just twenty-four hours after they had received the correspondence asking them to remove their gear, Pat and his son Jonathan were arrested at sea on board separate vessels at about 7 a.m. Pat O'Donnell was injured in the boarding by the Garda water unit, and was taken ashore by Ballyglass lifeboat and to hospital in Castlebar by ambulance.

Jonathan appeared on public order offences at Westport district court. He was refused bail and remanded in custody to Castlerea prison in County Roscommon. The O'Donnell vessels were detained by gardaí at Ballyglass pier under the Maritime Safety Act. By that time, the 300-metre *Solitaire* had arrived off Mayo to start its work. A fleet, involving two Naval Service patrol ships, *LE Orla* and *LE Emer*, and rigid inflatables crewed by the Naval Service, the Garda water unit and I-RMS, was on standby to escort the ship into Broadhaven Bay. In spite of this sea escort, five Shell to Sea kayakers managed to come within 100 metres of the vessel, before being apprehended by I-RMS.

The O'Donnell legal team of McGarr Solicitors secured a High Court hearing to challenge Jonathan's detention the day before. The striking dark-haired young fisherman was brought into court with his wrists shackled, to the great distress of his father. Solicitor Edward McGarr said that the gardaí had had a pre-determined plan to unlawfully seize the young man's vessel and had conspired to abuse their powers to have the waters cleared. The arrest was related to his client's objection to the pipelaying ship *Solitaire* arriving into north Mayo, he said. Mr Justice John Hedigan ruled that he had been detained legally, but consented to bail. However, the fisherman had to travel back to Castlerea to be signed out.

Back west, Eddie Diver of the Erris Inshore Fishermen's Association was 'gravely concerned' at the government's refusal to clarify the legal situation: the O'Donnells had a right to fish, while the developer had a temporary foreshore licence. This had resulted, Diver said, in 'one young man being kept in Castlerea prison, one boat sunk, two boats detained, and the Garda and Naval Service . . . in an impossible situation'. Gear belonging to the O'Donnells had also been removed and towed away to an unknown location. 'It's not good enough for a legislative authority to refuse to take responsibility for this project once again, as it has done from day one,' Diver added. It also transpired that a 500-metre exclusion zone, cited by the gardaí, had not been approved by the Department of Transport. The department had issued a 'marine notice', which advised all shipping of *Solitaire*'s presence. However, a 'marine information notice' had been issued separately by the Corrib gas developers, stating that 'mariners are advised to forgo any activity within a 500m radius of the *Solitaire* to avoid damage to fishing equipment and vessels and to prevent the endangerment of lives'.

Pat O'Donnell never did get a satisfactory answer as to why his vessels had been detained at Ballyglass pier. The Department of Transport told *The Irish Times* that vessels involved in the Corrib gas project were inspected under port state control or flag state inspections, and were normally carried out 'according to a targeting system based on a vessel's safety compliance record' or on the basis of an 'overriding priority' inspection following 'a report of an alleged breach of regulations or following an accident/dangerous occurrence'. The O'Donnell-owned *John Michelle* was one of two 'overriding priority inspections' carried out during the summer of 2009, the other being a cargo ship registered in Sierra Leone, which was acting as a Corrib guard vessel when it ran aground in Broadhaven Bay in August. However, in 2010 the department's statement and the legal basis for detention would be questioned in a subsequent report by the FrontLine human rights organization.

In late August, a Naval Service diving team found *Iona Isle*. The gardaí said that, due to the depth of water, she would not be salvaged. Several months later, the Garda closed its files on the investigation. A file had been sent to the Director of Public Prosecutions, which had directed that no charges be brought, gardaí told *The Irish Times*.[8] On 7 December, the public order charge brought against Jonathan O'Donnell was withdrawn in court with no explanation. By that time, the Corrib gas offshore pipeline had been laid.

Come midsummer 2009, Ireland was officially in recession, along with much of the developed world, apart from Norway, so the experience of a fisherman and his family in north Mayo who were objecting to a project billed as vital to Ireland's future energy needs did not cause any great political storm – even though the fisherman was a significant employer in his area, and his rescue efforts in 1997 had been recognized by the state. The questions raised in certain press reports, such as that of *The Sunday Times* on 5 July 2009, about *Iona Isle*'s sinking – similar in tone to those raised about Willie Corduff's assault – reflected a more fundamental issue.

'A deeply unlikeable lot' was how the *Irish Independent* television critic John Boland described the north Mayo representatives, interviewed for a TV3 documentary, fronted by crime journalist Paul Williams, in early June. Maura Harrington and Dublin Shell to Sea activist Niall Harnett were the principal objectors interviewed. The documentary incorporated footage of protests, including shots of Harnett flailing, and one of Harrington leaning provocatively against a line of gardaí as she smoked a cigarette for the cameras. Boland said he would have preferred to make his own mind up. He had expected Williams's 'hostility towards the protesters to be supported by some damning facts about their activities. However, instead of providing an exposé, he contented himself with innuendo and abuse, some of it so vehement that the viewer almost felt sympathy for the die-hard fanatics and professional agitators who have latched on to this bitter campaign.'[9]

Vincent McGrath of Pobal Chill Chomáin confirmed that TV3 had not sought an interview with the two groups proposing a compromise. *The Irish Times* environment correspondent Frank McDonald filed a complaint, which was turned down by the Broadcasting Complaints Commission. McDonald described it as 'one of the most tendentious pieces of television' he had seen in a long time. Shell confirmed to the *Phoenix* magazine that Williams had been on its 'stakeholder engagement list', receiving an invitation from the company to a historic Ireland–England rugby match in the GAA's national stadium, Croke Park, in 2007.[10]

Watching at home, Willie Corduff was deeply upset.[11] The programme had suggested that his father had opposed the arrival of electricity in Rossport. Father Kevin Hegarty had suggested the same in a recent column in the *Mayo News*. Corduff wrote a letter to the newspaper to defend his father's good name:

> The facts are that, at that time, he was not in a position to pay
> the extra connection fee required because his house was at the
> end of a road. However, his refusal did not affect the provision
> of electricity to any other house in Rossport. I'm deeply hurt by
> the manner in which this aspect of my father's life has been

used to damage me and my family's principled stand against the present Corrib gas project. Criticize me, by all means, on the issues, but, please, treat my family and community with respect.[12]

His hurt was still apparent when he was interviewed by RTÉ journalist Mick Peelo in a subsequent documentary for the RTÉ Television *Would You Believe* series, broadcast later in 2009. Peelo's sensitive portrayal of a small mixed farm on the Rossport peninsula, of Willie Corduff doting on his grandchild, his horse and his geese, and of his wife, Mary, outlining their concerns, painted a very different picture from that portrayed by the TV3 documentary. It showed a couple who had reared six children, who were deeply attached to their home and felt let down by the state. Ironically, both programmes had one common denominator: wide camera angles of a stunning Erris seascape.

Sunday Tribune journalist Justine McCarthy encapsulated the dilemmas in a comment piece in June 2009. She was familiar with the subject. Like Peelo, and like Kay Sheehy of RTÉ, whose four-part radio documentary in 2006, *From Rossport to the Niger Delta*, had explored the deeper issues surrounding the controversy, McCarthy had spent time in the area. Describing Harrington as the 'anti-Madonna' of north Mayo, she said that her 'abrasive personality has damaged her cause'. However, she added that she feared the country would wake one day to news of a 'terrible tragedy' in Erris:

If it happens, the news media will have to examine its role in ratcheting up the volatility of the confrontation. Seldom has the fourth estate exhibited such an appetite for spin and such a distaste for establishing facts.

If it is the case, as most of the media propound, that the Shell protest is being used by anarchists to destabilize the state, the media itself is complicit. Distorted reporting of the events in

Mayo serves to deepen the suspicions between protesters and gardaí. What is happening there is redolent of the [Northern] Troubles when ordinary northern nationalists came to regard the police as the enemy. A very dangerous game is played out along the beautiful north Mayo coastline. Like most games, there are two sides – at least.[13]

Sister Majella McCarron and independent observer Dónall Ó Mearáin spent three days in late June in the area, representing the Table human rights monitoring group established in the mid-1990s under the auspices of Afri. FrontLine, based in Dublin and with a distinguished international reputation in working with human rights defenders, had by this stage commissioned its own report on human rights aspects of the Corrib dispute from barrister Brian Barrington.

Former garda and human rights observer Benny McCabe also travelled to Erris on behalf of Afri, and subsequently filed a complaint to the Garda Síochána Ombudsman Commission on his treatment by gardaí on 28 June – as did photojournalist Eamonn Farrell. Farrell, who subsequently made details of his complaint public, had been present with McCabe when five protesters staged a 'lock-on', where they chained themselves together in a star shape, with reinforced concrete casing as added protection on their arms.

A team of gardaí arrived and erected a tent around the five to start cutting them free. McCabe asked for permission as an independent observer to supervise the cutting. As he waited for a response to his request, another group of gardaí, who had been involved in removing one of the Rossport Solidarity Camp members, Eoin Ó Leidhin, from a tripod – a set of scaffolding poles blocking a road – arrived at Glengad.

The atmosphere, which Farrell described as 'passive' until then, changed: one of the officers in this group started shouting at McCabe and Farrell, ordering them to leave. They were pushed out, and minutes later heard a 'loud scream' of protest. The noise of the angle grinders then began to 'pierce the night'. When Farrell tried to photograph one of the young women, who was carried out on a stretcher, he was confronted by a garda who shouted at him, fist raised, forcing him on to the ground. When he recovered, he went around to the front of the tent to see Benny McCabe 'being pushed back by members of the Garda public order unit, several of whom were not wearing identification numbers'.

Afri's Joe Murray had maintained a constant interest in the situation and, with the support of artists like Dublin actor Donal O'Kelly, he had hosted fundraising events for the Rossport campaign. However, the organization felt it was coming under pressure for its stance. MidWest Radio opted not to run an advertisement for an Afri-hosted Rossport fundraising event in Dublin in May, explaining that it had informed the Broadcasting Commission of Ireland (BCI) of its decision, which it said the BCI 'supported'. This prompted RTÉ to refuse a similar advertisement request, because of concerns about a reference to 'Rossport' in the copy.

The Table report – the first in a series by Sister Majella and Dónall Ó Mearáin – focused on three events: it reported on a 'tripod' road protest and a 'lock-on' protest at Glengad on Sunday, 28 June, and court hearings resulting from these actions on 29 June. *Solitaire* had been in the bay for three days, and Sister Majella and Ó Mearáin's report described garda filming of some residents as 'provocative or intimidatory'. They also remarked on the large security presence – a reported figure of 300 gardaí, along with 160 I-RMS staff, two Naval Service vessels and a police helicopter: 'Given the population of Kilcommon parish (circa 2,000) and the numbers at the Rossport Solidarity Camp ranging from four to 100, and that protest appears to have been mild in nature, the need for such numbers for the maintenance of order is questionable,' the Table report said.

Afri's chairman, Andy Storey, a lecturer in political economy and international development at University College Dublin, and journalist Michael McCaughan published a report on the oil and gas revenue issue in the early summer. Spending cutbacks and extra levies associated with the recession could be alleviated if the state took a greater stake in offshore oil and gas resources, Storey and McCaughan maintained.[14] Storey expanded on this in *The Irish Times* on 19 June, which attracted a response from Fergus Cahill of the Irish Offshore Operators' Association. Cahill argued that the chances of striking oil off the Irish coast were 'slim'.[15]

Residents spent much of the latter part of 2009 in the courts or in jail. A former I-RMS security guard also appeared before Ballina district court on a public order offence; Judge Mary Devins referred to the reported stringent vetting of I-RMS staff, as outlined by the Private Security Authority of Ireland's chief executive at the government's forum, and asked him how he had got his job.

Vincent McGrath was still working away tirelessly for Pobal Chill Chomáin, arranging a meeting with the Norwegian ambassador to Ireland, pursuing the OECD complaint, and waiting for the Bord Pleanála decision on the pipeline. Glengad resident Colm Henry had decided to take his own legal advice. At a hearing before Judge Raymond Groarke in Galway circuit court on 10 July, he sought an injunction to stop further works by the Corrib gas developers at the landfall valve installation, down the shoreline from his house.

Henry also sought an injunction restraining the defendants, Shell E&P Ireland, from conducting further works at two specified locations on the pipeline route and restraining the defendants or its agents from videotaping or harassing him at or near Pollathomas, close to his home. And he sought an order directing the defendants to restore the beach at Glengad to its original condition, and restraining them or their agents from committing 'further acts of nuisance' at or near the landfall valve installation at Pollathomas.

After several hours of discussion among and between the lawyers for Henry and for Shell, Shell agreed to permit Henry to have up to two inspections of the landfall site. Application for an interlocutory injunction sought by Henry would be adjourned until October. It allowed work at Glengad to continue that summer, but Henry was determined that stalling tactics within the courtroom, as he perceived it, would not put him off.

Eddie Diver, for the Erris fishermen, was also wise to delaying tactics. Shell had written to individual fishermen to state that it would honour its 'undertaking' of a year before in relation to minimizing harmful emissions from Ballinaboy into the marine environment. However, Diver and his colleagues had raised this in their meeting with the two Eamons after the arrest of the O'Donnells, the impounding of their two boats and confiscation of their gear. They had been advised by the ministers to seek a meeting with the Environmental Protection Agency. The fishermen said that they had been largely happy with their discussions. However, they noted that Ryan and Ó Cuív's cabinet colleagues, Minister for Transport Noel Dempsey and Minister for Justice Dermot Ahern, still had to explain 'why such directions were given under the Maritime Safety Act' to impound the O'Donnell boats.

Even as Royal Dutch Shell appointed a new head, Peter Voser, who would start a round of cost- and job-cutting in the group, Amnesty International published a report on the Niger delta, which accused Shell and other oil companies of responsibility for decades of environmental and human rights abuses in the region. Quoting a United Nations report of 2006, Amnesty described the region as suffering from 'administrative neglect, crumbling social infrastructure and services, high unemployment, social deprivation, abject poverty, filth and squalor, and endemic conflict'. Shell and the Nigerian government disputed the report's claims.[16]

Several days after the report's publication, Pat O'Donnell and Willie Corduff travelled south across the county border for the annual Galway Film Fleadh. *Sweet Crude*, a documentary by US film director Sandy Cioffi on the Niger delta, was preceded by a clip of Pat O'Donnell at sea, filmed by Richie O'Donnell (no relation) for his forthcoming documentary on the Erris controversy. It earned a warm round of applause from the audience.

Cioffi, a Seattle-based film and video artist, outlined how she, her producer, film crew and guide had been arrested and held for five days in Nigeria during filming in 2008; their material had been confiscated. While she was not fully familiar with the situation in north Mayo, she said that she was aware from discussions with human rights interests of allegations that a 'similar blueprint' was being worked from. In Nigeria, where she filmed the development of the Movement for the Emancipation of the Niger delta (MEND) over several years, US military intelligence had offered its services to the Nigerian military to quell dissent, she said, and communities involved in resistance were infiltrated.[17]

As the last sections of offshore pipeline were laid out towards the wellhead, Ballina town council in north Mayo debated a motion calling for royalties to be reintroduced on oil and gas production. In early August, it was reported that there had been several pollution incidents during construction work at Glengad. Environmental consultants EirEco of Carron, County Clare, had been appointed by Minister for Energy Eamon Ryan to monitor work, and had submitted reports, which were put up on the department's website.

There had been a 'minor' oil spill in late June at low tide, when contamination had been confined primarily to the causeway and an oil spill contingency plan was initiated immediately. On 29 July, there had been an accidental spillage of an estimated twenty litres of chemicals, comprising a 'mixture of liquids, including corrosion inhibitor dye' at Glengad. The chemical additive was being used as part of 'hydrotesting' of the offshore pipeline, which had been laid on an eighty-three-kilometre route from the landfall out to the Corrib gas field manifold. Both spills had 'no ecological impact'.

The EirEco reports noted a significant number of sightings near shore of five different whale and dolphin species, and noted that there had been a reduced number of sand martins in June. On 3 July, the consultancy recorded that the birds were breeding and this suggested that the site works were 'not having a negative impact on the colony at Glengad'. The reports recorded meetings with Shell officials at their offices in Belmullet, but deleted their names, and also deleted the names of those company representatives accompanying the consultant on site. A spokeswoman for the minister said that this was in line with 'standard practice' applied by consultants working on its behalf.[18]

On 16 August, a Corrib 'guard' vessel, the twenty-three-metre *Flamingo*, ran up on rocks off Erris Head in unexplained circumstances. Eddie Diver, still in negotiations with Shell about compensation for loss of fishing gear before the *Solitaire* arrival, and still holding the developers to their commitments on the discharge pipe, called for 'marine vigilance'. An Irish Coast Guard helicopter survey confirmed that there was no pollution, but also reported that the weather was not severe when the vessel grounded.

That same month, the *Sunday Tribune* carried a report in which gardaí denied authorship of a disparaging poem about Maura Harrington, circulated on mobile phone texts.[19]

On 4 September, Monica Muller was in court before Judge Mary Devins in Ballina. Muller argued that an order issued by Judge Devins almost two years before, prohibiting Shell or its agents from trespassing on the Rossport commonage, had been breached. Her application was successful. Four years after the jailing of the Rossport Five for ignoring a court order, Shell now also found itself in contempt of court – a ruling which it would go to some lengths to overturn the following year.

17

'UNACCEPTABLE' ON
SAFETY GROUNDS

Late autumn 2009, and the sun was draped over Westport, County Mayo, with Croagh Patrick, the 'Reek', almost close enough to touch under a clear blue sky. A family law hearing set for a Friday morning had finished early in the courthouse. Judge Mary Devins opted to hold the next hearing an hour earlier than planned. The court was empty, apart from one young garda on duty, as legal teams and reporters filed in, sitting at either side of a carpeted square box below the judge's bench. A chair on top of the box was for the defendant. It was a curious arrangement: the defendant's chair was set at such an angle that he or she could accidentally kick a solicitor or a journalist, one of the reporters quipped.

In this case, the 'respondent' was Shell E&P Ireland and the hearing before Judge Devins related to her ruling a month earlier when she had found the developers to be in contempt of court. The case had received minimal attention in the media, but it was significant enough now for RTÉ Television's western editor Jim Fahy to be there with several print reporters. After all, it was on the basis of a contempt of court ruling that the Rossport Five had been committed to jail indefinitely in late June 2005.

The judge appeared in a smart jacket and her trademark large earrings. She hadn't robed, she explained, as she had been involved in the family law hearing not long before. She was well familiar with the Corrib gas project, having sat before many a hearing involving people arrested by the Garda at various protests since 2006. Married to Fianna Fáil politician Dr Jimmy Devins TD, a former junior health minister who had resigned the party whip for a period over cuts to cancer services in Sligo in 2009, she was respected for her essential fairness in interpretation of the law. At times the hearings had been controversial: on one occasion in late June of that year, she had refused bail and denied legal aid to seven of nine people arrested at Glengad, stating that the courts couldn't be 'giving out legal aid like Smarties'.[1] Her decision was subsequently overturned by the High Court.

This was 8 October, and Judge Devins was due to give sentencing in relation to her contempt verdict of 4 September against Shell, which had breached an order secured by Monica Muller and more than twenty of some sixty-two shareholders in the Rossport commonage on 14 November 2007, prohibiting the company from entry or from carrying out 'site investigations or otherwise'. Subsequently, Shell had bought into the commonage, but it hadn't returned to court to vacate the order.

Shell's solicitor, John Gordon, began in a conciliatory tone. His clients took the contempt finding 'very seriously' but were unhappy with it, he said. Three conditions set by the judge were being met, in that a donation of €3,000 had been made to An Taisce's planning unit, and he was in 'active negotiation' with Monica Muller's legal representative on costs. His clients were also seeking to vacate the original November 2007 order banning them from the commonage. This would be applied for at a court hearing the following week.

However, his clients had compelling reasons to have the case clarified in the High Court (by way of a Case Stated), as the applicant and others 'may rely on this judgment into the future'.[2] They had two questions that required responses. Mairead Casey, counsel for Muller, said that in her view the two questions were not relevant to Judge Devins's ruling. Judge Devins said that she had difficulty in linking the questions to what was decided: the High Court opinion was being sought in a 'very wide way', and had no relevance to her order. Gordon argued that if the High Court responded in a 'negative way' to either of the questions, the finding of contempt of court 'must fall'. The questions raised a 'fundamental question of law', he pointed out.

There was some protracted legal argument, and a break to allow Judge Devins to get a legal textbook from the boot of her car. After some further discussion, Gordon quietly dropped his bombshell: he had found a precedent for referring a civil case of this nature to the High Court. It was a ruling made in 2008 in relation to a small claims court case in Cork on payment of waste charges.

He hadn't given Mairead Casey a copy beforehand. Casey didn't have much time to respond. However, as she argued, there was a precedent for her case in domestic violence legislation. A spouse prohibited from a family home by a barring order did not automatically resume co-ownership if he or she moved back again at the invitation of a partner. An application had to be made to the court to have the barring order vacated. Shell's share in the Rossport commonage did not automatically negate the judge's ruling of November 2007 unless it went back to court – which it had not – to have it vacated. The central point, therefore, was that the respondents were in contempt of court for breaching a court order, Casey contended.

Judge Devins appeared to agree with her, but her hands were tied. She suggested that Shell's two questions for the High Court be amalgamated into one. There was another adjournment.

By this time, there was company in the courtroom, and an interlude that underlined the nature of the legal system. A slightly dishevelled young man in beige jacket and denims had been led in by three gardaí, accompanied by a superintendent. Could the judge hear an application for remand? the superintendent queried.

Judge Devins agreed. The young garda who had arrested the man took the high chair. He described how he had arrested the man that morning and was seeking to have him remanded to Castlerea prison in County Roscommon, pending a court hearing in Galway. The man had been due before a previous court hearing there and had failed to turn up.

The man stood motionless, looking the judge straight in the eye. He had been asleep on a park bench, he explained. He was in Castlebar seeking social welfare. He was doing his Leaving Certificate exams and trying to get his life back together. 'Where did you spend last night?' the judge wanted to know, looking at the charge sheet and noting that there were several public order issues listed. 'In a tent at Loughatalia in Galway,' he replied. She looked at him in disbelief. 'It's very comfortable,' he added.

'I don't want to spend the weekend [in prison] with heroin addicts,' the man pleaded, as the judge addressed the gardaí. She remained unmoved and granted the gardaí their request. The man sat down, threw his head back against the wall and looked up at the ceiling in a gesture of absolute despair and resignation. It seemed that he was beyond tears. On his budget, he wouldn't have been able to buy a solicitor a cup of coffee, never mind pay for half an hour of the legal representation that large companies could afford to engage.

The lawyers returned, and there was some further discussion with Judge Devins. The case was set for mention in Ballina district court on 27 October. Both parties were aware that An Bord Pleanála was due to give its verdict by then on the application for a modified route for the onshore pipeline – part of that route being through the same commonage. Gordon had been sent in to buy more time for the developer, and he had won.

Shell E&P Ireland's application to vacate the original court order and lift the prohibition on it entering the commonage would come before Belmullet district court on 14 October, six days after the Westport hearing. Judge Devins granted the application. Monica

Muller's solicitor, Marilyn McNicholas, made it clear that her client had no problem with this, but asked that the company give an undertaking to comply with legislation on notifying other commonage shareholders. Gordon was unwilling to concede.[3]

In early October 2009, the two Eamons announced that a sum of €750,000 had been secured from Éamon Ó Cuív's department, Mayo County Council and the national tourist board, Fáilte Ireland, to pay for 'community improvements' in Chill Chomáin parish, including Rossport village. The improvements would focus on tourism, recreation, roadworks and 'village infrastructure', and would include extending the north Mayo coastal walk from Belmullet to Ballycastle, linking existing loop walks and developing a cycle route. An extension had also been approved for Coláiste Chomáin, Rossport's secondary school. The ministers said that the measures had been identified in their ministerial forum.[4] Micheál Ó Seighin noted wryly that walks had already been developed in the area, way back: was there some element of repackaging? Shell to Sea in Dublin said that the issue was not about 'loot', but about health and safety concerns relating to the Corrib gas project.

Shell contractors had already moved in on land at Aughoose, between Ballinaboy and Pollathomas, erecting fencing and lighting, when Raidió na Gaeltachta was first with the apparent 'news' in late October 2009: An Bord Pleanála had 'approved the new pipeline route' – or so it seemed. It was the list of conditions that gave the hoax away: during the removal of 75,000 tonnes of peat from Rossport commonage, 'endangered species of frog are to be removed by hand, not by JCB bucket as has been general practice', and Shell would be required to built sports and leisure facilities in Rossport and Pollathomas, with 'free access to residents living within the 400-metre kill-zone along the route of the pipeline. In the unlikely event of a pipeline rupture, people within the kill-zone only have thirty seconds to avoid thermal radiation, and regular access to exercise facilities will minimize evacuation time.' And if Shell was to use its outflow pipe to discharge untreated waste into Broadhaven Bay, this was 'only to occur at night, and with prior collusion of the Mayo County Council and National Parks and Wildlife Service'. If the local fishermen noticed the waste discharge, Shell 'will be forced to have the fishermen given more money or otherwise "dealt with"'.

A speed limit of 25 k.p.h. would be imposed on each of the four road crossings over the pipeline and children would 'not be allowed to jump around in the back of the car while driving over the pipeline'. Shell would be required to fund construction of a Garda barracks on top of the pipeline, 'adjacent to the landfall valve which is 345 bar pressure', to 'encourage their continued facilitation of the project' and to 'reward them for their hard work'. Shell would be required to provide a dolphin sanctuary at Inver, where dolphins could 'look through the glass on to their former home, Broadhaven Bay, as it slowly deteriorates'. The email was from 'Farah Marie'. Whoever he or she was, they had provided a few brief moments of laughter.

Back in the real world, An Bord Pleanála deferred its key decision for a second time in late October, but it was imminent as the two Eamons headed again for north Mayo in early November. This time, the forum would meet in the heart of the community affected by the project – rather than in Belmullet, miles away. The venue was the small health centre at Corrán Buí. Once again chaired by Joe Brosnan, it had a low-key Garda presence, but there was some Garda interest in its proceedings. The Garda Síochána Ombudsman Commission (GSOC) had been invited to speak on policing issues relating to the project. Several days before, Mary Corduff had received confirmation from the ombudsman that there was an outcome to her complaint about the handling of events at Pollathomas pier in June 2007, when landowner Paddy McGrath had sought to protect his property from forced access by Shell contractors under Garda escort.

Curiously, the first letter to Mary Corduff informing her of this had been posted on 14 July by the Garda Ombudsman. She had never received the letter; neither had at least five others among fourteen complainants whose submissions had been deemed 'admissible' for investigation.

The Garda Ombudsman said it would reissue the letters – the initial batch had not been sent by registered post – and confirmed that a file in the case had been 'received for a decision to be made by the Garda commissioner'. In fact, as the letters to Corduff and others would confirm, the monitoring body had recommended that disciplinary action be taken by the commissioner against a named senior garda who had been serving in the area at the time. He was no longer based in Erris.

The recommendation suggested that a 'less serious breach of discipline be considered' under the Garda Síochána (Discipline Regulations) Act 2007, regarding the garda at the centre of the investigation. Sanctions applying to a 'less serious' breach range from a cut in pay of not more than two weeks, a reprimand, a warning, a caution or 'advice'. The Garda Ombudsman had initially asked the minister for justice whether it could investigate the complaints under section 106 of the Act, which allows for examination into practice, policy and procedures. This had been turned down by the minister, Dermot Ahern.[5]

Some sixty-eight gardaí were contacted by the investigation team to speak about the events at Pollathomas, which had led to twenty civilians and two gardaí being injured and which – as Mary Corduff had noted – were so dangerous that it was a 'miracle' nobody had been killed. Landowner Paddy McGrath had been hospitalized afterwards. A young woman in the area, whose husband had tried to protect his neighbours when the JCB had continued to drive down a private access road to the pier, would also remember it for a long time. She suffered a miscarriage in the immediate aftermath.

*

Maura Harrington was out of jail. She moved quickly when she heard the forum's agenda. She contacted the Garda Ombudsman to ask if it would address a separate people's forum, which she had established as an alternative to the ministerial initiative. Graham Doyle of the Garda Ombudsman's office travelled at Harrington's invitation to the Broadhaven Bay Hotel in Belmullet, where he fielded a variety of questions from a packed audience. There were many expressions of frustration. 'We've been wounded and brutalized, and we feel we are wasting our time now talking to the Garda Ombudsman,' Winifred Macklin explained, in her strong Scottish accent, as Doyle listened patiently.

Several members of the Rossport Solidarity Camp explained that complainants had become the subject of undue Garda attention, with warrants even being issued after complaints had been made. Doyle said that a survey his employer had undertaken on the reasons why people would not make complaints to it had identified two factors: people believed nothing would be done, or that it would make matters worse for them in relation to the Garda. 'That's exactly what I'm hearing here today,' he said. The session ran right up until lunchtime, with Doyle happy to stay for as long as it took.

Over in Corrán Buí, his colleague Kieran FitzGerald, the commission's head of communications, had arrived to give his presentation to the official forum. He described the Garda Ombudsman's work, its powers, and then ran through some statistics. A total of 111 complaints had been received about policing at the Corrib project since the Garda Ombudsman was established in May 2007, and this was 'unique' in that the commission had not dealt with any other situation that had elicited so many submissions.

At this juncture, the forum was on camera – Maura Harrington, Terence Conway and several members of the Rossport Solidarity Camp and Shell to Sea had arrived at Corrán Buí, having concluded their own people's forum at lunchtime in Belmullet. They bore a banner, which proclaimed that '€540 billion worth of resources' had been 'robbed' from Irish waters. The group of less than a dozen had taped over their mouths. Harrington positioned herself behind the two ministers and chairman Joe Brosnan, while the rest of the group stood silently across the room from her. Conway held up his video-camera to film the proceedings.

The colour drained from several forum participants' faces, including that of Shell managing director Terry Nolan, but otherwise there was little reaction. Joe Brosnan recognized the opportunity for what it was: it was the first time that any of the groups objecting to the terms of reference had attended the actual forum, as distinct from the abortive direct talks between Shell and the two Pobals. He issued a warm welcome. Several minutes later, he was passed a note by a department adviser. He paused again to remind participants that it had been agreed there would be no electronic media – no filming or tape recording. Terence Conway ignored him at the back of the hall. Fine Gael councillor Gerry Coyle broke the impasse: the meeting should continue – 'There's nothing here we want to hide,' he said. Brosnan looked around the room for any dissenting voices. He then invited RTÉ's western editor Jim Fahy to bring in his cameraman out of the pouring rain.

In his presentation, Kieran FitzGerald outlined how some 75 per cent of complaints relating to policing aspects of the Corrib gas project had been admitted for investigation. Around fifty-five files were 'closed' as the investigation had been completed, and twenty-three were still 'open'. Significantly, seven files had been sent to the Director of Public Prosecutions (DPP) in relation to possible legal proceedings. However, the DPP would not accept evidence for criminal prosecution of gardaí in all seven cases, and six of those cases were now 'closed'. The seventh case was being pursued in 'another investigative form', FitzGerald said. There were few enough questions, and no mention of the report about Pollathomas. In any case, FitzGerald had made it clear that he could not discuss specific instances.

The silent protest continued as the forum heard about the recent approval of €750,000 in state spending on various projects, including roads and walking routes. Shell's Denise Horan gave a PowerPoint presentation on another new funding initiative, involving support for local businesses. The Erris business development initiatives programme would offer financial assistance to companies seeking to 'optimize' their presence on the Internet, it would run marketing workshops, and it would give money to 'new and early start' businesses. As the funding couldn't be given directly, it was being channelled through two state agencies, the Mayo County Enterprise Board and Údarás na Gaeltachta.

Less than twenty-four hours later, An Bord Pleanála issued its decision – this time it was the real deal, and no spoof email, and it was unexpected. In its four-page letter, An Bord Pleanála said it was 'provisionally' of the view that it would be 'appropriate to approve the proposed onshore pipeline development, should alterations be made'. However, the applicant did 'not present a complete, transparent and adequate demonstration' that the high-pressure pipeline 'does not pose an unacceptable risk to the public'. The impact of construction on a designated rural area in Rossport would 'seriously injure residential amenities', and the development potential of lands there. It noted that part of the pipeline route onshore was omitted from the application – this being the part that was deemed 'exempt' the previous year by the Department of the Environment, and which Bríd McGarry and Brendan Philbin had tried to clarify at the oral hearing.

It found that up to half of the proposed new route was 'unacceptable' on safety grounds, due to proximity to housing in Rossport and between Glengad and Aughoose. The letter also noted that Ireland had not adopted a risk-based framework for decision-making on major hazard pipelines and related infrastructure. It said that British health and safety risk thresholds, and a standard for allowing hazard distances in the event of a pipeline failure, should be applied to the route and design. It outlined fifteen 'alterations', but also suggested that the developers explore another route, up the Sruwaddacon estuary; it gave three months for further information to be submitted on the route, design and safety of the high-pressure pipe.

The fifteen alterations included modifying the route again, providing new design and risk-assessment information, addressing problems with the landfall valve installation at Glengad, providing details of hazard distances, building burn distance, and escape distances. The developers were also asked to provide an assessment of the societal risk for Glengad, and given until 5 February 2010 to respond. However, Shell consultants RPS had ruled against Sruwaddacon almost two years before on environmental and technical grounds, when undertaking the research for a modified pipeline route. The consultants had noted in December 2007 that Sruwaddacon's estuarine and inter-tidal approaches were EU Habitats Directive Annex 1 habitats, and the bay was an 'integral part of the Glenamoy river salmonid fishery'. Specialized construction methods would also be required for Sruwaddacon, RPS had said.

The Corduffs were delighted. Their phone hadn't stopped ringing for several days after Mick Peelo's sensitively filmed documentary about them for the RTÉ *Would You Believe* series. Now it was ringing again, and Mary was determined to take that afternoon to celebrate – knowing well it was far from over. This was the outcome of 'project splitting', she said. 'In 2003, An Bord Pleanála inspector Kevin Moore described the Ballinaboy site for the refinery as "the wrong site" from a strategic planning perspective,' she said. 'Mr Moore should have been listened to. Shell should go back to the drawing board now.' Minister for Energy Eamon Ryan and his department had 'serious questions to answer' in relation to its endorsement of the safety of the proposed modified pipeline route at the oral hearing, Corduff added. A spokeswoman for Ryan countered that 'his first priority has and always will be the safety of the affected community'.[6] Former Bord Gáis engineering manager Leo Corcoran said the decision showed that the planning application had represented 'an attempt to retrofit a failed design to meet the required codes of practice', and the landfall valve installation at Glengad, just below Colm Henry's house, was 'socially unsustainable' and didn't comply with internationally acceptable codes. Maura Harrington called for a 'total overhaul', but also took issue with the board's provisional approval if alterations were effected.

Justice and peace group Afri welcomed the acknowledgement of 'legitimate safety concerns of local people'. A 'crazy approach to planning' was how its chair, Andy Storey of UCD, put it. 'Shell built the refinery in the wrong place, and laid an offshore pipeline, and they can't connect one to the other,' he said.

Shell said it would give 'detailed consideration' to An Bord Pleanála's difficulties with the Corrib gas onshore pipeline, but it still believed the current design was 'safe' and met 'all international standards'.[7] Several days later, an oil and gas industry website quoted the company as welcoming the fact that the Corrib pipeline had been 'approved – partly'.[8]

The decision received extensive coverage in the Mayo media, but elsewhere RTÉ's Jim Fahy was among a minority of national journalists to highlight the fact that the decision vindicated the position taken by the community. James Laffey, editor of the *Western People*, placed the decision in the wider context of severe flooding, which had affected much of the west coast in the latter part of 2009:

> Events in Ireland in recent weeks would remind one of the final scenes of Shakespeare's *King Lear*. Even the gods of nature have turned against us as we are made to learn the bitter lessons of our imprudent past. Not only have we discovered that the homes and apartments we bought were overpriced but now it appears many of them were built on swamplands that are susceptible to flooding.

He referred to a debate about the flooding on RTÉ Television's *Frontline* programme with Pat Kenny that previous week.

And the question everyone kept returning to during the course of the evening was how could planning permission have been granted for commercial and residential developments in flood plains?

Once upon a long time ago I was as naïve as them when it came to planning matters. That was back in the early part of this decade when I was reporting on the early stages of the Corrib Gas project. The original plan for the terminal at Ballinaboy in North Mayo was so ridiculous it defied parody. [Described then as] the largest piece of infrastructure in the history of the state, [it was] to be constructed on an unstable bog, within close proximity of private residences and in an area that was utterly cut off from basic emergency services . . .

I sat through those early public hearings into the Corrib project with my mouth agape. I had gone there with pre-conceived prejudices: I believed the local residents, who were against the project, were essentially harebrained idealists with no concept of the real world, and were holding up badly needed development in the west. But after hearing the plans for the terminal and pipeline I went away wondering who were the harebrained idealists and who were the realists. By the end of the second day I had made up my mind.

Almost a decade later I can still recall my utter astonishment when officials from the Health and Safety Authority told the hearing that dumping 500,000 tonnes of peat on the side of a hill in an area with a history of landslides was not a risk to public safety . . . But the officials didn't even stop to blink. They tipped their forelocks to the oil executives, gave a two-finger salute (metaphorically speaking, of course) to the local residents and told the hearing they were entirely satisfied with the project. The rest, as they say, is history.

Several of those protesters who lodged the early objections to Corrib now possess criminal records. They were right about one thing: Corrib was bad for their health and safety. It has to be said that some protesters have been lost in the fog of war and have done things that no law-abiding citizen could condone, but it should equally be acknowledged that they were treated with utter disdain from the first moment they lodged objections that were rooted in common sense . . . I only hope the sins of bad planning in Ballinaboy do not come back to haunt us as they have done in other parts of the country. Such a scenario is just too awful to contemplate.[9]

Fine Gael leader Enda Kenny told the *Mayo News* that he believed 'mistakes were made' over the Corrib gas project, confirming comments he had made to this effect in response to a question at a conference in Cork the previous week. He stressed that he was referring to the early years of the project, when it was being developed by Enterprise. 'My view was clear from the beginning. I always said that the Corrib field should be harvested, with due regard to the highest standards of safety and the environment,' he told the newspaper. 'Six years ago, I did say, "Put the pipe up the bay [Sruwaddacon estuary]",' Kenny added, in relation to An Bord Pleanála's advice to the company.[10]

Farmers who had raised concerns about a pipeline through their farmland were in the British news that month, November 2009. The *Guardian* reported that British Petroleum (BP) faced damages over claims lodged in the High Court in London by some ninety-five Colombian farmers who alleged that the oil company had caused serious damage to their lands, crops and animals. They cited breach of contract and negligence, claiming they had experienced harassment and intimidation from Colombian para-militaries employed by the Colombian government to guard the pipeline.[11]

*

The security issue in relation to Corrib was never far from minds, even as the number of gardaí in north Mayo was scaled down considerably in the aftermath of the 'pause button' pressed by An Bord Pleanála. Marilyn Horan, a North American with Irish links who had volunteered as an observer during contentious parades in Northern Ireland, produced a report in late 2009 in which she said that the legal rights of residents had been 'trampled upon'.

Chief Superintendent Tony McNamara retired from his post as head of the Mayo Garda Division before Christmas, as did Mayo county manager Des Mahon from the local authority. In a subsequent interview with *The Irish Times*, McNamara compared policing at Corrib to that at the international climate change conference in Copenhagen, Denmark; he had watched the television shots of confrontations between police and environmental activists, and believed his force had handled the situation in north Mayo 'far better'. However, he also said that he believed lack of adequate consultation by developers with residents in the early stages of the Corrib gas project had caused a 'fundamental breakdown in trust. That's been a failure of politics at local level in many instances in Ireland,' he told the newspaper. 'People get frustrated when they don't get adequate information. Eventually, dialogue will have to provide the solution to the Corrib gas dilemma.' The gardaí had dealt with 'very challenging' situations where he believed that a 'small group of people' had influenced others.

He was also keen to take issue with some media reporting. 'There was one occasion [in autumn 2006] where it was reported that there was a baton charge on protesters,' he said. 'In fact, there was technical deployment of batons by a sergeant and seven gardaí.' A number of people had been injured in that incident, but in his view his colleagues were 'still held in high regard in Mayo, in spite of all that has happened, and I was told this on my retirement by a large number of people from Erris'.[12]

The Department of Justice's figure for the Garda overtime cost for Corrib was updated to €14 million in a Dáil reply from

Minister for Justice Dermot Ahern to Sinn Féin TD Aonghus Ó Snodaigh on 20 January 2010.

Former garda and human rights observer Benny McCabe took issue with some of McNamara's points. Garda handling of the Corrib gas project security had been 'anathema to the spirit of community policing', he told *The Irish Times*. McCabe said that he had worked on developing community relations when he was based in the Garda deputy commissioner's office more than two decades before. He had also worked as a detective, before resigning and training as a psychologist.[13]

McCabe believed that the gardaí were 'working with impunity' in north Mayo. 'I have worked as a human rights observer with the UN, the EU and the OSCE [Organization for Security and Cooperation in Europe] in Cambodia, the Balkans, South Africa and in many post-conflict situations, but I have never been treated the way I was in Glengad in late June last year,' he said. McNamara rejected McCabe's assertions, and said that McCabe had not presented himself as a human rights observer beforehand when he attended the protest at Glengad. McCabe countered that he had presented himself to Superintendent Michael Larkin at Belmullet Garda Station some hours before, and this was 'on station records'.[14]

Afri initiated a petition in mid-January 2010, which called on Taoiseach Brian Cowen and the government to 'suspend all work on the Corrib gas project, pending a full investigation by an independent body qualified to assess the economic, environmental, safety and human rights aspects of the project'. Early signatories were former UN assistant secretary general Denis Halliday, who attended the press conference, Senator David Norris, Sinn Féin TD Aengus Ó Snodaigh, broadcaster and journalist Vincent Browne, NUIG Galway lecturer and poet Louis de Paor, Lelia Doolan, artist Jim Fitzpatrick and writers Rita Ann Higgins and John Arden. Afri called for prioritization of the security and policing operation in such a review.

At this stage, the *Guardian* had published a series of reports on policing in Britain, which indicated that environmental activists were regarded as a threat, and their groups were often infiltrated. A British justice ministry memo on extremism listed British environmental campaigners alongside dissident Irish republican groups, loyalist paramilitaries, far-right activists and al-Qaeda-inspired extremists.[15]

In February 2010, the Garda Síochána responded to the Garda Ombudsman's recommendation of the previous July. It said it found 'no breach of discipline' by the senior garda at Pollathomas pier in 2007. The Garda Ombudsman wrote to the fourteen complainants, informing them that it had sought an explanation from the gardaí for this decision. Mary Corduff and Betty Schult were disappointed, but not surprised. 'We have been told time and again by judges that we should take every legal course of action in relation to our concerns about the Corrib gas project, which we have done over the past ten years,' Mary Corduff said. 'There were people who were hurt both physically and emotionally at Pollathomas who never filed a complaint, because they felt it would go nowhere. It seems now that they were right.'

Betty Schult, Pollathomas hostel owner, had a similar reaction. 'If something like this can happen with no sanction in the civilized world, it is very shocking. But in the context of how we have been treated here, I am not surprised.'

The Garda Síochána had no comment to make.

In the meantime, An Bord Pleanála had come under some pressure over its direction on the onshore pipeline, half of which it had found unsafe. On 15 January 2010, RPS consultants wrote to the appeals board on behalf of the Corrib gas developers, seeking more time beyond the 5 February deadline for further information, but also raising a number of queries about the board's decision. This was formalized in a subsequent letter of 27 January, where it said it intended to submit a formal response before the end of May. This was granted by the appeals board.

On 20 January, Bob Hanna, chief technical adviser to the minister for energy and the energy installations inspector, wrote to the board taking issue with its risk-assessment methodology, and pointing out that this was based on the consequences of an accident, and not on the 'likelihood of occurrence'. 'This is different from international best practice in this area,' Hanna said, underlining this phrase. 'There are very significant potential consequential implications arising from this approach,' which would have the effect of 'prohibiting all significant infrastructural developments'.

To illustrate this, he said, 'a "consequence only" approach meant one would have to design and build an aircraft 'that would protect its passengers from harm when it crashes'. It was a misleading analogy. People chose to fly, but did not choose to live beside a high-pressure gas pipeline.

'The current competent authority for upstream gas safety, TAG [the Technical Advisory Group], has concluded that the design and proposed construction, installation and commissioning of the onshore section of the Corrib gas pipeline meet or exceed all relevant safety standards and codes,' Hanna concluded. The appeals board acknowledged his letter and said that the contents 'will only be considered by the board upon receipt of any additional information from the applicant'. It sent the correspondence out to the appellants, giving them an insight into the pressure the appeals board had come under.

Up in north Mayo, residents were angry but not surprised – yet again. A group including Ed and Imelda Moran, Éamon Ó Murchú from Glengad, Niall King, Kevin Healy and St John Ó Donnabháin from the Rossport Solidarity Camp sought, and were granted, a meeting with the minister, Eamon Ryan, who denied that Bob Hanna's letter constituted interference in the board's work.

On 28 January, Ryan and party colleague Senator Niall Ó Brolcháin had paid homage to Rossport when introducing legislation in the Seanad that aimed to transfer responsibility for pipeline safety from the energy department to the Commission for Energy Regulation. This had been promised at least three and a half years before by Ryan's predecessor, Noel Dempsey, when he published the Advantica safety review on the Corrib onshore pipeline.

In mid-February, Castlebar circuit court held a special sitting to deal with appeals lodged by Maura Harrington, Niall Harnett, Terence Conway and Pat O'Donnell against Corrib protest-related district court convictions the previous year. Harnett lost his appeal, as did Harrington and O'Donnell. Harnett apologized in court to Garda Hugh Egan, whom he had assaulted, and Judge Raymond Groarke agreed that he would be assessed for 240 hours of community service in lieu of a six-month prison term. Harrington was described by the judge as engaging in 'vigilante-type activity' to enforce a 'law she believed in'.

Harrington had described herself in court as a 'specialist', and said that as a citizen she felt abandoned by the state since a large police presence had appeared in Erris in 2006. She said that there was a 'degree of societal dysfunction in Erris where Shell contractors and supporters are given a degree of impunity not given to those who are trying to exercise their rights'.

Judge Groarke said that he recognized the right of communities to protest where they perceived unfairness, wrongdoing or in-equality, but believed Harrington and her supporters were acting like secret police, 'chasing after people trying to stop them'. He adjourned her sentencing on several cases for a year, while also dis-qualifying her from driving for two years and fining her €200 for an incident at Glengad. The adjourned sentencing was curious – but it would make sense as the summer wore on.

O'Donnell was given three-month and four-month consecutive

prison sentences for his part in a breach of public order on 13 September 2008, and incidents at Doolough, Erris, on 14 September when four gardaí who had been monitoring a Shell to Sea protest cavalcade were surrounded in their vehicle. Garda Seán McHale told Judge Groarke that he and his colleagues were trapped in the vehicle while the group hurled abuse, and said O'Donnell had opened the car door and asked him to hand over a video-camera that was 'not being used', he said.

Judge Groarke said that the gardaí had 'effectively been set upon by a large group in a most cowardly fashion' and O'Donnell, whom he described as a 'thug' and a 'bully', was clearly the leader. The description was not one that fitted with the O'Donnell that many in his community knew, or with the people whose lives he had helped to save at sea – gardaí serving up in Erris felt differently. The fisherman who had first raised concerns about the project's potential impact on the marine environment would miss the 2010 fishing season. One of his sons, Patrick, was studying for his Leaving Certificate and elder daughter Rachel was studying for final third-level exams. His youngest, Aisling, was just nine and due to make her first communion that May.

Residents on both sides of the Sruwaddacon estuary were deeply upset at the jailing of fisherman Pat O'Donnell, as were residents in Ballinaboy. However, there was also an unease about meetings, billed as 'community gatherings', in support of O'Donnell, but with an agenda set by Maura Harrington of Shell to Sea. When Willie and Mary Corduff and other residents travelled to Dublin for a protest in support of O'Donnell, they were dismayed to see a toilet bowl, symbolizing the state's 'begging bowl' attitude, which had been transported there by Naoise Ó Mongáin, Maura's husband.

Three of the Rossport five – Willie Corduff, Vincent McGrath and Micheál Ó Seighin – were granted leave to visit O'Donnell in Castlerea, but were asked to leave by one prison officer in a confusion over their status. The Irish Prison Service subsequently apologized. On their return to Castlerea in late March, they were asked by the fisherman to support a rally outside the prison the following day. Mary Corduff travelled with her husband to the rally, and Vincent McGrath agreed to play music. Again, to their dismay, the gathering of around a hundred people was hijacked. Members of the dissident republican socialist organization Éirigí, comprising disaffected Sinn Féin supporters who opposed the Northern Ireland institutions, turned up with banners, demanding that the fishermen be set free. They marched alongside Maura Harrington.

The significance of Bob Hanna's letter to An Bord Pleanála, on energy department-headed notepaper, was not picked up at national level. Ryan's Green Party colleague, environment minister John Gormley, was embroiled in a public row over the Ringsend incinerator. A week after an explosion at a gas-fired power plant under construction in Middletown, Connecticut, which claimed the lives of five workers and injured twenty-seven others, Fintan O'Toole of *The Irish Times* cast his eye over Hanna's intervention, and didn't like what he saw:

> Instead of accepting that the residents have reasonable grounds
> for their fears and pushing for a genuine dialogue, the department
> is yet again lining up entirely with one side in the conflict. Instead
> of acting as an honest broker, the Green Minister is continuing
> with the disastrous approach of his Fianna Fáil colleagues. This
> is both economic madness and political folly . . . What Corrib is
> telling us is that Irish politics, at either a local or a national level,
> is incapable of generating a decent compromise even when vital
> economic interests are at stake.[16]

O'Toole was the subject of an attack published in *The Irish Times* a week later. The respondent, film-maker Gerry Gregg, who was producer/director of the Paul Williams documentary for TV3, accused him of 'dodgy details' and of misleading his readers. The

Gregg/Williams documentary had been broadcast again by TV3 in the second week in February.

Gregg referred to the recent circuit court hearings. 'The truth is that there has been suppression of the popular will in north Mayo and the agents of the suppression have been Shell to Sea campaigners, not the Garda,' he wrote. 'But the only acceptable narrative to *The Irish Times* is that the bad guys (i.e. Fianna Fáil, hand in glove with a foreign multinational) are despoiling the land and seas and trampling on the people's rights and dreams.'

Gregg described An Bord Pleanála's request that the pipeline route be re-examined on safety grounds as 'an exercise that seems to be more about optics and PR than engineering', and he said that more than 90 per cent of the ten thousand people on the peninsula supported the political parties that backed the gas field exploitation. Most objectors, bar Maura Harrington, had always backed it. Their political affiliations were not the issue here.

Gregg wrote of the 'sinister side' of the Corrib campaign, and claimed that the Garda Ombudsman Commission had dealt with at least 110 complaints against the police and 'deemed most of them vexatious or without foundation'.[17] This claim took some facts from but was at odds with the Garda Ombudsman's own presentation to the ministerial forum in Erris the previous November, where it was stated that 75 per cent of complaints had been deemed admissible and seven files on possible further action against gardaí had been sent to, and returned from, the Director of Public Prosecutions.

Over at the *Irish Independent*, Kevin Myers steered clear of personal attacks, but described the Bord Pleanála philosophy, based on revival of the 'donkey-in-the-bog image so beloved of John Hinde postcards', as 'a recipe for poverty'.[18]

Journalist and broadcaster Vincent Browne followed up the issue on his nightly programme on TV3, inviting both Shell and Gerry Gregg to appear with Fintan O'Toole and a representative from Erris, Pobal Chill Chomáin spokesman John Monaghan. Shell and Gregg withdrew, but the programme went ahead – it had been

Browne's second attempt to debate the Gregg documentary with its producer/director and Shell. Former senior Labour Party adviser Fergus Finlay, chief executive of Barnardo's and also contributor on RTÉ Radio, had been invited to discuss issues relating to state protection of children. Finlay said he knew the Erris area very well, visited at least once a year, spoke to many people and had not yet met anyone who was enthusiastic about the Corrib gas project.

Even as Shell appeared to be making a case for staying with its modified pipeline route, it applied for a foreshore licence to drill up to eighty boreholes in Sruwaddacon estuary, and sought tenders to design and build a tunnel, pipeline and accompanying services. The company said this related to design investigations 'with a view to addressing, as fully as possible, all matters raised by An Bord Pleanála in its request for further information'. In its application to Minister for the Environment John Gormley for a foreshore licence for the drilling work, it said that there would be minimal adverse impact on the special area of conservation.

Gormley was by now responsible for foreshore legislation, apart from aquaculture, under a transfer of government responsibilities. Betty Schult, the Corduffs and their neighbours feared the worst. The drilling would degrade the sensitive area, they feared; once degraded, it could lose its special area of conservation status.

And so in mid-February 2010, as continuing frost and ice and sub-zero temperatures marked one of the coldest winters in fifty years, Terence Conway got into his van and drove to the department's office in County Wexford to hand in 139 objections, and there would be more in the post, including one from Pobal Chill Chomáin and one from shellfish company owner Tony McGrath who noted that his oyster farm in Sruwaddacon wasn't mentioned by the developers in its foreshore licence application. The North Western Regional Fisheries Board submitted an observation on the movement of wild salmon, and on the need for strict compliance with an environmental management plan. The foreshore licence was approved by Gormley with twelve conditions in June 2010. One of Gormley's conditions on protecting overwintering birds from work disturbance was notably more liberal than that recommended by the National Parks and Wildlife Service district conservation officer Denis Strong ten years before, when the agency was consulted about a foreshore licence for subsea pipelines in Broadhaven Bay. In a memo obtained under the Freedom of Information Act, Strong said 'yes' when asked if the proposed development would have an 'adverse effect on a habitat of high conservation value', and advised that all work should be carried out at a time of least disturbance to breeding, migrating and wintering birds – from 1 August to 30 September only.

Sister Majella McCarron had always maintained that the objecting Kilcommon residents should keep a 'victory list', to tick off achievements, which would sustain them through their darker days. On 4 March 2010, there was one such list entry when Bríd McGarry and Brendan Philbin heard the outcome of their long-running counterclaim against Shell E&P Ireland. Ms Justice Mary Laffoy ruled that they were entitled to have the issue of ministerial consent and various compulsory acquisition orders for the onshore pipeline determined by the court, in spite of a bid by Shell and the state to stop the action. Responding outside the court, Bríd said that she was 'delighted that we can continue with our counterclaim'.[19]

Shell pointed out that it was a 'notice party' only in the case, and had no comment to make. Two weeks later, the state gave notice of its intention to appeal Ms Justice Laffoy's decision.

Mairead Corduff, Willie and Mary's daughter, and Bríd Ní Sheighin were also up in court in the same month – but this time in Belmullet district court, as two of some twenty-seven residents summoned to appear on various charges relating to Corrib. The two women had been among a group of six who had taken down nets put up by Shell contractors at Glengad in April 2009 to stop sand martins nesting. They had folded the nets carefully and put them into the boot of a jeep. Several of the group had already witnessed birds dying in the nets – it was the nesting season – and there was video evidence of this. The group, who had planned to hand the nets into Belmullet Garda Station, were charged with theft, and the jeep driver with not having insurance.

Solicitor Marilyn McNicholas served a witness summons on the National Parks and Wildlife Service to appear at the court hearing. Under section 65 of the Wildlife Act, the gardaí had a duty to investigate threats to wildlife, but had not done so. On 27 March, the case was struck out by Judge Gerald Haughton and no reason was given, although McNicholas was told by a garda duty officer that the case would not be revisited. She had intended to argue that the netting was in breach of the EU Habitats Directive, but the validity of works at Glengad would inevitably have come under scrutiny. She applied for costs for her clients, but this was rejected.

Labour Party president Michael D. Higgins, who had raised the issue of the netting at Glengad as far back as 2008, and had mentioned it to Statoil when in Norway, was incensed. He welcomed the fact that the case wasn't proceeding, but was very concerned that 'a further step to vindicate the protection of the birds, as required under EU law' had not been afforded a hearing.[20] 'Ireland leaves itself open to legal challenge in a case where nesting habitats are interfered with without approval,' he said. The six people should now appeal a decision not to award their costs, he said. 'This trampling over environmental considerations in pursuit of some greater project is an outrageous provocation to environmentalists everywhere.'

On the same day as the district court hearing took place, the Environmental Protection Agency confirmed that it had received Shell's submission to review its emissions licence relating to discharges into Broadhaven Bay – eighteen months after its promise to the fishermen as a 'goodwill' gesture.

18

COUNTERPOINTS

He might have left it, but the Corrib gas field had never quite left Brian Ó Catháin. After the Shell takeover of Enterprise, he had attempted to brief his successor, Andy Pyle, on the situation with residents living close to the proposed onshore pipeline route. The meeting 'never happened', Ó Catháin recalled later. Pyle was 'more focused on other project issues'.

Ó Catháin moved to Irish company Tullow Oil, and from there to PetroCeltic, where he was appointed managing director. In a letter to *The Irish Times* in September 2005, however, he expressed some of the frustration he might have felt about the Mayo project. Responding to an opinion column by Fintan O'Toole on the state's attitude to offshore resources, he wrote:

> Enterprise Oil, Marathon, Statoil and, more recently, Shell have
> invested hundreds of millions of euro in the appraisal and
> development of the Corrib gas field, with no revenue shown to
> date. The state's inability to keep its side of the bargain, and
> to deliver due process to the developers of this asset, allowing

them to develop it in a timely fashion, augurs extremely badly
for the future of Irish offshore oil and gas exploration.

He couldn't imagine that multinational investors, such as
computer giants Dell and Intel, would have been so keen to invest
in Ireland if the 'approval process for a new factory was subject to
the same degree of arbitrary, ad-hoc policy-making and the
invention of additional hoops through which prospective
developers have to jump – which has been the fate of the Corrib gas
field's developers over the past few years'. The Irish state was 'not
willing to bear the risks which the Norwegian state has by creating
and funding a national oil company to explore for oil and gas.
Without private sector investment, we would not have an offshore
natural hydrocarbon resource.'[1]

The Rossport Five were still in jail when Ó Catháin's letter was
published – but had he still been with the project, he said he would
not have taken that route. He knew some of those men. Micheál Ó
Seighin had sent him a letter during the first oral hearing into the
Ballinaboy terminal. It was 'vitriolic', but Ó Catháin wished he had
kept it for its 'beautiful *cruinn* Irish'. In his view, the decision to jail
the five men was 'a huge mistake, and turned what had been a
broadly supportive local population 100 per cent against the
project'.

The government had not given any specific guarantees to
Enterprise, and neither did the company donate to political parties,
Ó Catháin told this author in early 2010. 'It would be unusual for
oil and gas exploration companies to make political donations (outside of North America), and Enterprise Oil Ireland certainly did
not.' However, he still held firmly to the views articulated in his
2005 letter on the state's approach, which, he believed, had denied
the developer 'due process'. For this reason, he said, he would have
sued the state if he were running the project still.

'There are two interesting points about Corrib gas which never
seem to get into the public debate,' Ó Catháin explained. 'The first

is that Irish gas has the potential to substantially reduce the cost of energy in Ireland. If a second Corrib were discovered, the Irish market would be full, and gas would be exported, rather than imported, via the interconnector to Britain. Irish gas prices would reduce to a British rate, minus the interconnector charge which we are currently paying. Thus, indigenous gases could reduce Irish energy costs (gas and electricity) by 15 to 20 per cent. This should be a national policy objective, of interest to all political parties. I don't know why it is not.

'The second interesting point is that all of the delay caused by endless obstruction and oral hearings has the interesting impact of substantially increasing the cost of the development of Corrib. This means that the point where the developers make a profit – and therefore are liable to tax on that profit – is pushed far into the future, if ever. The objectors are causing a major loss to the Irish exchequer, to tax revenue, and ultimately to the Irish people. It would be useful to have this point more widely understood for a more informed debate.'[2]

Fergus Cahill of the Irish Offshore Operators' Association concurred: 'The state could have been benefiting by up to €1 billion in tax if Corrib came on stream when forecast,' he told this author. Only two applications had been made for exploration licences in 2009 under the Rockall licensing round, compared to 193 and 46 applications respectively for similar rounds in offshore areas off Britain and Norway. Serica Energy had reported oil in 2009 – the first off the west coast in thirty years – but the company had subsequently played down reports that it would be commercial, and said it was re-evaluating the area to determine the potential of other prospects. Finding oil in sufficient quantities to prove commercially viable was always a challenge for exploration companies in a hostile Atlantic environment, Cahill stressed.

Norway 'refunded 78 per cent of the cost of unsuccessful exploration wells' and this was one of several key reasons for its attractiveness, along with a greater potential success rate. Cahill had made this case repeatedly at industry seminars and conferences. Asked about his advocacy of a Norwegian Statoil model for Ireland back in 1973, he recalled that there was great excitement in Europe at the time due to the discovery of the Ekofisk field by the Norwegians. There was a 'queue of oil and gas company executives at Minister Justin Keating's door', and Cahill was concerned that the state should not 'give it all away'.

Cahill cited the case of New Zealand, at the end of a long supply line and with only one producing gas field until recently. It had undertaken its own seismic work to identify hydrocarbon prospects. 'This is something that Ireland should undertake,' he said. He also emphasized the considerable employment benefits Corrib had already provided, not just in Mayo but also in Killybegs. The south Donegal port had developed a logistical service base for the exploration on the Rockall, Donegal, Erris and Slyne basins.

Its €50-million deep-water harbour had proved to be a blessing – initiated by former marine minister Frank Fahey in 2002 and completed several years later just as the full impact of shrinking EU quotas and rising fuel prices hit the port's fishing fleet. Fishing crews had found alternative employment with Jim Parkinson, the diver who had founded his own company, Sinbad Marine Services, in 1982. His lucrative contracts with Shell, Statoil, Allseas and others had led to work in British waters, and there was potential to service the offshore renewable sector.

Minister for Energy Eamon Ryan was convinced that the government was making every effort to ensure Corrib was developed in a safe manner, and said that his department was relying on the Advantica and Cassells reports as templates. The government's forum was 'not working as well as it should', he acknowledged, but transfer of safety functions to the Commission for Energy Regulation would be significant, he believed – even though it had been recommended back in 2006.

Ryan was enthusiastic also about north Mayo's potential as a wave energy location. His forecast was that Ireland could exceed its renewable energy target of 40 per cent dependence by 2020. Yet wind energy projects already drawn up for Mayo were caught in a planning limbo, due partly, it was said, to problems with the grid connection. Wave energy promoters were also voicing concern about infrastructural and bureaucratic impediments, even though their technology was still at trial stage. Erris fishermen facing further potential disturbance were also keen to ensure there was adequate consultation about wave energy plans, and that 'Corrib-type mistakes' over full stakeholder involvement were not repeated.

Ryan was convinced that there was 'no one view' in Erris on how Corrib should be developed, and he had been told at one point that the state should ensure the developer would stick to the original pipeline route 'because the Advantica report says it is safe'. It would be 'symbolic' if the state should live up to its responsibilities and show that a project could be built, and that the community would benefit, he said. 'The hardest bit is the legacy of hurt and the divided communities – our job is to try and undo some of that.'[3]

Fine Gael leader and Mayo TD Enda Kenny believed that Corrib would serve as a learning curve. 'There won't be the like of this again – mistakes were made. But it has taken less to divide communities. In my thirty-five years as a TD, I've never come across more bitterness than that associated with the Land Commission, which took land away and subdivided farms in the early decades of the new state, and with the rod licence dispute of the 1980s over state control of angling activity,' Kenny told this author in his Castlebar constituency office. 'Land and water issues can evoke very strong passions.' His constituency and party colleague Michael Ring was more specific. 'Mayo County Council bears a huge responsibility for what happened with this project. It opened every door, but it should never have approved that site at Ballinaboy. I was always in favour of the gas, as were most of the residents, but it was the wrong site then and it is still the wrong site now. I never agreed with breaking the law, but the Garda Síochána was put in an impossible situation. People have to live with the law of the land. But the "law of the land" has to live within the law of the land – statutory bodies have a responsibility too.'

Debates over issues such as peak oil, low-carbon futures and climate change were a constant theme of international news in spring 2010 – and, not surprisingly, the debate about resource exploitation versus environmental impact was the theme of the Oscar-nominated James Cameron film *Avatar*, which was a spectacular box-office success. Cameron subsequently received an appeal from eastern India, where the Dongria Kondh tribe feared the impact of bauxite mining on its sacred Niyamgiri mountain. In Argentina, there were rumblings over British plans to drill for hydrocarbon deposits off the Falkland Islands, while Greenpeace called on the European Commission to restrict the import of 'carbon-intensive' petroleum products, such as diesel and petrol derived from tar sands in Alberta, Canada.

As gas prices remained low in early 2010 due to a glut on the international market, Bord Gáis seemed upbeat about Ireland's supply. According to its website, interconnectors to Britain gave 'adequate capacity', with 87 per cent imported through the inter-connectors from Britain coming from 'diverse' sources, including Norway's Ormen Lange and Statfjord fields, Dutch, Danish, Russian, Norwegian and Algerian sources, and liquefied natural gas imports from Africa and the Middle East. Its management stressed the importance of Corrib, but its chief executive, John Mullins, who was keen to steer the semi-state into wind farm investment, said he was 'a believer in peak oil, but not peak gas'.[4]

Oil was a different matter, but as one senior Norwegian industry executive noted, 'Oil is not a word that goes down well with communities living close to the resource any more.' In other words, if Shell and its partners believed that their west coast base might eventually service oil finds, they were certainly not going to high-light this. A 2008 industry press overview of activity on Irish waters noted that Ireland's favourable tax rate made it attractive for the oil majors, along with high fuel prices, improved deep-water technology and a 'safe political environment'.

Providence Resources chief executive Tony O'Reilly said there were 'huge potential rewards off the west coast of Ireland, but also a higher risk', while University College Dublin geology professor Patrick Shannon said tests showed that any oil around the Atlantic margin might lack sulphur, making it easier to refine and therefore more valuable. The report referred to the Indecon analysis for government of 2007, which had estimated a potential reserve of 10 billion barrels of oil off the Irish Atlantic seaboard.[5]

Statoil had run into its own difficulties with its Norwegian constituents over involvement in tar sands oil extraction in Alberta, and plans to drill for oil near the Lofoten islands off the Norwegian west coast, best known for the seasonal cod fishery and a beautiful but fragile environment. A 'state within a state' was how an influential former Labour junior minister, Arne Strand, described the company in *Dagsavisen*, the newspaper he edited, in late May 2010.[6] The Norwegian government was too weak in its dealings with Statoil, still 70 per cent state-owned, he intimated, in a comment on a row over postponement of plans for a promised carbon capture facility in Mongstad. Strand had served with a Labour party that had virtually created Statoil a decade before his term as junior minister, so his criticism was particularly significant.

Norwegian non-governmental organizations had made contact with Vincent McGrath and his Pobal Chill Chomáin colleagues over several years; in late May 2010, McGrath addressed the People's Campaign for an Oil-free Lofoten, Vesteralen and Senja. 'For more than ten years, we have experienced conflict, anxiety and continual misrepresentation in public,' McGrath told the Lofoten conference. 'When Shell, Statoil and the Irish government failed to buy the support of our community with false promises of jobs and prosperity, they embarked on a campaign to terrorize and criminalize us . . . We have always considered that Norway has a key role to play in resolving the Corrib conflict, now in its tenth year . . .'

Employment at Ballinaboy peaked at 1,500 in 2009, according to company figures. As construction work began scaling down, there was renewed impetus to highlight Shell's investment in the community, through its local grants programme, its scholarship programme, which was extended for another three years and opened up to third-level students of all disciplines, and its long-term Erris development fund. By early 2010, some €4.4 million had been allocated, five community liaison officers had been appointed, and bi-monthly newsletters were being distributed to 3,500 homes.

Letters sent to residents invited them to 'small group' or 'face-to-face' talks. 'This is how Enterprise tried to talk to us over the pipeline,' Mary Corduff sighed. 'Will they ever learn?' In early April, some 409 letters were delivered by residents to Shell's offices in Belmullet, criticizing the company for bombarding them with 'propaganda' such as the company's community update. Signatories like John Monaghan and Bríd Ní Sheighin objected to the 'feedback form' enclosed, which, they said, would be 'misused as a statistic in future promotional propaganda' if returned.

'We think it is most unacceptable that Shell should try and use people in this way,' they wrote. 'Seeking opinions on the Corrib project from people who live outside the danger zone is simply dishonest.' In Shell's response, it said that while it respected the objectors' decision not to accept the invitation to talks, the door would remain open for dialogue.[7]

Over in Belmullet, home, temporary or otherwise, to dozens who had found construction work on the project, one might be forgiven for forgetting that the country was in deep recession. Cars spilled out of the large car park at the Broadhaven Bay Hotel: there was a function on, and there wasn't a seat to be had in the lobby, lounge or bar. However, the talk there was of the recently reported threat to the local hospital, due to a moratorium on recruiting staff. If it was downgraded, the nearest hospital would be some eighty kilometres away in Castlebar.

An editorial in the *Western People* put it in context, referring also to the long wait for a new sewerage scheme for Belmullet, and the fact that thirty-nine of the town's street lights were out of order.

> It is remarkable that the people of Erris should find themselves in a battle to protect the facilities at their local hospital at a time when they are extending the hand of welcome to one of the largest infrastructural projects in the history of the state ... Other parts of the world that have become centres for oil and gas exploration – including parts of Scotland – have witnessed dramatic improvements in public services.
>
> Yet that is clearly not happening in Erris, and it is quite worrying. If an area does not see a long-term economic and social lift after the discovery of a multi-billion euro gas field, what hope is there for the rest of rural Ireland?[8]

The wily solicitor Gordon Urquhart, protagonist of the 1983 Bill Forsyth film *Local Hero*, would never have let that happen to his community when US oil magnates expressed an interest in buying up the coastline.

Several weeks later, *Western People* editor James Laffey posed a related question, in a tongue-in-cheek response to reports that Irish company Providence Resources, owned by the O'Reilly newspaper family, was about to start drilling for 'oil' on the Kish Bank basin off the south Dublin/Dalkey coast. 'Let's see if Ballinaboy rules apply in Dalkey,' he wrote, describing how U2's Bono and his fellow Dalkey residents, including singer Van Morrison and film-maker Neil Jordan, might be very perturbed to find their vast back gardens used for a pipeline linked to a terminal built near Dalkey village. And if 'heavily armed gardaí have to beat up Van Morrison then so be it; and if Neil Jordan has the audacity to film the scenes, he'll face the full rigours of the law in Dublin Circuit Criminal Court . . ."[9]

That same month of April, it was confirmed that Shell was also sponsoring the GAA's Comórtas Peile na Gaeltachta 2010 tournament for Gaeltacht teams in Erris over the June bank holiday weekend. Pobal Chill Chomáin wrote to all thirty-two county boards of the GAA to inform them of the 'hurt' this had caused within their community, without specifically asking teams to stay away. Vincent McGrath's letter cited the founding principles of the GAA, which extolled parish and pride of place. The previous year, Shell and its Corrib gas partners had given €450,000 to Belmullet GAA club as part of a €1.8 million development, the club's chairman Sean Ó Gallchóir pointed out, and he valued the company's support. He did not believe the decision to accept sponsorship had caused hurt 'within this club or this area'. The finals were 'the biggest ever since the competition was started in 1969'. The tournament earned the local nickname 'Shell-Peil'.

In mid-April 2010, the Labour Party unveiled several new candidates for the next general election – including former independent TD for Mayo, Dr Jerry Cowley. He had played a prominent role in the Corrib gas issue, but told journalists that An Bord Pleanála's ruling of the previous November on the onshore pipeline had 'vindicated' the residents on the Sruwaddacon estuary. It was time to 'move on', he said. He believed the pipeline would proceed safely. If residents had safety issues, he would 'certainly be glad to support them' but he was 'never one for doing anything outside the law . . .'[10]

Willie and Mary Corduff were confused by his press interviews. Cowley had only visited their home the previous Christmas. They felt he was now trying to distance himself for political ends. Vincent McGrath was sanguine: 'Dr Cowley did more than any politician for us and this community. We won't forget that.'

The Corduffs were boosted by a spring 2010 bulletin from the

Goldman Environmental Prize. It led with the Bord Pleanála decision on the onshore pipeline, which it described as a 'stunning victory' for the local campaign, and it published an interview with Willie. 'Obviously we cannot return to any kind of normality until this dark cloud hanging over us is finally removed,' he said. 'The health and safety of present and future generations and the long-term sustainability of our community depend on the project being delivered safely.'[11]

Pat O'Donnell was into his third month in Castlerea prison when Shell to Sea campaigner and Clareman Niall Harnett was also jailed. Harnett had received a five-month sentence on 23 March for assaulting three senior gardaí in a courtroom on 11 March 2009. The court was told that Harnett launched himself 'rugby style' at the officers; Harnett pleaded not guilty and said that he had been pushed, 'flung around' and 'manhandled' when he went to the assistance of a man he knew. He was also fined and banned from driving for two years on a separate charge of obstructing traffic on a roadway at Pollathomas.

The following month, he tried unsuccessfully to overturn the jail conviction in Ballina circuit court. The court heard that he had declined to undertake community service because he believed his protests in north Mayo were 'a form of community service'. Several days after his jailing, members of the British Earth First group staged a protest outside the Bacton gas terminal in Norfolk in solidarity with the Corrib gas campaign.[12]

Harnett had been the only one convicted among twenty-seven defendants called before the special court sitting in Belmullet from 23 to 26 March in relation to various public order charges, several cases of alleged criminal damage, one trespass and one vehicle insurance issue in Kilcommon parish between April and June 2008. In the case of nine Shell to Sea support activists, the judge found them to have been unlawfully before the court due to the manner of their arrest and a time lapse in their access to the legal system.

A subsequent report by Sister Majella McCarron and fellow observers in the Table group was critical of handling by the gardaí of the arrests and said 'The general demeanour of a number of gardaí in the court was disrespectful.' It noted that twenty-three charges were withdrawn out of a total of twenty-seven, and yet several defendants had incurred significant expense and disruption to travel for district court appearances over several months 'throughout Mayo'. The report recommended that the Director of Public Prosecutions review the outcome of the special court sitting in relation to Garda and court practice. Sister Majella criticized Ireland's delay in fully implementing the Aarhus Convention on access to environmental information: 'Were the convention in place, people in Mayo would not find themselves in a position where they were forced to protest to gain information which should be their right,' she said.

The government's record here in relation to Corrib was analysed in a damning independent report by barrister Brian Barrington for the FrontLine human rights organization, published on 27 April 2010. FrontLine had asked Vincent McGrath to speak as a human rights defender at its international conference in February of that year. The report 'headline' was that gardaí from outside Mayo should reinvestigate the assault on Rossport farmer Willie Corduff at Glengad the year before, but that was just one of a series of recommendations in the eighty-six-page document.

As an international foundation for the protection of human rights defenders, founded in Dublin in 2001 and supported by the government's Irish Aid programme, FrontLine explained that it had sent a small delegation of observers to monitor protests in north Mayo in 2006, and found it could not respond in an 'ad-hoc' way to the escalation of incidents in the following two years. It referred to its invitation to Vincent McGrath to speak, and it explained that it had commissioned research from Barrington, son of a respected member of the judiciary and a lawyer in his own right with extensive experience relating to policing and human rights work in Northern Ireland. He had worked as special adviser to Northern deputy first ministers Seamus Mallon and Mark Durkan.

The research's remit was to examine whether those engaged in protests could be considered as 'human rights defenders', and to ascertain whether or not there were legitimate human rights concerns about the policing of the dispute. The organization took no position on the safety of or the route for the Corrib gas onshore pipeline, still before An Bord Pleanála at that stage. Barrington made four visits to north Mayo in 2008–9.

Barrington's detailed analysis was critical of Shell, the Garda, I–RMS security, the Department of Energy, the Private Security Authority of Ireland and a minority of protesters. He found that residents protesting peacefully on health and safety grounds about the gas project had the right to be described as 'human rights defenders' under United Nations guidelines, but a minority engaging in unlawful activity did not. He published medical records and an independent medical assessment, which disputed claims by Shell's security contractor, I–RMS, that 'not a finger' was laid on Corduff on the night of 22/23 April 2009. Hospital records stated that Corduff had 'been kicked all over the body' and experienced loss of consciousness, headaches, nausea and vomiting. All X-rays and a CT brain scan proved normal, and he was discharged the following day.

Barrington sought the independent opinion of Dr John Good on the medical reports. Good had worked for the International Committee of the Red Cross, and regularly assessed asylum seekers who had made claims of assault or torture. He found that Corduff's injuries were 'totally consistent with a history of assault'. The Health Service Executive West refused Barrington's request to interview the ambulance drivers on duty that night.

Barrington was critical of the subsequent handling of the Garda investigation into Corduff's assault, and for this reason he urged a reinvestigation by gardaí outside Mayo. He was unable to come to a conclusion about the 2009 sinking of Pat O'Donnell's boat, *Iona Isle*, but said that detention of O'Donnell's other boat appeared to be 'unlawful', and he criticized the government's failure to clarify the legal situation at sea in relation to competing rights.

He examined complaints of surveillance, including Betty Schult's experiences at Kilcommon Lodge Hostel in Pollathomas, and those of Colm Henry and Father Michael Nallen in Glengad. He noted that members of Pobal Chill Chomáin, established to seek a compromise onshore solution for refining gas, had also claimed that they were placed under Garda surveillance.

'Given that some people with paramilitary backgrounds have attended the days of action, it is understandable that the gardaí may be anxious to know who is involved in the different organizations,' Barrington wrote, but Pobal Chill Chomáin had not organized any such days of action, he noted. He said that any such monitoring should be 'proportionate'.

He examined issues relating to Shell security firm I-RMS, including surveillance of residents, and licensing and vetting of staff, such as the former staff member who travelled to Bolivia with Irishman Michael Dwyer in late 2008. He recommended that the Irish authorities press their Bolivian counterparts to carry out a full independent inquiry into all the circumstances of Dwyer's shooting.

He found no evidence of republican direction of protests, although individuals associated with various republican organizations did attend 'days of action'. He noted that 'protesters in Mayo appear to have had little control over who attended, although it is clear that the attendance of republican groups like Éirigí has been encouraged by some', he said. Maura Harrington had addressed Éirigí's conference in 2009. She told Barrington that she would accept any invitation to speak: she had spoken at the National Women's Council of Ireland, and 'if Fianna Fáil wanted to invite her to speak at their Ard-Fheis [national conference], she would attend there too . . .'

Barrington examined reports of intimidation but found most related to non-verbal communication. He was unable to substantiate allegations of criminal damage to property owned by or associated with Shell, but noted that some protesters did cause damage to a Shell fence on the night of Corduff's assault in 2009. He criticized Garda handling of the protest at Pollathomas pier in 2007, when twenty people were injured, and highlighted lack of information supplied by the Department of Energy, which had led to a situation in which thirteen people were arrested in 2008 over their concerns about unpublished authorization for work.

He was also critical of misleading or incorrect statements by several senior gardaí in relation to the handling of a court order obtained by Monica Muller preventing Shell or its staff from trespassing on commonage. Some gardaí had told him they were unaware of the court order. Given that Muller had always taken such care to remain within the law, in spite of her concerns, Barrington found the treatment of her to be unconscionable.

Barrington recommended that a human rights observer be assigned to Erris if the onshore pipeline was given approval by An Bord Pleanála. He also recommended that the Garda Síochána should employ a trained lawyer as a human rights adviser to review policies and practices and advice on policing; and he said that the Garda Síochána Ombudsman Commission (GSOC) should seek permission again to investigate policing of the Corrib gas dispute – having applied unsuccessfully to do so to then justice minister Brian Lenihan in 2007. The Corrib gas dispute had been the 'single greatest cause' of complaints to GSOC since its foundation, he said.

Barrington recommended that Shell intensify its existing efforts to ensure regulatory compliance, in the light of previous breaches by the developers. He noted that Shell and its agents were employing a former Mayo county secretary, a former Garda Síochána chief superintendent and a former editor of a Mayo paper – giving rise to 'the appearance that Shell is seeking to influence those who regulate them, rather than to comply with those who regulate them'. He noted that 'the announced role' of the former chief superintendent was to liaise with the community, but one leading community group had told him that the former senior garda had never contacted it.[13]

Shell defended its approach in a statement issued several days later, on 29 April.

> Shell E&P Ireland has always applied the highest professional standards in its approach to the Corrib gas project. The regulations relating to the Corrib gas project are enormously complex, involving multiple consents under different legislation, from several government departments, multiple statutory agencies, and the local authority in Mayo. We devote an enormous amount of time and resources to complying with the regulations relating to all aspects of the development of the pipeline and the terminal site. We are committed to doing everything possible to fully comply with these regulations at all times. Whenever an unforeseen issue has arisen, we have moved quickly to ensure full compliance as quickly as possible.

Jim Farrell of I-RMS was more forthright, and issued his statement within several hours of the report's release. He rejected 'outright' what he described as a 'central conclusion' in the report that Corduff was assaulted. He sought a 'retraction of any insinuations that I-RMS was involved in any assault on Mr Corduff' and said that the FrontLine report must be amended to include recent findings by the Director of Public Prosecutions (DPP) that there were no grounds for criminal prosecution in relation to the alleged assault.

Farrell said that he had three reasons for believing Corduff was not assaulted. He said he was 'personally present at all times' during Corduff's 'presence under and adjacent to the truck' and, as a professional, my duty of care was to Mr Corduff. Mindful of his welfare and the deteriorating weather conditions, our concern was to keep Mr Corduff warm and dry, that food and water was provided to him, that medical support was in place, and that for his own welfare, the situation was safely resolved. At all times, our first consideration was the health and safety of Mr Corduff. The DPP has ... found that I–RMS has no case to answer. Furthermore, the DPP concluded that I–RMS actions were both legal and justified and that I–RMS operated within the remit of the law at all times.

He said that the report's failure to mention this was a 'staggering omission', and his company would be consulting its lawyers. The DPP did not generally make public its reasons for its decision not to prosecute except in cases of murder; investigating gardaí were informed. FrontLine questioned how the company could have this level of detailed information. Deputy director Andrew Anderson said that the organization stood by its recommendation of a reinvestigation by the gardaí.

Curiously, Barrington had been told by I–RMS on 22 October 2009 that Corduff had 'made no complaint to the gardaí and that the gardaí had never investigated them in connection with any complaint'. He had reiterated this in writing on 8 December 2009. Corduff had in fact reported the incident to the gardaí on 31 August 2009 – having delayed doing so due to a 'lack of confidence' in the local force. Superintendent Larkin of Belmullet had told Barrington on 2 November 2009 that the investigation was at an advanced stage, and that I–RMS 'had provided statements'.[14]

Energy minister Eamon Ryan welcomed the FrontLine publication and said his department would 'take all points on board' relevant to its involvement on disseminating information. 'Independent oversight is always welcome, and we take this report seriously. The more neutral observers the better, if this helps to give people confidence,' he noted.

Minister for Justice Dermot Ahern said that the report represented a 'contribution to the debate on all the issues surrounding the opposition by some to the Shell gas project' and its recommendations addressed to state agencies would 'be considered by them'. However, Ahern said that policing of Corrib gas protests had been 'a difficult one for the gardaí' and the force would continue to uphold key principles and 'defend the human rights of all involved'. A number of protesters had been convicted of offences and some had been imprisoned in relation to protest actions, Ahern noted.

At that stage Corduff had not received a copy of the report. However, he told *The Irish Times* that he knew what had happened to him and 'the truth will come out eventually'.[15] In subsequent correspondence, the DPP assured him that 'Mr Farrell [of I–RMS] was not given the direction of this office.'

The Garda Síochána offered no immediate response. Off record, however, the force was stung by some of Barrington's criticisms. At a time of increasing pressure to tackle violent crime, Mayo had proved to be a drain on resources and on morale – even if some young officers keen for overtime opportunities had nicknamed the Ballinaboy bridge dawn protest route as the 'golden mile'. Some senior gardaí expressed disillusionment with the way the force had been used by the state and Shell. 'The same old story – we always get sent in to clean up the mess . . .'

19

ENDGAME

A harmless-looking black pole protruded above a farm building at Glengad. On closer inspection, it was a mount for a security camera, which rotated at 360-degree angles diagonally opposite Colm and Gabrielle Henry's home. Echoing the lyrics of the Police's 1983 hit single, the camera seemed to watch every breath they took, every move they made. In the first six months of 2010, there wasn't much else for a security camera to catch. The courts might have been busy with Corrib-related cases, but there was a lull in activity around Sruwaddacon estuary and Broadhaven Bay, while the developers prepared to respond to An Bord Pleanála over the prospective pipeline route.

Henry was determined to continue with his legal action initiated in the year before. There might be no substantial security presence for some months, but inevitably he and Gabrielle feared that the construction equipment, the fencing, the diggers would all return – and perhaps 24/7 security, even after the engineers moved on, at the critical pipeline landfall point.

On 8 May 2010, the government's forum met for the sixth time since December 2008 under the chairmanship of Joe Brosnan.

Newly appointed Gaeltacht minister Pat Carey, promoted in a cabinet reshuffle, had his first taste of the controversy when Maura Harrington and her husband Naoise Ó Mongáin and four Shell to Sea supporters turned up. With Carey were Minister for Energy Eamon Ryan, newly promoted Minister for Social Protection Éamon Ó Cuív, who was handing over to Carey, and junior minister and Mayo TD Dara Calleary. When proceedings opened, Harrington attempted to disrupt the gathering by slowly reading out over a loud-hailer the number of days that Pat 'the Chief' O'Donnell and Shell to Sea activist Niall Harnett had spent in prison. She continued with extracts from the recently published FrontLine report.

Brosnan asked Harrington to desist. She ignored him. The forum adjourned and resumed in a smaller meeting room, where Shell confirmed it was planning to run the onshore pipeline up Sruwaddacon estuary, subject to permissions. It outlined its various spending initiatives, confirmed that construction employment would begin to fall from the 750 currently engaged to between 250 and 200 by the end of the year. There was an update on state investment from new Mayo county manager Peter Hynes, and from Údarás na Gaeltachta's Séan Ó Coisdealbha.

There was some discussion on the forum's future; Pobal Chill Chomáin, Pobal le Chéile and Shell to Sea were still declining to participate, and Eamon Ryan had indicated earlier that week that the forum was 'flawed', and needed to take a 'bottom-up approach'. Several proposals were discussed. A family who had booked the room for a first communion function were patiently waiting outside. Mindful of this, Joe Brosnan suggested that 'focus groups' might provide scope for further discussion on the future.

And then Ballinaboy resident Jacinta Healy spoke. Earlier, during one of the unscheduled breaks, she had asked the new county manager if there was a safety plan. It was a question she had first asked back in 2002 during the Bord Pleanála oral hearing into the planning application for the terminal. Over the past month, bog and forest fires had caused extensive damage in the west, mainly in Connemara, parts of Sligo, Roscommon, Clare and Mayo. Coillte had estimated that more than 520 hectares of mainly coniferous woodland, with a replacement value of €2.5 million, had been lost. There had been a gorse fire at Attycunnane on the Belmullet road – several weeks later, Belmullet's part-time fire service was called out to forest fires close to the terminal at Glenamoy, and at Bellagelly on the old pipeline route opposite the McGarrys' farm.

Hynes told Healy at the forum break that he would 'get back to her' on the subject, but she wasn't too satisfied with the response. When it resumed one could have heard a paperclip drop as she spoke, facing the three senior ministers and one junior, and with senior gardaí, Terry Nolan and senior executives from Shell and a number of state officials right behind her in the small room. Healy recalled how the members of her residents' association had been virtually 'on their own' when they first appealed the terminal's planning approval by Mayo County Council, how they had been successful then, and how they had accepted the final Bord Pleanála decision as they had 'neither the energy nor the money' to continue.

She recalled how the group had opted to remain within the law, declining to participate in the dawn protests outside their doors – but how she and her neighbours still harboured serious fears. An Bord Pleanála had deemed half of the modified onshore pipeline route to be unacceptable on safety grounds. So if it wasn't safe near housing at Rossport, Aughoose and Glengad, how 'in God's name' could it be safe near housing at Ballinaboy now?

Healy was passionate, articulate, clear. She described the 'hurt' that was still embedded in the community, a 'hurt' that had been exacerbated by the jailing of fisherman Pat O'Donnell. She did not mean to criticize the judicial process, she said, but O'Donnell would not be in jail if it were not for the gas project, and he would not be at risk of missing his youngest daughter's first communion. Her husband was O'Donnell's first cousin, and her brother-in-law had rescued him on the night that his boat sank.

'He's not a thug,' she continued, taking issue with the judge's description of the fisherman at his sentencing. Her husband worked in the terminal, she explained, but 'We'd rather not see the gas come in here . . . This will go on for another eleven years . . . and we really need closure.'

Joe Brosnan said it was 'refreshing' to hear her; this was what the forum was about. Eddie Diver of the Erris Inshore Fishermen's Association endorsed her comments afterwards. His association, which was still awaiting an Environmental Protection Agency decision on its efforts to extend the terminal discharge pipe beyond Broadhaven Bay, had written to the Castlerea prison governor. It had asked that O'Donnell be released on compassionate grounds, given the onset of the fishing season; O'Donnell was granted temporary release for a few hours in mid-May to attend his daughter's first communion.

Diver's members were due to negotiate another summer, or two, of compensation for lost fishing time from Shell. 'The fishermen only ever got what they were entitled to,' Diver told this author. 'It was never "bribery", as some people unfairly said.' His association had stuck to its aims, and its remit up to the high-water mark; and it had achieved no small victory in pressurizing Shell to apply to amend its marine emission plans. 'However, it's a pity that we didn't have a combined strategy here,' Diver added. The words of Shetland's John Goodlad some years earlier had proved to be prophetic – 'Hang together or you will hang separately.'

Journalist Arthur Reynolds, who had extensive contact with coastal communities during his twenty-seven-year tenure as editor of *The Irish Skipper* fishing industry journal, could empathize with Diver and his members. 'The economic structure of these harbours, particularly the smaller ones, is so dependent on few alternatives that changes and intrusions have to be treated with extreme caution. It was a pity that the diplomatic approach taken by Statoil in the early days of its development in Norway wasn't reflected to a greater degree in the approach taken by the Corrib gas partners.'

Jacinta Healy experienced a mixed response when word got out about her intervention. The Ballinaboy residents had been criticized by objectors for receiving some compensation for the disturbance associated with works on the project, and some had accepted work themselves. However, as TD Michael Ring observed, the company had not been easy to deal with on the disturbance compensation issue. As one resident remarked, 'Children who haven't even been born yet won't be able to speak to each other, long after the gas is gone.'

Micheál Ó Seighin had been impressed by Healy: 'The Ballinaboy residents were there with us until such time as they made a decision not to continue their campaign. There was many a time when they could have come out publicly to welcome this gas project. They didn't, and we should always be grateful for that,' he said.

Vincent McGrath also paid tribute to Healy: 'I've had to revise my views of eight hundred years of our nation's glorious struggle against colonization,' he said. 'It was always just a few heads like ours, sticking above the parapet, and waiting to be shot down by our good neighbours . . .'

And Mary Corduff was impressed with her courage. Healy's words reminded her of something she had forecast in the direct talks with Shell back in spring 2009. 'Shell managing director Terry Nolan kept saying that the project had the support of the majority of the community,' she recalled. 'My response was "Never under-estimate the people of Erris." It's a pity he didn't listen.'

On an unseasonably warm and humid May morning, staff were busy with the last stages of building the gas terminal at Ballinaboy. Two weeks before, a Glengad resident had joined a bus tour of seventy people to see the plant, having applied three to four years before. She described afterwards how her fellow visitors, some of whom had made several previous visits, had been enthralled by the financial benefits, the jobs, the PowerPoint presentation on the terminal's operation and how the children present were 'all en-couraged to run up to Shell staff to get their presents', while there were caps and fridge magnets for the adults. When she had asked several technical questions, the 'entire assemblage of the room' had turned to stare at her, as if to say how dare she 'disrupt the party atmosphere . . .'

A smaller tour for the press in late May opened with a detailed PowerPoint presentation by Shell project site manager Brendan Butler, who had more than twenty-four years' experience with oil and gas production and had worked with Shell on its Sakhalin project in Russia before moving to Corrib in 2007. He explained that testing of gas fed from the Bord Gáis pipeline to the west would take place over three months.[1] The Ballinaboy plant would then be filled with nitrogen to protect against corrosion, while the developers awaited the outcome of An Bord Pleanála's decision, Butler's colleague and project construction manager Paul Hughes of CMC explained. The system had already been subject to testing over the past two months to ensure pipework was 'leak–tight'.

Some 750 people currently employed would be reduced to 200 by the end of 2010. This included up to half of some 120 staff work-ing for Longford firm Gilmore Security, who were not happy with their employer's parting terms – and had threatened to protest during the press visit. Shell anticipated that up to eighty-five permanent staff would be employed at Ballinaboy, with 130 'long-term' positions overall, once construction at the terminal was complete in July.

Butler stressed the local involvement. As of mid–May, some 23 per cent of 773 people employed directly or through various con-tractors were from Erris, while 22 per cent were from other areas of Mayo and 55 per cent from elsewhere. A group of fifteen local third-level students had spent time on site during the summer of 2009, and interviews were in progress for similar paid placements this year.

Paul Hughes, along with Corrib press staff Christy Loftus and

Denise Horan, escorted the small group, including this author, on a walkabout. First stop was the 'slug catcher', which would remove the first stage of liquid from gas piped in from the wellhead. Condensates extracted would be sent by truck to the Whitegate refinery in Cork. Water would be treated on site. The methanol still, recycling methanol injected as a type of 'anti-freeze' into the gas, and the 'flare' stack were among the tallest structures on the sixteen-hectare terminal 'footprint'.

'Flaring', where gas is released into the atmosphere and ignited to release pressure, would 'rarely' take place, the company team said. However, a 'small amount of gas' would be 'flared' during start-up until export gas composition met the required Bord Gáis specification. The press group was shown fire-water ponds, and told how measures had been taken to ensure there would be no repeat of the earlier high discharges of aluminium – Shell's press team in Dublin had vehemently denied the company was responsible for elevated discharges when the issue was first high-lighted by residents in 2006.

There were impressive statistics – 440 kilometres of cabling, 40 kilometres of piping, 1,800 tonnes of steel on site, and equipment delivery generating 20,000 road trips. Further statistics on safety, including a recent 'safety milestone', were cited. It was the third time to date on the project that one million man hours had been worked without a single 'lost time' injury. The plethora of safety signs extended to internal stairways, urging staff to keep one hand constantly on banisters.

The press group was told how wildlife protected under habitats directives was transferred to the dormant wooded area on the extensive site, and how additional trees had been planted. There had been 'a pine marten or two, a badger set was relocated, there were plenty of frogs and a colony of bats'. At one stage, a number of cats had found their way on site from the neighbouring animal testing laboratory.

Young staff working with the various contractors explained how they took pride in their work, and how optimistic they were about their career prospects. Opportunities like this did not often come this way in north Mayo, they pointed out, and the company offered an invaluable career path in terms of training. Adrian Deane from Gortbrack in north Mayo was one of twenty-five site operators trained for permanent posting, with another eighteen due to be recruited, he said. Deane, originally hired as an electrician, had opted to take a course in operations at Middlesbrough in Britain and had then worked at the Shell gas plant in Bacton, Norfolk, before returning to Mayo.

His brother Brendan was one of the first locals to be hired by Dutch company Hertel for the Corrib contract three years before. Along with colleagues Christopher Reilly from Ballinaboy and John McAndrew from Carramore, Brendan Deane had worked on scaffolding, insulation and painting. The Mayo staff concurred with their project manager Andrew Canning, from Blackpool, that they were 'very lucky' to have found jobs on the project during the construction downturn.

Canning had rented accommodation near Carne golf club, loved the landscape, and said he had experienced very little negative reaction locally during the height of protests in 2006–7. Had he availed himself of a Garda escort to work? 'No.' Had reports of intimidation of staff been overblown? He nodded. Out of earshot, several other staff paid an unusual tribute to the protesters – 'Well, if it wasn't for them, how many Mayo people would have got work here? And if it wasn't for the delays, we wouldn't still be working now!'

The dense coniferous shield created a cocoon effect. Most staff coming and going from gate two could miss the road signs declaring, 'No Consent', and the nine crosses planted in memory of Ken Saro-Wiwa and his eight Ogoni colleagues opposite gate one.

An explosion and leak at a British Petroleum (BP) oil well on 20 April in the Gulf of Mexico killed eleven staff on the Deepwater

Horizon rig, owned by Transocean. Weeks later, as oil continued to spew from the sea floor, it was described as the worst environmental disaster in US history. Seabirds and wildlife were photographed coated with crude, fishermen feared for livelihoods, and the vast slick spread across the Louisiana marshland. Meanwhile, in the North Sea in late May, a 'well-control incident', in which pressure built up, was recorded on a Statoil platform – the third in a series.

Paraic Cosgrove of Pro-Gas Mayo believed the Deepwater Horizon incident vindicated the arguments put forward by supporters of the Corrib gas developers on the risks of processing gas offshore. Offshore rigs just weren't safe, and Shell to Sea should accept this, he said. However, community groups Pobal Chill Chomáin and Pobal Le Chéile already had done so when they had sought to compromise two years before.

The industry, including BP, initiated a public relations offensive to reassure the North American public that offshore drilling was safe; just a month before, US President Barack Obama had agreed to open up new areas to offshore drilling, including Alaska. After Deepwater Horizon, Obama announced that no new licences would be issued until an investigation was complete. Official statistics appeared to show a steady improvement in terms of accidents and spills, but non-governmental organizations, like the Canadian campaign group Alberta Wilderness, pointed out that, in many cases, governments could not be relied on to provide independent and rigorous monitoring and oversight. *The Wall Street Journal* noted on 16 June 2010 that BP was 'not the only spill master – check Shell in Nigeria' and referred to a leak on the Trans-Niger pipeline in April 2009 which caused a 'giant fire'. The newspaper referred to a contractor's report which found that the fire was due to 'rusty, damaged and [leaking] pipes'. Unlike the Gulf of Mexico, 'the remote wetlands of Nigeria's Niger Delta' were 'far from the eyes of the western public'. The line in particular struck a chord for Mary Corduff up in her remote part of north Mayo.

President Obama shone a spotlight on the culture of information control within the oil and gas industry at a White House press conference on 27 May 2010. He admitted that he had been wrong to believe oil companies were prepared for a catastrophic spill. He also acknowledged that he had underestimated the 'scandalously close relationship between oil companies and the agency that regulates them'. Pobal Chill Chomáin's John Monaghan hoped Eamon Ryan was taking note, as the minister proceeded with plans to transfer upstream pipeline safety to the national energy regulator. The legislation for this was enacted on 3 April 2010, but Ryan's junior minister, Fianna Fáil TD Conor Lenihan, said that the department would 'still have a role'. On 14 June 2010, Lenihan announced 'special oversight measures' to step up monitoring of drilling rigs in Irish waters – two days before he opened up the entire Atlantic margin in a new licensing round designed to boost exploration. Padhraig Campbell of SIPTU's offshore oil and gas committee questioned the independence of the measures. On 28 June, Lenihan told *The Irish Times* that a BP-Gulf-of-Mexico-type spill 'could never happen here' because of the strict regulation.

Commenting on the new Atlantic margin licensing round in the *Irish Mail on Sunday*, business journalist Colm Rapple described it as 'scandalous'. There has 'long been a suspicion that oil has already been found in commercial quantities off the west coast, and that it suits oil companies to underestimate the prospects while they hoover up rights to more of our hydrocarbon resources,' Rapple wrote. 'Mr Lenihan says that his decision was reached after consultation with the industry and is designed to boost the level of exploration activity. Applications don't close until May [2011], so maybe he is also hoping for some good pre-election publicity,' Rapple observed.

The Sedco 711 submersible rig – owned by the beleaguered Transocean still embroiled in the row over responsibility for the Gulf of Mexico pollution – was booked by Shell for further drilling on Corrib in the summer of 2010. A brief entry on the Irish Offshore Operators' Association website referred to drilling well 18/20-G – in fact, this was the separate Corrib North, for which the company had an exploration licence, and seismic results looked very promising. There could be another substantial quantity of gas, perhaps oil. If it proved commercial, would it be pumped to Ballinaboy? At that point, the company couldn't say. It depended on what was there, and its constitution. Shell managing director Terry Nolan said it was 'early days yet' with Corrib North in an interview with *The Sunday Times* on 20 June 2010. 'I want to get Corrib finished first,' he told the newspaper's business journalist Mark Paul. Nolan said he was 'surprised and disappointed' at the Bord Pleanála decision on the last section of the Corrib pipeline the previous November. The decision would delay the project by another year and add €100 million to a final bill that could reach €2.5 billion, the report said. 'I don't foresee any circumstances where this project is not going to go ahead,' he said – even as Bord Pleanála was still considering the new route application. Mr Nolan said he conveyed his 'regrets' when asked for an interview by this writer in March 2010.

There was one certainty for the residents: the Ballinaboy site was big enough for several more terminals, and a tunnelling route under Sruwaddacon estuary could provide scope for several more pipelines, Vincent McGrath believed. On 31 May, Shell submitted details of the Sruwaddacon estuary pipeline route option to An Bord Pleanála. There was some uneasiness among some of the Rossport landowners who had consented to the original pipeline route on their land, and who now risked losing that compensation. One of the landowners asked this writer 'how much it would be worth him' to give an interview, and refused the request when it was made clear that there would be no payment for same. Several weeks later, a meeting of Rossport residents about their water scheme discussed the option of Shell paying for an upgrade – those residents who had campaigned against Corrib's methodology couldn't believe their ears. Shell's continued strategy of buying support was having a corrosive effect.

Even before it had been laid in the ground, the Corrib gas pipeline had not only rent a community asunder but had become a metaphor for a continuum of struggles – habitat and environment versus resources and lifestyle, right around the planet.

On 6 July 2010, Shell applied to revise its extant plan of development for developing the Corrib gas field. The Irish Whale and Dolphin Group subsequently wrote to Minister for Environment John Gormley and his Cabinet colleague, Minister for Energy Eamon Ryan, reminding them both of a section of the original plan of development which, the group said, had still not been complied with, eight years after it had been approved. Seismic surveys, drilling and Corrib field-related construction could continue offshore, but there had still been no baseline survey of marine mammals beyond Broadhaven Bay.

Fisherman Pat O'Donnell was released from Castlerea prison early on a wet Saturday morning of 17 July 2010, having served just over five of a seven-month sentence. Shell to Sea had organized a welcoming rally for 2 p.m. that day. It was shortly after 7 a.m. when he stepped out into the rain, with a bus ticket to Belmullet and no phone to contact his wife, Mary. Two gardaí parked nearby in an unmarked car gave him a lift to a phone box.

The fisherman pledged to continue his opposition to the Corrib gas project, before being piped home in a cavalcade to a bonfire at Glenamoy cross, where he was plied with hot cups of tea. O'Donnell said that his 'worst fears had been realized' in prison, when the Deepwater Horizon rig exploded in the Gulf of Mexico on 20 April. He felt for fellow fishermen affected by the pollution which US scientists described as the worst ever in those waters. 'They were depending on "fail safe" valves which didn't work . . . the same thing could happen here,' he said.

He had met some of 'the most decent people ever' in prison – men from 'difficult social backgrounds who had committed crime at a young age' and 'were unable to put their lives back together'. He recalled that on his return to prison from day release for his daughter's first communion, he brought back lobster and twenty pounds of crabmeat for fellow inmates. 'This is a prison, not a hotel!' he was told, as the seafood was confiscated. 'Well, I suppose the officer was just doing his job…'

Just over a week before O'Donnell's release, Stein Bredal, a former trade union representative on Statoil's board, was in Ireland for the first screening of Richie O'Donnell's film on Corrib, entitled *The Pipe*, which won best documentary award at the Galway Film Fleadh. Bredal was on the board when allegations of bribery to gain contracts in Iran in 2002/3 had come to light, leading to multi-million fines being imposed on the company, and the resignation of Statoil top management.

In an *Irish Times* interview, Bredal said that Minister of State for Natural Resources Conor Lenihan was 'naïve' to state that a BP Gulf of Mexico-type incident could not happen in Irish waters. An economics graduate and Stavanger resident, Bredal had spent twenty-five years working on offshore rigs, and had taken a very keen interest in health and safety after the capsize of the Alexander L Kielland semi-submersible drilling rig on the Norwegian Ekofisk oil field in March 1980, which killed 123 people. He had been due to fly out to start a new shift on the rig when the capsize occurred.

'Accidents do happen, even in Norway with our experience and tight regulation,' he said. A Gulf of Mexico-type accident had been narrowly avoided on Statoil's Snorre A platform when a gas blowout occurred in November 2004, he pointed out.

Bredal had also unsuccessfully opposed semi-privatization of Statoil, as he believed semi-privatization would dilute emphasis on social responsibility. 'Statoil's approach in Norway was to ask the community first what it wanted from a project, and to listen,' he said. 'It was only when it joined with BP to work in other countries that it moved away from this model. The Norwegian authorities were naïve about this partnership. BP had the contacts but no money, and Statoil had the money but no contacts.'

Bredal said that one could not be 'totally against' oil and gas exploitation if it brought wealth to countries, but the history of resource exploitation was not always happy. 'The reality is that good economics is often seen as more important than good relations with communities.'

Oil and gas companies tended to act totally on behalf of shareholders, were in a position to hire the best legal expertise, and liked to hire journalists to work for them, he said. And, he said, a common public relations strategy was to present critics of projects or project methodology as 'crazies or fundamentalists'. When oil was first found in the North Sea, Norway was 'very very lucky to have had some strong politicians who planned for the future', he said.[2]

Bord Pleanála, which had received Shell E&P Ireland's revised pipeline application for Sruwaddacon, confirmed that it would re-open its oral hearing of 2009 on 24 August 2010. Objections lodged by the community included one joint submission signed by 300 people, to save 15,000 euro on appeal fees. Up on Sruwaddacon estuary, several members of the Rossport solidarity camp were injured and there were arrests after protests on water over Shell's 'investigative' drilling of boreholes in the SAC.

On 3 August, the OECD representatives in Ireland and Holland released their 'final statement' on the Corrib gas issue, two years after a complaint submitted by Pobal Chill Chomáin. It found no grounds for mediation, given the differences over the location of the Ballinaboy terminal. It found that while Shell admitted to problems in the early stages, it had, since 2005, 'shown a willingness to address health and safety concerns, of which the revised route for the onshore part of the pipeline seems the clearest proof'. It pointed out that government compliance with EU legislation on Corrib consents was an issue for the judicial system and not within its remit. Pobal Chill Chomáin was very disappointed, and Afri, which had supported the complaint, described the OECD's response as 'lethargic'. It had only held two meetings with the complainants, and had not visited the area once, Afri pointed out.

Former energy minister Justin Keating passed away in his sleep on New Year's Eve 2009, having never changed his views on 'state take'. He hadn't followed the Corrib gas controversy closely, he said, in one of his last interviews, but what he had read pained him greatly. Speaking to film-maker Richie O'Donnell, he recalled how he had established the terms for oil and gas exploration in 1973, during the oil crisis, and how he had admired the Scandinavian social democrat model.

He said he had also been impressed by *Limits to Growth*, by two members of the Club of Rome in the 1970s, 'who essentially pointed out that the human population is going up, the standard of living of each individual is going up, therefore consumption of earth resources is exploding and resources are finite'. They had 'got their time base wrong', but their warning had been borne out by events, he said.

The geopolitics of mineral resources had always fascinated him, Keating explained. 'If you look at the British Empire, coaling stations for its navy were dotted around the world strategically. Likewise, when oil became important, countries which supplied it became extremely important also. Both oil and politics have always been mixed. I read histories of the big companies, not very charming, a bit rough, a bit ruthless, a bit inclined to buy people if they had to . . .'

So if oil and gas were lying off the west Irish coast, and had not yet been fully exploited, his heart was not broken, he said. The value of the finite resource would rise, not fall, and the technology of deep-water ocean drilling would inevitably improve – making its exploitation much more possible, once the state had not 'sold the pass' entirely on licensing terms.

'We weren't ready back then,' he said. 'If you have a windfall like oil and gas, and if it comes before you are ready to use it prudently, you squander it. We didn't have the political support or under-standing back then to build a good infrastructure with it – we would have consumed it. We've seen since then that the gap between rich and poor has got wider, and money didn't get spent on infra-structure, health, education – not good.

'We had a "Celtic Tiger" for reasons that were not much to do with us, and now it is gone,' Keating noted. 'We will desperately want a source of income that will rescue us . . .' Exploitation of hydrocarbon resources, and development of tidal and wind power and biomass growth in the west could offer this, he stressed, but only if managed properly and sensitively for and by the state.

'I hope that it is all done in a social context, building a community rather than giving out money. And I am glad the oil and gas is still there. I am glad it is not gone . . .'[3]

TIMELINE

1996–2001

1996: Gas discovery reported off Mayo by Enterprise Energy Ireland (EEI).

October 2000: Bord Gáis announces gas pipeline from Mayo to Galway.

November 2000: Enterprise applies for planning permission for gas processing plant at Ballinaboy, north Mayo.

3 August 2001: Mayo County Council approves permission for Ballinaboy.

15 November 2001: Government petroleum lease granted for Corrib gas field.

2002

February 2002: An Bord Pleanála appeal hearing into Ballinaboy terminal opens.

8 April 2002: Minister Fahey publishes Marine Licence Vetting Committee (MLVC) report, used as basis for signing consents.

9 April 2002: Independent general election candidate in Mayo, Dr Jerry Cowley, calls for a renegotiation of the state's deal with Enterprise on Corrib, following takeover by Shell.

April 2002: Minister for the Marine, Frank Fahey, approves plan of development for Corrib gas field.

May 2002: Fahey signs thirty-four compulsory acquisition orders for access to private lands on the pipeline route.

June 2002: An Bord Pleanála seeks further information on Corrib terminal.

25 November 2002: Bord Pleanála Ballinaboy oral hearing resumes, until 10 December.

2003

April 2003: An Bord Pleanála turns down planning permission for Ballinaboy terminal.

19 September 2003: Landslide at Dooncarton Hill in Pollathomas/ Glengad. Taoiseach Bertie Ahern and two ministers meet Shell president, Tom Botts, and senior management in Dublin.

December 2003: Shell submits new planning application for Ballinaboy.

2004

April 2004: Mayo County Council approves new Ballinaboy terminal application.

17 October 2004: New planning application for Corrib onshore terminal given final approval by An Bord Pleanála, with forty-two conditions.

December 2004: Judicial review sought of Bord Pleanála decision.

2005

January 2005: Shell agents attempt to gain entrance to land on onshore pipeline route.

April 2005: Shell obtains High Court injunction restraining six named parties from interfering with preparation, construction and installation of the pipeline and ancillary works.

18 June 2005: Landowner Bríd McGarry reports to Belmullet Garda Station that Shell is carrying out unauthorized work on the onshore pipeline.

29 June 2005: High Court jails Willie Corduff, Brendan Philbin, Philip McGrath, Vincent McGrath and Micheál Ó Seighin for contempt of court over breaches of the April 2005 injunction.

30 June 2005: Minister for the Marine, Noel Dempsey, states in the Dáil that Shell has consent for preparatory work, but not to lay the pipeline.

13 July 2005: Noel Dempsey commissions a new safety review, and says

Shell has agreed to suspend work on the onshore pipeline pending its publication.

26 July 2005: High Court turns down judicial review application over Bord Pleanála Ballinaboy decision.

30 July 2005: Noel Dempsey orders Shell to dismantle almost three kilometres of unauthorized welded pipeline onshore, citing a 'prima facie breach' of consents.

2 August 2005: Minister approves offshore pipeline consent.

4 August 2005: Shell says it will delay laying offshore pipeline until next year to 'allow for a period of discussion'.

8 August 2005: Shell agrees to dismantle welded onshore section of pipeline, and announces formal suspension of work at Ballinaboy.

15 August 2005: The Rossport Five offer to hold direct talks with Shell and its 'government partners' if injunction is lifted. Shell declines the offer. Mayo residents plan a visit to Norway to highlight men's case and Statoil involvement.

30 September 2005: The Rossport Five are released after Shell drops injunction.

October 2005: Advantica consultants begin work on onshore pipeline safety review.

November 2005: Peter Cassells is appointed as government mediator on Corrib gas pipeline. Centre for Public Inquiry (CPI) publishes report on Corrib project.

2006

May 2006: Safety review by Advantica consultants recommends pressure be limited to 144 bar.

July 2006: Mediator Peter Cassells recommends that the pipeline route be modified.

September 2006: Shell announces immediate return to work at Ballinaboy terminal.

3 October 2006: Some 170 gardaí are deployed to police protest at the terminal gates. Two people are injured.

5–6 November 2006: Shell to Sea calls for commission of inquiry into optimum development concept for Corrib – call rejected by Minister for the Marine Noel Dempsey.

8 November 2006: Eight people, including four gardaí, are injured after batons are drawn on protestors in Shell to Sea 'day of action'.

2007

January 2007: EPA gives preliminary approval for IPPC (emissions) licence for Ballinaboy.

16 January 2007: Shell states that it will not be reliant on current compulsory acquisition orders for lands, as new consents will be needed for modified onshore pipeline route.

28 March 2007: Shell denies responsibility for highly elevated aluminium levels in Carrowmore lake.

29 March 2007: Dr Jerry Cowley TD and five Shell to Sea supporters complain to the Minister for Justice over phone tapping.

16 April 2007: EPA hearing opens into IPPC licence for Ballinaboy terminal.

April 2007: Rossport farmer Willie Corduff wins Goldman Environmental Award in US.

25 April 2007: Nine TDs meet Statoil in Dublin on project.

11 June 2007: Twenty people injured when Shell contractors attempt to enter privately owned land in Pollathomas.

13 June 2007: Greens vote at Mansion House to enter government with Fianna Fáil.

28 July 2007: European Commission's environment directorate deems as admissible a complaint lodged by Dr Mark Garavan.

1 August 2007: Revised oil and gas licensing terms announced by Minister for Energy, Eamon Ryan, from 2008.

3 August 2007: Ryan says the government cannot commit to proposed review of entire Corrib gas project.

September 2007: Shell announces local investment fund, and publishes a shortlist of onshore pipeline corridors.

October–November 2007: Minister for the Environment, John Gormley, orders the restoration of special area of conservation at Glengad by RPS Consultants.

13 November 2007: EPA grants emissions licence for Ballinaboy terminal.

14 November 2007: Three Kilcommon parish priests propose an alternative site for the gas terminal at Glinsk. Judge Mary Devins prohibits Shell or its agents from conducting site investigations on Rossport commonage.

2008

February 2008: Fire at a storage compound near Ballinaboy.

February–April 2008: 125 Pollathomas residents sign a petition objecting to road-widening plans for Corrib project.

9–11 April 2008: Concerned residents and politicians highlight Corrib conflict on a visit to Norway.

28 April 2008: Shell submits modified pipeline route application to An Bord Pleanála.

Late April 2008: Pobal Chill Chomáin, formed by seven former Shell to Sea supporters, backs compromise terminal at Glinsk and drops demand for refinery at sea.

May 2008: Andy Pyle retires and Terry Nolan is appointed managing director of Shell E&P Ireland.

14 May 2008: US Global Community Monitor report is critical of gardaí in north Mayo.

19 May 2008: Pobal Chill Chomáin lodges a formal complaint to European Commission.

July 2008: Shell begins excavation work at Glengad landfall. Thirteen people are arrested after questioning authority for work. Minister for Energy, Eamon Ryan, apologizes that approvals were not published.

24 July 2008: An Bord Pleanála seeks further information on modified onshore pipeline route.

6 August 2008: Shell proposes alternative method of discharge from terminal in talks with Erris Inshore Fishermen's Association, and offers compensation.

21 August 2008: Gardaí in Mayo call for protesters to pull back, following a near-accident at Glengad. Eight are arrested.

23 August 2008: Pobal Chill Chomáin lodge a complaint about Shell with Organization for Economic Co-operation and Development (OECD)

28 August 2008: EPA confirms that Corrib gas partners will have to reapply to change the marine dimension of emissions licence for Ballinaboy.

9 September 2008: Maura Harrington goes on hunger strike over the entry of pipelaying ship *Solitaire* into Broadhaven Bay.

10 September 2008: Pat O'Donnell and son Jonathan are arrested in Broadhaven Bay, and later released. Shell withdraws *Solitaire* over 'technical' difficulties.

16 September 2008: Device found outside the Shell headquarters in Dublin is detonated by the army.

18 September 2008: Offshore pipelaying is delayed until the following year.

19 September 2008: Maura Harrington calls off her hunger strike.

29 September 2008: Ministers Eamon Ryan and Éamon Ó Cuív make the first direct government intervention since 2006, with talks in Ballina.

6 November 2008: Ministers Ryan and Ó Cuív announce forum for Corrib gas project.

12 November 2008: Four men, held by gardaí over device at Shell offices in September, are later released with files sent to Director of Public Prosecutions.

5 December 2008: The first meeting of government's North West Mayo Forum. Pobal Chill Chomáin, Pobal Le Chéile and Shell to Sea decline participation.

2009

Late January 2009: New Corrib long-term investment fund is announced.

3 March 2009: New planning application to revise plans for Ballinaboy terminal is advertised in local press.

11 March 2009: Maura Harrington is jailed for twenty-eight days and ordered to undergo psychiatric assessment.

12 March 2009: Gardaí inform Pat and Jonathan O'Donnell that the Director of Public Prosecutions has recommended no prosecution as a result of their September 2008 arrests.

20 March 2009: Direct talks between government ministers, Shell and Pobal Chill Chomáin are chaired by Joe Brosnan.

Early April 2009: The direct talks collapse. Minister for Energy, Eamon Ryan, approves the environmental management plan for offshore pipeline work. Maura Harrington is released from prison.

16 April 2009: Michael Dwyer, graduate and former security guard at Glengad, is shot dead in Bolivia.

22–23 April 2009: Shell begins work at Glengad. Willie Corduff stages protest under a truck and is hospitalized after an alleged assault.

27 April 2009: Pobal Chill Chomáin and Pobal Le Chéile meet OECD representatives in Dublin.

30 April 2009: Ministers Ryan and Ó Cuív attend heated public meeting in Inver.

May 2009: Shell-commissioned report by Arup consultancy dismisses Glinsk as a terminal option.

5 May 2009: Over twenty recipients of the Goldman international environmental award appeal to President Mary McAleese, Taoiseach Brian Cowen and Norwegian Prime Minister Jens Stoltenberg to intervene in the Corrib gas conflict.

9 May 2009: There are seven arrests during a protest at Glengad.

13 May 2009: The preliminary Bord Pleanála oral hearing on modified onshore pipeline is adjourned due to the funerals of Garda Terence Dever and Stephen Conway, both killed in a collision on 10 May outside Belmullet.

18 May 2009: Case taken by Monica Muller over an alleged Shell trespass on Rossport commonage opens.

19 May 2009: Bord Pleanála full hearing on modified pipeline route opens in Belmullet.

10 June 2009: A record number of nineteen Corrib gas protesters appear before Belmullet district court.

11 June 2009: Pat O'Donnell and crewman Martin McDonnell are rescued after O'Donnell's boat, *Iona Isle*, sinks off Erris Head.

22 June 2009: The government's forum opens to press for the first time.

25 June 2009: On the last day of the Bord Pleanála oral hearing, Pat O' Donnell and son Jonathan O'Donnell are arrested at sea. Pipelaying ship *Solitaire* returns to the Mayo coast, under escort by two Naval Service patrol ships.

28 June 2009: Seven of nine arrested at the Glengad protest are remanded in custody and refused bail, and seven are denied legal aid.

1–2 July 2009: The High Court overturns Judge Devins's decision to refuse bail to seven Corrib protestors.

10 July 2009: Glengad resident Colm Henry makes an injunction application.

30 July 2009: Maura Harrington is jailed for four months and Niall Harnett for eight months for Corrib gas-related offences. Both are released pending appeal.

6 August 2009: The sale of Marathon's Corrib gas field shareholding to Vermilion Energy is confirmed. Offshore pipeline is laid.

Early September 2009: Table human rights observer report is published.

4 September 2009: Shell is found in contempt of court over Rossport commonage.

9 September 2009: Maura Harrington is sentenced at Belmullet district court to two three-month sentences, to be served concurrently.

1 October 2009: Seven Corrib gas protesters are convicted at Belmullet district court.

30 October 2009: GSOC confirms it recommended earlier in the year that disciplinary action be taken against a senior member of the Garda for the handling of the Pollathomas protest of 11 June 2007.

2 November 2009: GSOC confirms at government's forum that some 75 per cent of over one hundred complaints about Garda actions in relation to Corrib are deemed admissible for investigation.

3 November 2009: An Bord Pleanála finds up to half of the modified pipeline route 'unacceptable' on safety grounds, and suggests the developers look at Sruwaddacon estuary.

December 2009: Shell applies to Mayo County Council for further amendment of permission for Ballinaboy.

7–11 December 2009: The sentencings of Shell to Sea supporters for various public order offences are heard at Belmullet district court. Public order charge against Jonathan O'Donnell is withdrawn.

7 December 2009: Gardaí confirm the completion of an inquiry into the sinking of Pat O'Donnell's boat *Iona Isle* in June 2008 and say that DPP has directed that no charges be brought.

7 December 2009: Maura Harrington jailed for the fifth time in a year.

2010

18 January 2010: Former UN assistant secretary general, Denis Halliday, launches Afri petition, calling for suspension of all work on the Corrib gas project pending an independent investigation.

20 January 2010: Garda Corrib policing costs from autumn 2005 are quoted at almost €14 million.

2 February 2010: Shell applies for a foreshore licence for investigation works in Sruwaddacon Bay.

3 February 2010: Shell to Sea group meet Minister for Energy, Eamon Ryan, and seek resignation of his technical adviser, Bob Hanna, for 'gross interference' in the planning process.

9 February 2010: Pat O'Donnell is given four- and three-month sentences consecutively in Castlebar circuit court. Maura Harrington's sentence is deferred by a year.

18 February 2010: Shell informs residents that several key stages will be postponed until the following year.

25 February 2010: There are at least 139 objections to Shell's foreshore licence application for investigative work on new pipeline route in Sruwaddacon estuary.

4 March 2010: Ms Justice Laffoy in the High Court rules that Brendan Philbin and Bríd McGarry are entitled to have claims determined.

3 April 2010: The government enacts legislation transferring upstream pipeline safety to Commission for Energy Regulation.

20 April 2010: BP Deepwater Horizon rig explodes and sinks, with a loss of eleven lives and widespread pollution in the Gulf of Mexico.

27 April 2010: Front Line calls for the re-investigation of assault on Willie Corduff in a human rights report on Corrib by barrister Brian Barrington.

31 May 2010: Shell E&P Ireland submits an application for Sruwaddacon estuary pipeline route to An Bord Pleanála.

June 2010: The Sedco 711 rig arrives to drill at Corrib and at satellite Corrib North field.

14 June 2010: The Minister of State for Energy, Conor Lenihan, announces a 'step-up' of 'oversight' of Irish oil and gas well operations in wake of the BP Gulf of Mexico spill.

16 June 2010: Conor Lenihan opens up the entire Atlantic Irish margin for petroleum exploration.

18 June 2010: The High Court overturns finding that Shell acted in contempt of court over entry on Rossport commonage.

20 June 2010: Shell E&P Ireland MD Terry Nolan tells *The Sunday Times* that he doesn't foresee 'any circumstances where this project is not going ahead'.

17 July 2010: Fisherman Pat O'Donnell released from prison.

3 August 2010: OECD final statement finds no grounds for mediation between Shell and Pobal Chill Chomáin.

24 August 2010: Bord Pleanála hearing into last section of pipeline re-opens.

NOTES

PROLOGUE: FROM SRUWADDACON TO SANTA CRUZ

1   An Bord Pleanála: inspector's report PL.16.126073, April 2003.

Chapter 1: BEDROCK

1   Ireland's application to the UN Law of the Sea Convention for further rights off the west coast may extend the seabed area to an estimated twelve times the land mass.

2   Peadar McArdle, *Rock Around Ireland: A guide to Irish geology* (Science Spin Discovery Series, 2008).

3   *Irish Skipper*, May 1973.

4   Interview with Justin Keating by Richie O'Donnell, 2008.

5   Ibid.

6   *The Great Corrib Gas Controversy* (Centre for Public Inquiry [CPI], 2005).

7   Dáil record, Minister for Energy Ray Burke, 8 April 1987.

8   In documentation supplied to Mayo County Council on 12 August 2005, subsequent energy minister Noel Dempsey said that his department had found 'nothing to support' claims that the changes were 'somehow linked to, or are a result of, Mr Burke's period as minister' (Centre for Public Inquiry [CPI] report, p. 60).

9   World Bank (Centre for Public Inquiry [CPI] report, p. 65). Terms were changed again by Minister for Energy Eamon Ryan in 2007, shortly after his Green Party had formed a coalition government with Fianna Fáil, under which an additional resource tax of up to 15 per cent on the oil and gas sector would apply to all new licences issued from 1 January 2008.

10 *Western People*, 18 August 1999.

11 Ibid., 9 August 2000.

12 Eoin Neeson, *Celtic Myths and Legends* (Mercier Press, 1998).

13 Seamus Heaney, 'Belderg', from *North* (Faber & Faber, 1975).

14 Uinsionn Mac Graith and Treasa Ní Ghearraigh, *Logainmneacha agus Oidhreacht Dhún Chaocháin/The Placenames and Heritage of Dún Chaocháin in the Barony of Erris, County Mayo* (Comhar Dún Chaocháin Teo, 2004).

15 Anne Chambers, *Granuaile: The Life and Times of Grace O'Malley c. 1530–1603* (Wolfhound, 1998).

16 Bernard O'Hara, *Davitt* (Mayo County Council Library, 2006).

17 John M. Synge, *Collected Plays* (Penguin, 1952).

18 Mac Graith and Ní Ghearraigh.

19 Dr Mark Garavan (ed.), *Our Story: the Rossport Five* (Small World Media, 2006).

20 Ibid.

21 *Western People*, 20 September 2005.

Chapter 2: THE WAKE

1   Garavan (ed.), *Our Story: the Rossport Five*.

2   *Irish Independent*, 15 October 2007.

3   *The Irish Times*, 2 August 2000.

4   Garavan (ed.), *Our Story: the Rossport Five*.

5   Ibid.

6   Harry Corduff, *The Irish Times*, 3 May 1957.

7   Garavan (ed.), *Our Story: the Rossport Five*.

8   Ibid.

9   In June 2009, Shell agreed to pay a US$15.5-million settlement to end a lawsuit that claimed it was complicit in the execution of Nigerian activist Ken Saro-Wiwa. The company, which had

consistently denied that it had advocated violence and had said it lobbied Nigerian officials to grant Saro-Wiwa clemency, paid the sum to end a court action in the US district court in New York and because it hoped it would aid a 'process of reconciliation'. Associated Press calculated on 9 June that the sum amounted to less than one-hundredth of a per cent of Shell's annual revenue.

10   Garavan (ed.), *Our Story: the Rossport Five*.
11   *The Irish Times*, 14 August 2000.

Chapter 3: SEASCAPE

1   Lorna Siggins, *Mayday! Mayday! Heroic Rescues in Irish Waters* (Gill and Macmillan, 2004).
2   Dr Alex Rogers, *Initial comments on the Bellanaboy Bridge Terminal and Corrib natural gas field environmental impact assessments by Enterprise Oil Ireland* (School of Ocean and Earth Science, University of Southampton, February 2001).
3   *The Irish Times*, 12 February 2001.
4   *The Great Corrib Gas Controversy* (Centre for Public Inquiry [CPI], 2005, p. 30); and Dáil record, 8 February 2005.
5   *The Irish Times*, 12 February 2001.
6   Dr Mark Garavan, *Patterns of Environmental Activism* (2003); and *Sunday Business Post*, 27 May 2001.
7   *Western People*, 12 June 2001.
8   Michael Viney, 'Another Life', *The Irish Times*, 17 February 2001.
9   *The Irish Times*, 17 April 2001.
10   Mike Cunningham, *Western People*, 6 June 2001.
11   *The Irish Times*, 23 January, 18 and 28 February 2001.
12   Ibid., 21 May 2001.

Chapter 4: THE BOGONI

1   *The Great Corrib Gas Controversy* (Centre for Public Inquiry [CPI], 2005, p. 30).
2   Garavan (ed.), *Our Story: the Rossport Five*.
3   *Sunday Business Post*, 27 May 2001.
4   *The Irish Times*, 2 July 2001.
5   Ibid.

6   *Mayo News*, 27 June 2001.
7   Shay Fennelly, *The Irish Times*, 21 July 2001.
8   The press releases referring to the plan of development were dated 16 January and 27 January 2001, with Frank Fahey speaking to the Claremorris Chamber of Commerce on the latter date.
9   Garavan, *Patterns of Environmental Activism.*
10 Report of the Department of the Marine and Natural Resources public information seminar on the Corrib gas field development, Teach Iorrais, Geesala, County Mayo, 25 July 2001.
11 Under a change in legislation in July 2000, which intended to transpose an EU gas directive that effectively liberalized the energy market, the then minister for public enterprise permitted private companies to acquire compulsory purchase acquisition orders for pipelines over private land. In September 2000, then Taoiseach Bertie Ahern approved transfer of regulatory power over 'any upstream pipeline network' from the minister for public enterprise to the minister for the marine and natural resources. This minister was Galway West TD Frank Fahey, whose brief already included Coillte, the state forestry company, and Bord na Móna, the peat board.

Chapter 5: THE BERTIE BYPASS
1   Department of the Marine, press release, 23 October 2001.
2   Green Party TD Eamon Ryan also lodged an objection to Mayo County Council.
3   Dáil report, *The Irish Times*, 16 October 2001.
4   *The Irish Times*, 18 February 2002.
5   Ibid., 19 November 2001.
6   Dr Mark Garavan, 'Seeking a Real Argument', from Mary P. Corcoran and Michael Peillon (eds), *Uncertain Ireland: A Sociological Chronicle, 2003–2004* (IPA, 2006).
7   *The Irish Times*, 11 June 2002.
8   Ibid., 26 June 2002.
9   Conor Newman and Ulf Strohmayer, *Uninhabited Ireland: Tara, the M3 and Public Spaces in Galway* (Arlen House, 2007).
10 *The Irish Times*, 21 August 2002.

11   Ibid., 25 October 2002.
12   Ibid., 25 November 2002.

Chapter 6: DANCING ON THE BEACH

1   Sam Smyth, *Irish Independent*, 26 November 2002.
2   *The Irish Times*, 29 November 2002.
3   De Facto, *Mayo News*, 18 December 2002.
4   Garavan, 'Seeking a Real Argument'.
5   *The Irish Times*, 1 May 2003.
6   Ibid., 24 April 2003.
7   An Bord Pleanála, inspector's report, PL.16.126073, April 2003.
8   *The Irish Times*, 6 May 2003.
9   Ibid., 7 May 2003.
10   Ibid., 8 May 2003.

Chapter 7: THE SLIDE

1   *The Irish Times*, 22 September 2003; and Geological Survey of Ireland, Irish Landslides Working Group, *Landslides in Ireland* (Geological Survey of Ireland, 2006).
2   *The Irish Times*, 24 October 2003.
3   *Western People*, 24 September 2003.
4   *The Irish Times*, 24 September 2003.
5   *Irish Examiner*, 23 October 2003.
6   *The Irish Times*, 26 September 2003.
7   Contractors Ascon Ltd and ESB International Engineering Services were subsequently convicted of pollution as a result of the Derrybrien landslide, and fined €1,200 each. The European Commission found in 2004 that there were 'deficiencies' in the environmental impact statement for the project in failing to provide for adequate geophysical risks.
8   *The Irish Times*, 22 October 2003.
9   Ibid., 24 October 2003.
10   Ibid., 15 October 2003.
11   Ibid., 20 November 2003.
12   Ibid., 21 November 2003.
13   Ibid., 2 February 2004.

14   Ibid.

15   Ibid., 11 February 2004.

16   An Amsterdam court of appeal decision in May 2009 cleared the way for just over US$350 million in compensation to be paid out to non-US Royal Dutch Shell shareholders, following the 2004 shares scandal.

17   The Planning and Development (Strategic Infrastructure) Act 2006 came into force in 2007 to fast-track certain projects, bypassing local authorities for planning approval. Its first oral hearing on 21 January 2008 dealt with a €500-million terminal for liquefied natural gas at the Shannon estuary in County Kerry.

18   *The Irish Times*, 25 October 2004.

Chapter 8: 'WHY NOT JUST HAVE ALL OF THEM COMMITTED?'

1   Garavan (ed.), *Our Story: the Rossport Five*.

2   *The Irish Times*, 15 January 2005.

3   Ibid., 24 January 2005.

4   Ibid., 7 March 2005.

5   Garavan (ed.), *Our Story: the Rossport Five*.

6   *The Irish Times*, 2 March 2005.

7   Ibid., 7 March 2005.

8   Ibid., 15 March 2005.

9   Ibid., 25 May 2005.

10   Ibid., 28 May 2005.

11   Ibid., 30 May 2005.

12   Ibid.

13   In 2007 negotiations between Fianna Fáil and the Green Party on forming a new coalition government, a commitment was secured by the Greens to fully implement the Aarhus Convention. Green Party environment minister John Gormley told *The Irish Times* of 2 April 2008 that he was 'looking' to transpose it into Irish law. As of summer 2010, it had not been. The department attributed the delay to the fact that every consent procedure in every department had to be reviewed.

14   *The Irish Times*, 1 June 2005.

15   Ibid., 3 June 2005. Beverly Flynn was readmitted to Fianna Fáil in April 2008.

16   Ibid., 17 June 2005.

17   Extracts from the 7, 8 and 10 June 2005 memos in relation to Shell's discussions with lawyers were read at a rally in Castlebar on 3 July 2005 and reported in *The Irish Times* on 5 July 2005.

18   *The Irish Times*, 29 June 2005.

19   *Mayo Echo*, 6 July 2005.

20   *The Irish Times*, 2 July 2005.

21   Ibid.

22   Garavan (ed.), *Our Story: the Rossport Five*.

Chapter 9: 'LIKE CATCHING A FOX . . .'

1   Garavan (ed.), *Our Story: the Rossport Five*.

2   Ibid.

3   *The Irish Times*, 1 July 2005.

4   Ibid., 11 July 2005.

5   Ibid., 6 July 2005.

6   Ibid., 11 July 2005.

7   Ibid., 14 July 2005.

8   Ibid., 18 July 2005.

9   Ibid., 27 July 2005.

10   Ibid., 20 July 2005.

11   *Western People*, 16 February 2000.

12   Garavan (ed.), *Our Story: the Rossport Five*.

13   *The Irish Times*, 11 and 13 October 2005.

14   Ibid., 26 July 2005.

15   Ibid., 29 July 2005.

16   Ibid., 1 August 2005.

17   Ibid., 30 July and 2 August 2005.

18   Ibid., 3 August 2005.

19   Ibid.

20   Ibid., 5 August 2005.

21   Ibid., 9 August 2005.

22   Ibid., 19 September 2005.

23   Ibid., 22 September 2005.

24   Ibid., 12 September 2005.
25   Ibid., 14 September 2005.

Chapter 10: 'SPACE SHUTTLE SYNDROME'
1   Garavan (ed.), *Our Story: the Rossport Five*.
2   Ibid.
3   *The Irish Times*, 1 October 2005.
4   Ibid.
5   *Western People*, 18 October 2005.
6   *The Irish Times*, 22 October 2005.
7   Ibid., 25 and 29 October 2005.
8   Ibid., 17 November 2005.
9   Eamon Ryan statement, 22 November 2005; and *Mayo News*, 5 August 2008.
10   *The Irish Times*, 24 November 2005.
11   Conor O'Clery, *The Billionaire Who Wasn't: How Chuck Feeney Secretly Made and Gave Away a Fortune* (Public Affairs, 2007).
12   *The Irish Times*, 9 December 2005.
13   Ibid., 27 December 2005.
14   Ibid., 26 January 2006.
15   Ibid., 27 January 2006.
16   Ibid., 3 February 2006.
17   Harry Browne, *Hammered by the Irish: How the Pitstop Ploughshares disabled a U.S. war-plane – with Ireland's blessing* (CounterPunch and AK Press, 2008). Ciaron O'Reilly and four colleagues, who became known as the 'Pitstop Ploughshares', were found not guilty in 2006 of charges relating to an estimated US$2.5 million damage to a US Navy transport plane at Shannon.
18   *The Irish Times*, 6 May 2006.
19   Ibid., 4 May 2006.
20   Ibid., 5 May 2006.
21   Ibid., 6 May 2006.
22   Ibid., 8 May 2006.

Chapter 11: NO 'MARTYRS'

1   *The Irish Times*, 5 May 2006.
2   Ibid., 8 May 2006.
3   Ibid., 20 May 2006.
4   Ibid., 9 May 2006.
5   Ibid., 2 June 2006.
6   Ibid., 30 June 2006.
7   Ibid., 14 July 2006.
8   Ibid., 29 July 2006.
9   Ibid., 27 July 2006.
10 Ibid., 23 May 2006.
11 Ibid., 4 August 2006.
12 Bríd McGarry said that later her fears were realized when Shell's response was to amend the pipeline route to facilitate distance requirements. Her stance has always been that an inland refinery in association with the pipeline complex was in the wrong location as a totality and was 'unacceptable as a whole'.
13 *The Irish Times*, 11 September 2006.
14 Ibid., 22 September 2006.
15 *Garda Review*, November 2006.
16 *The Irish Times*, 19 October 2006.
17 *Irish Independent*, 29 September 2006.
18 *The Irish Times*, 27 September 2006.
19 Ibid., 29 September 2006.
20 Dáil written reply, 1 October 2006, from Minister for Justice Michael McDowell to Dr Jerry Cowley TD.
21 *The Irish Times*, 27 November 2006.
22 *Garda Review*, November 2006.
23 Roy Greenslade, *Daily Mirror* editor at the time of some of the reports, apologized in print to Arthur Scargill twelve years later.
24 *The Irish Times*, 17 October 2006.
25 Ibid.
26 Ibid., 24 October 2006.
27 Ibid., 27 October 2006.
28 Ibid., 6 November 2006.
29 Ibid.

30   Ibid., 7 November 2006.

31   Ibid., 13 November 2006.

32   RTÉ Radio and Television news, 10 November 2006.

33   *The Irish Times*, 13 November 2006.

34   Ibid., 22 November 2006.

35   *Mayo News*, 22 November 2006.

36   *Irish Independent*, 24 November 2006.

37   *The Irish Times*, 27 November 2006.

38   Ibid., 28 November 2006.

39   Ibid., 23 December 2006.

Chapter 12: THE SPLIT

1   *The Irish Times*, 31 January 2007.

2   Ibid., 17 February 2007.

3   Ibid.

4   Brian Barrington, BL, *Breakdown in Trust: A Report on the Corrib Gas Dispute* (Front Line, 2010); and *The Irish Times*, 28 March 2007.

5   *The Irish Times*, 27 January 2007.

6   Ibid., 15 March 2007.

7   Ibid., 21 March 2007.

8   Ibid., 29 March 2007.

9   Ibid., 21 March 2007.

10   Ibid., 21 April 2007.

11   Ibid.

12   Barrington, *Breakdown in Trust*; and *The Irish Times*, 28 March 2007.

13   *The Irish Times*, 13 June 2007.

14   Ibid.

15   Ibid., 14 and 18 June 2007.

16   Ibid., 30 October 2009.

17   Ibid., 7 June 2007.

18   Ibid., 2 August 2007.

19   Ibid., 3 August 2007.

20   Ibid., 14 August 2007.

21   Ibid., 22 September 2007.

22   Ibid., 14 September 2007.

23   Ibid., 8 October 2007.
24   Ibid., 11 October 2007.
25   Ibid., 22 September 2007.
26   Ibid., 10 November 2007 and 16 March 2009.
27   Barrington, *Breakdown in Trust*.
28   *The Irish Times*, 22 September 2007.
29   Ibid., 20 November 2007.

Chapter 13: 'FIVE "GREEN" MEN . . .'
1   *The Irish Times*, 29 November 2007.
2   Ibid., 7 May 2008.
3   Ibid., 22 September 2007.
4   Ibid., 25 February 2008.
5   Paddy Briggs, *The Tragedy of Corrib Gas* (self-published, March 2008).
6   *The Irish Times*, 9 and 14 April 2008.
7   Ibid., 18 April 2008.
8   Ibid., 6 May 2008.
9   Ibid., 28 April 2008.
10   Ibid., 23 August 2008.
11   Ibid., 26 May 2008.
12   Ibid., 12 May 2008.
13   Ibid., 5 June 2008.
14   Terje Nustad, SAFE website, July 2008.
15   *Mayo News*, 5 August 2008.

Chapter 14: OPERATION GLENGAD
1   *The Irish Times*, 23 July 2008.
2   Ibid., 29 July 2008.
3   Ibid.; and Bord Pleanála report, 3 November 2009.
4   Arthur Boland, *The Irish Times*, 1 August 2008.
5   *The Irish Times*, 2 August 2008.
6   *Irish Mail on Sunday*, 17 August 2008.
7   *The Irish Times*, 8 August 2008.
8   Ibid., 22 August 2008.
9   Ibid., 26 August 2008.

10 *Irish Independent*, 4 August 2008.

11 *The Irish Times*, 1 September 2008.

12 *Upstream*, 22 August 2008; *Mayo News*, 2 September 2008; and *Irish Independent*, 29 August 2008.

13 *Mayo News*, 2 September 2008.

14 The offshore pipeline laying from Broadhaven Bay out to the well-head was completed in autumn 2009. By early 2010, Shell said it was still preparing its application to the Environmental Protection Agency in relation to the emissions licence and the discharge pipe.

15 *The Irish Times*, 30 August 2008.

16 Ibid., 16 September 2008.

17 *Mayo News*, 16 September 2008.

18 *The Irish Times*, 27 November 2007.

19 Lelia Doolan, *The Irish Times*, 29 November 2007.

20 Harrington listed this author, without permission, as one of the contact points for the *Solitaire* captain to verify his departure. She subsequently amended this.

21 *The Irish Times*, 17 September 2008.

22 Ibid., 3 October 2008.

23 Ibid., 27 October 2008.

24 Ibid., 7 November 2008.

25 Ibid., 19 November 2008. Ms Justice Laffoy allowed Bríd McGarry and Brendan Philbin to pursue their claim in March 2010 but the state signalled it would appeal her ruling.

26 *The Irish Times*, 29 December 2008.

Chapter 15: NO QUARTER

1 *The Irish Times*, 27 January 2009.

2 Ibid., 4 February 2009.

3 Ibid., 7 April 2009; I-RMS subsequently commented on this when responding to the Front Line report on Corrib in April 2010. It also told this writer that the term "hostile environment" is an internationally recognized term used to describe unsafe or high-risk locations. The term is often used by NGOs, medical professionals, peacekeepers, media, army and security personnel. Given the security risk in many hostile environments, advance training is generally

undertaken and I-RMS has provided such training for NGOs, media, medical teams and others.

4 Ibid., 10 June 2009 (Associated Press report).
5 Barrington, *Breakdown in Trust*.
6 *Western People*, 5 May 2009.
7 *The Irish Times*, 14 October 2009.
8 Ibid., 17 October 2009.
9 *Sunday Tribune*, 12 July 2009.
10 *The Irish Times*, 24 April 2009.
11 *Mayo News*, 5 May 2009.
12 *The Irish Times*, 2 May 2009.
13 Ibid.
14 *The Irish Times*, 19 May 2009.
15 An Bord Pleanála subsequently ruled that the onshore section of pipeline, which had been deemed exempt from development, was in fact subject to planning, as McGarry and Philbin had both argued.
16 *The Irish Times*, 19 June 2009.

Chapter 16: 'A VERY DANGEROUS GAME'

1 Pat O'Donnell had been arrested on his boat during the summer of 2008 on two occasions while protecting his gear, ahead of the off-shore pipelaying by *Solitaire*. On one of two occasions he was released from Belmullet Garda Station two minutes before his lawyers in Dublin were due to present papers to the High Court for an inquiry into his arrest under Article 42.4.2 of the Constitution.
2 *Mayo News*, 16 June 2009.
3 *The Irish Times*, 12 June 2009; Royal Dutch Shell became embroiled in controversy in 2004 for overstating oil reserves. It led to a £17-million sterling fine by the Financial Services Authority and ousting of the group's then chairman Sir Philip Watts. In May 2009, an Amsterdam court of appeal ruling cleared the way for more than £218 million sterling to be paid out to non-US shareholders.
4 *The Irish Times*, 13 June 2009.
5 Ibid., 23 June 2009.
6 *Western People*, 16 June 2009.
7 *The Irish Times*, 26 June 2009.

8   Ibid., 10 December 2009.
9   *Irish Independent*, 6 June 2009.
10   *Phoenix*, 19 June 2009.
11   The documentary for TV3, made by Gerry Gregg of Praxis Pictures, was nominated for an Irish Film and Television Award the following year. It was broadcast again a week before the award ceremony in February 2010 – at the end of a week of special court sittings relating to Shell to Sea cases. It was one of four programmes nominated in its category. The award went to the RTÉ Television series *Front Line* presented by Pat Kenny.
12   *Mayo News*, 9 June 2009.
13   *Sunday Tribune*, 14 June 2009.
14   Andy Storey and Michael McCaughan, *The Great Gas Giveaway: How the Elites have Gambled our Health and Wealth* (Afri, 2009).
15   *The Irish Times*, 26 June 2009. On the day that Cahill's opinion piece was published in *The Irish Times*, the newspaper reported that Marathon was selling its interest in Corrib and the agreement valued the field at over US$540 million before it produced any gas.
16   *Nigeria: Petroleum, Pollution and Poverty in the Niger Delta* (Amnesty International, June 2009); and *New York Times*, 6 July 2009.
17   *The Irish Times*, 10 July 2009.
18   Ibid., 12 August 2009.
19   *Sunday Tribune*, 23 August 2009.

Chapter 17: 'UNACCEPTABLE' ON SAFETY GROUNDS

1   *Mayo News*, 30 June 2009.
2   Gordon's application to Judge Devins to state the case on 17 September 2009 was made under the 1857 Summary Jurisdictions Act and the 1961 Courts Supplemental Provisions Act.
3   *The Irish Times*, 15 October 2009.
4   Ibid., 6 October 2009.
5   Ibid., 30 October 2009.
6   Ibid., 4 November 2009.
7   Ibid.
8   Offshore247.com, 5 November 2009.
9   *Western People*, 8 December 2009.

10   *Mayo News*, 17 November 2009.
11   *Guardian*, 12 November 2009.
12   *The Irish Times*, 5 January 2010.
13   Ibid., 7 January 2010.
14   Ibid.
15   *Guardian*, 27 January 2010.
16   *The Irish Times*, 16 February 2010.
17   Ibid., 23 February 2010.
18   *Irish Independent*, 18 February 2010.
19   *The Irish Times*, 5 March 2010.
20   The Table human rights group monitored this special court sitting in March 2010 and produced a report that highlighted the denial of the right to a fair trial among other issues.

Chapter 18: COUNTERPOINTS

1   *The Irish Times*, 1 September 2005.
2   Interview with the author, February and March 2010.
3   Interview with the author, 4 May 2010.
4   *Irish Independent*, 13 May 2010.
5   *Offshore Technology*, 17 September 2008.
6   *Dagsavisen*, 22 May 2010.
7   *Mayo News*, 20 April 2010.
8   *Western People*, 23 February 2010.
9   Ibid., 13 April 2010.
10   *Irish Independent*, 17 April 2010.
11   *Ouroboros*, newsletter of the Goldman Environmental Prize, Spring 2010.
12   *The Irish Times*, 24 March 2010; and *Western People*, 27 April 2010.
13   Barrington, *Breakdown in Trust*.
14   Ibid., p. 57.
15   *The Irish Times*, 28 April 2010.

Chapter 19: ENDGAME

1   *The Irish Times*, 10 May 2010.
2   *The Irish Times*, 12 July 2010.
3   Justin Keating, interview with Richie O'Donnell, 2008.

BIBLIOGRAPHY

Amnesty International, *Nigeria: Petroleum, Pollution and Poverty in the Niger Delta*, London: Amnesty International Publications, 2009.

An Bord Pleanála, Inspector's Report PL.16.126073, Dublin: April 2003.

Barrington, Brian, *Breakdown in Trust: A Report on the Corrib Gas Dispute*, Dublin: FrontLine, 2010.

Briggs, Paddy, *The Tragedy of Corrib Gas*, Middlesex, UK: March 2008.

Browne, Harry, *Hammered by the Irish: How the Pitstop Ploughshares disabled a U.S. war-plane – with Ireland's blessing*, California: CounterPunch and AK Press, 2008.

Centre for Public Inquiry, *The Great Corrib Gas Controversy*, Dublin: Centre for Public Inquiry, 2005.

Chambers, Anne, *Granuaile: The Life and Times of Grace O'Malley c. 1530–1603*, Dublin: Wolfhound, 1998.

Creighton, Ronnie (ed.), *Landslides in Ireland: A Report of the Irish Landslides Working Group*, Dublin: Geological Survey of Ireland, 2006.

Garavan, Mark, (ed.), *Our Story: the Rossport Five*, Wicklow: Small World Media, 2006.

Garavan, Mark, 'Seeking a Real Argument', from Corcoran, Mary P. and Peillon, Michael (eds), *Uncertain Ireland: A Sociological Chronicle, 2003–2004*, Dublin: Institute of Public Administration, 2006.

Heaney, Seamus, *North*, London: Faber & Faber, 1975.

McArdle, Peadar, *Rock Around Ireland: A guide to Irish geology*, Dublin: Science Spin Discovery Series, 2008.

MacGraith, Uinsionn and Ní Ghearraigh, Treasa, *Logainmneacha agus*

Oidhreacht Dhún Chaocháin/The Placenames and Heritage of Dún Chaocháin in the Barony of Erris, County Mayo, Carrowteigue: Comhar Dún Chaocháin Teo, 2004.

Neeson, Eoin, *Celtic Myths and Legends*, Dublin: Mercier Press, 1998.

Newman, Conor and Strohmayer, Ulf, *Uninhabited Ireland: Tara, the M3 and Public Spaces in Galway*, Dublin: Arlen House, 2007.

O'Clery, Conor, *The Billionaire Who Wasn't: How Chuck Feeney Secretly Made and Gave Away a Fortune*, New York: Public Affairs, 2007.

O'Hara, Bernard, *Davitt*, Castlebar: Mayo County Council Library, 2006.

Praeger, Robert Lloyd, *The Way That I Went*, Dublin: Hodges, Figgis & Co, 1937.

Report of the Department of the Marine and Natural Resources public information seminar on the Corrib gas field development, Teach Iorrais, Geesala, Co Mayo, 25 July 2001, Dublin: Department of the Marine and Natural Resources, 2001.

Siggins, Lorna, *Mayday! Mayday! Heroic Rescues in Irish Waters*, Dublin: Gill and Macmillan, 2004.

Storey, Andy and McCaughan, Michael, *The Great Gas Giveaway: How the Elites have Gambled our Health and Wealth*, Dublin: Afri, 2009.

Synge, John M., *Collected Plays*, London: Penguin, 1952.

PICTURE ACKNOWLEDGEMENTS

Maps: Tom Coulson at Encompass Graphics. Thanks also to Francis Bradley and John Cassidy of *The Irish Times* studio for assistance in compiling the maps

Page 1: Willie Corduff at Broadhaven Bay: John Monaghan; Sedco 711 semi-submersible rig: Alan Betson, *The Irish Times*; model of the proposed terminal at Ballinaboy: Eamonn O'Boyle

Page 2: Bríd McGarry and Mary Corduff: Alan Betson, *The Irish Times*; a message written in turf: Peter Wilcock

Page 3: Eamon Ryan outside Leinster House: Shay Fennelly; rally outside the Dáil in Dublin, autumn 2005: John Monaghan; The Rossport Five following their release: Alan Betson, *The Irish Times*

Page 4: The Rossport Five near the terminal at Ballinaboy: Henry Wills; workers dismantle a section of pipeline: Dara Mac Dónaill, *The Irish Times*; the silver trailer at Ballinaboy: John Monaghan

Page 5: protestors stage a 'lock on': Eamonn Farrell; Gardaí scuffle with protestors, November 2006: Eric Luke, *The Irish Times*; Ray Corduff admonishes Shell's Corporate and Social Responsibility (CSR) Committee members: John Monaghan

Page 6: The garda presence at the Ballinaboy Terminal: John Monaghan; residents scuffle with Shell contractors: John Monaghan; *Solitaire* is towed into Broadhaven Bay: Peter Wilcock

Page 7: The 'stinger' on the *Solitaire*: Shay Fennelly; protestors in wetsuits at Broadhaven Bay: Peter Wilcock; Jonathan and Pat O'Donnell outside Belmullet Garda station: Peter Wilcock

Page 8: Pat O'Donnell after a protest at Ballinaboy: John Monaghan; unhappy residents at a meeting of the ministerial forum: Henry Wills; the Corrib gas terminal at Ballinaboy, spring 2010: Eamonn O'Boyle

INDEX

419